Business Analytics for Decision Making

Business Analytics for Decision Making

Steven Orla Kimbrough

The Wharton School
University of Pennsylvania
Philadelphia, USA

Hoong Chuin Lau

School of Information Systems
Singapore Management University
Singapore

 CRC Press
Taylor & Francis Group
Boca Raton London New York

CRC Press is an imprint of the
Taylor & Francis Group, an **informa** business

A CHAPMAN & HALL BOOK

First published in paperback 2024

First published 2016
by CRC Press
2385 NW Executive Center Drive, Suite 320, Boca Raton FL 33431

and by CRC Press
4 Park Square, Milton Park, Abingdon, Oxon, OX14 4RN

First issued in hardback 2019

CRC Press is an imprint of Taylor & Francis Group, LLC

© 2016, 2019, 2024 by Taylor and Francis Group, LLC

Library of Congress Cataloging-in-Publication Data

Names: Kimbrough, Steve, author. | Lau, Hoong Chuin, author.
Title: Business analytics for decision making / Steven Orla Kimbrough, Hoong
Chuin Lau.
Description: Boca Raton : CRC Press, [2016] | Includes bibliographical
references and index.
Identifiers: LCCN 2015037871 | ISBN 9781482221763 (alk. paper)
Subjects: LCSH: Decision making--Statistical methods. | Decision making--Data
processing. | Management--Statistical methods.
Classification: LCC HD30.23 .K3985 2016 | DDC 658.4/032--dc23
LC record available at http://lccn.loc.gov/2015037871

ISBN: 978-1-4822-2176-3 (hbk)
ISBN: 978-1-03-292271-3 (pbk)
ISBN: 978-1-315-37242-6 (ebk)

DOI: 10.1201/9781315372426

**Visit the Taylor & Francis Web site at
http://www.taylorandfrancis.com**

**and the CRC Press Web site at
http://www.crcpress.com**

Contents

Preface

Business analytics is about using data and models to solve—or at least to help out on—decision problems faced by individuals and organizations of all sorts. It is a broad, open, and expanding concept, with many facets and with application well beyond the prototypical context of a commercial venture. We take the term business analytics to apply in any situation in which careful decision making is undertaken based on models and data (including text, video, graphic, etc. data, as well as standard numerical and transaction processing data).

The two principal facets or aspects of business analytics are data analytics (associated with the popular buzz term "big data") and model analytics. These two facets, of course, interact and overlap extensively. Models are needed for data analytics and model analytics often, as we shall see, involve exploration of large corpora of data.

This book is mainly about model analytics, particularly model analytics for constrained optimization. Moreover, our focus is unremittingly practical. To this end, we focus heavily on *parameter sweeping* (a term of art in the simulation community, apt for but not much used in optimization) and *decision sweeping,* a new term we introduce here. Both are methods and conceptual tools for model analytics. The larger, governing principle is what we call *solution pluralism.* It is constituted by the collection of multiple solutions to models as an aid to deliberation with them.

Our primary topics emphasis is three-fold and distinctive in each case.

1. We focus on computationally challenging problems actually arising in business enterprises and contexts. This is natural for us because of where and who we teach, because of the research we do, and simply because there are very many such problems.

2. We dwell extensively (but not exclusively) on using heuristics for solving difficult constrained optimization problems. Modern metaheuristics—such as simulated annealing and genetic algorithms, which we discuss in detail—have proved their worth compellingly and have received enthusiastic uptake among practitioners. This book is unusual among business analytics texts in focusing on using heuristics for solving difficult optimization problems that are important in practice.

3. We emphasize throughout the use of constrained optimization models for decision making. In consequence, *post-solution analysis of models* as it contributes to deliberation for decision making is a main topic in the book. We take seriously the saying, commonly assented to among practitioners, to the effect that *after* a model has been specified, formulated, implemented, and solved, then the real work begins. This book is very much about that real work, undertaken after a model has been solved. The emphasis on post-solution analysis is another distinctive feature of the book, motivated by quite apparent practical needs.

Our emphasis on post-solution analysis is in distinction to the usual preoccupation with model formulation and solution theory. These are important topics and we certainly touch upon them in no small way, as is appropriate. Model formulation and model solution theory are topics that complement the study of post-solution analysis, which is a main concern in this book.

Our intended audience for the book consists of those who would undertake serious deliberative decision making with models, in short, practitioners of business analytics both present and future (that is, working professionals and university students). The subject is larger than can be comprehensively treated in a single book, and we mean this book to be introductory and accessible to a broad range of readers, having a broad range of mathematical and computational preparation. We aim to present advanced ideas of post-solution analysis and model-based deliberation in a way that is accessible to tyros and informative for veteran practitioners.

Our approach, our strategy, is to pick a part of business analytics that is important in itself, one that should be known by practitioners as an essential part of the background knowledge for analytics (whether or not the analyst is working directly in the area) and one that affords discussion of post-solution analysis of models. To that end and as already noted, we have chosen that part of business analytics that contains model analytics for constrained optimization models. Further to that end, the book dwells in large part on combinatorial optimization problems, and the use of modern metaheuristics to solve them and to obtain useful information from them for post-solution analysis. Programming is not required to understand the material, but we make available programming examples for those interested. We give examples in Excel, GAMS, MATLAB, and OPL. The metaheuristics code is available online at the book's Web site in a documented library of Python modules (`http://pulsar.wharton.upenn.edu/~sok/biz_analytics_rep`), along with data, material for homework exercises, and much else. For readers without programming skills, we hope to communicate useful information about modeling and model deliberation with the examples in the text. For readers chary of programming, we note that this material has been taught and tested with many others who share your outlook. We do our best to make valuable information for modern decision making accessible to all.

The book is organized into five parts, each containing one or more chapters that naturally group together.

Part I, "Starters", contains three chapters. Chapter 1, "Introduction," frames the entire book and introduces key concepts and terminology that are used throughout what follows.

Chapter 2, "Constrained Optimization Models: Introduction and Concepts," is an overview of the various kinds of constrained optimization models, categorized conventionally for ease of reference to the literature. A higher level classification divides these models into those that are *tractable* (relatively easy to solve, even for large instances) and those that are *intractable* (in theory very difficult to solve when scaled up). Our attention in this book is mainly directed at models in the intractable category and at how we can use heuristics and metaheuristics to solve them and gain useful information from them for purposes of decision making.

Chapter 3, "Linear Programming," presents an overview of linear programming models. These constrained optimization models are tractable. Large scale instances, with thousands and even millions of decision variables, are widely deployed. Our treatment of linear programming models is light and brief. We say what they are and give examples of how they can be applied. We touch on solution methods and formulations. The exposition is introductory, with no attempt to be comprehensive. We aim to provide a useful point of departure for exploration of post-solution analysis and deliberation, as well as for deeper exploration of model formulation and solution theory. This is also the pattern we follow in Part II of the book.

Part II, "Optimization Modeling," consists of seven chapters, each providing an introduction to an important class of (mostly) intractable constrained optimization models. As is the case with linear programming, our treatment is brief, light, and broadly accessible.

We aim to provide a useful point of departure for exploration of post-solution analysis and deliberation, as well as for deeper exploration of model formulation and solution theory. The seven classes of models are:

- Chapter 4, knapsack problems. These models are often used for R&D and for financial portfolio planning.

- Chapter 5, assignment problems. These models are often used for assigning workers to tasks in a factory or consulting firm.

- Chapter 6, traveling salesmen problems. These problems very often arise in transportation, logistics, and tourism applications.

- Chapter 7, vehicle routing problems. These problems occur ubiquitously in service firms having responsibilities that range geographically.

- Chapter 8, resource-constrained scheduling problems. These occur very often in project scheduling for construction and manufacturing ventures, as well as elsewhere.

- Chapter 9, location analysis problems. These arise ubiquitously in siting service and distribution centers, both for commercial ventures and governments. Related to these are zone design problems (Chapter 18) that seek to find good service areas, given demands. Designing sales and service districts are example applications.

- Chapter 10, two-sided matching problems. These are important for designing markets, typically of buyers on one side and sellers on the other, when what is exchanged is not a commodity.

Familiarity with and exposure to these seven classes of models, plus linear programming, belongs in the intellectual armamentarium of every business analyst. Whether or not any of these models are directly employed, familiarity with them affords a certain maturity of mind that can recognize and act upon opportunities to improve deliberation and decision making with models.

Part III, "Metaheuristic Solution Methods," is about modern metaheuristics, which in many if not most cases have become preferred methods for solving intractable optimization problems. A *heuristic* is a method thought to perform well in most cases. We give examples throughout the discussion in Part II of heuristics for solving constrained optimization problems. These traditional heuristics, however, are tailored to very specific model classes and in many cases metaheuristics perform better. A metaheuristic may be thought of as a general, parameterized heuristic. For example, genetic algorithms constitute a kind of metaheuristic. They are general in that they apply very broadly and do not require (much) problem specific information. Also, they are parameterized. For example, genetic algorithms will typically have a mutation rate parameter (inspired by biological evolution), which can be tuned for good performance on a particular class of problems. A genetic algorithm with all of its parameters set—"instantiated"—constitutes a particular heuristic. Thus, genetic algorithms in their various forms are said to constitute a metaheuristic because they may be instantiated in indefinitely many ways to create particular heuristics. Genetic algorithms are *types* of heuristics; instantiated as concrete procedures they are *tokens* of heuristics. The point applies to other kinds of metaheuristics.

Part III contains three chapters. Chapter 11, "Local Search Metaheuristics," presents, motivates, and discusses several metaheuristics in the broader class of local search metaheuristics, including greedy hill climbing, simulated annealing, and tabu search. Using Python code available on the book's Web site, the chapter also discusses how these methods may be used to solve exemplar problems discussed in Part II.

Chapter 12, "Evolutionary Algorithms," presents, motivates, and discusses several meta-heuristics in the broader class of population-based metaheuristics, including evolutionary programming and genetic algorithms. Again, using Python code available on the book's Web site, the chapter also discusses how these methods may be used to solve exemplar problems discussed in Part II.

Chapter 13, "Identifying and Collecting Decisions of Interest," bridges Parts III and IV, "Post-Solution Analysis of Optimization Models." The emphasis on metaheuristics is a distinctive feature of this book, well justified by current developments in research and trends in applications. Metaheuristics as we describe them are very commonly accepted and deployed in practice. Chapter 13 develops connections between these two topics by showing how metaheuristics may be used to generate and collect information that is valuable for post-solution analysis. Metaheuristics thus receive a boost. Besides being appropriate when exact methods fail and besides being useful for finding good starting solutions for exact methods, metaheuristics are invaluable for post-solution analysis.

Part IV contains five chapters discussing post-solution analysis. The presentation is example driven. Although we provide ample computational information for the interested reader, the discussion is designed to be very broadly accessible.

Chapter 14, "Decision Sweeping," is about using a plurality of solutions or decisions for a model in support of decision making. It discusses how these data—the plurality of solutions—obtained from metaheuristic solvers may be used to obtain rich corpora of decisions of interest, and how these corpora may be used to support deliberative decision making with models. Continuing to use a set of exemplar problems from Part II, the chapter presents actual data produced by metaheuristic solution of the exemplar models and discusses its use for decision making. An important point made by the examples is that the data so obtained are easily understandable and afford deliberation by any general stakeholder.

Chapter 15, "Parameter Sweeping," discusses how a plurality of solutions to a model, obtained by varying its parameters systematically, may be used to support deliberative decision making with the model. Again, the chapter presents actual data obtained from exemplar models and shows how the results are easily understandable and afford deliberation by any general stakeholder.

Chapter 16, "Multiattribute Utility Modeling," presents the very basics of utility theory and then discusses in detail how to build simple, practical, and useful models for comparing entities on multiple dimensions at once. The method is quite general and has many successful applications outside of model analytics. Our discussion of how to apply multiattribute utility modeling for post-solution analysis is the first we know of. Essential to the project is access to a corpus of solutions or decisions for the model(s) under consideration. The use of metaheuristics not only for solving optimization models, but for generating useful corpora of decisions, is a main theme in the book. Thus, the general method of multiattribute utility modeling is linked innovatively with model analytics.

Chapter 17, "Data Envelopment Analysis," shows how linear programming may be used to filter a large number of found solutions for a model, as were produced and used in Chapters 14 and 15, in order to distinguish efficient from inefficient solutions.

Chapter 18, "Redistricting: A Case Study in Zone Design," is a case study of the application of many of the post-solution analysis ideas found in this book. It describes work related to a contest to design councilmanic districts in Philadelphia, work which was recognized both in the original contest and in the INFORMS Wagner Prize competition.

Finally, Part V, "Conclusion," consists of a single chapter, Chapter 19. It wraps up the book and directs the reader to points beyond.

* * *

We wish explicitly to thank a number of people. Steven Bankes, Roberta Fallon, Margaret Hall, Ann Kuo, Peter A. Miller, Fred Murphy, and Ben Wise each read portions of the book and offered sage and useful comments. Nick Quintus did the excellent GIS work. Over the years we have benefitted greatly from many ideas by and conversations with Steven Bankes, Harvey Greenberg, Fred Glover, and Fred Murphy. Our heartfelt thanks to all.

List of Figures

List of Tables

Part I

Starters

Chapter 1

Introduction

1.1 The Computational Problem Solving Cycle

Business analytics is about using data and models to solve—or at least to contribute towards solving—decision problems faced by individuals and organizations of all sorts. These include commercial and non-profit ventures, LLCs, privately held firms, cooperatives, ESOPs, governmental organizations, NGOs, and even quangos. Business analytics is, above all, about "thinking with models and data" of all kinds (e.g., in the case of data, including text data). It is about using them as inputs to deliberative processes that typically are embedded in a rich context of application, which itself provides additional inputs to the decision maker.

Focusing on the general analytics at the expense of details of the governing context of any particular case, there are three kinds of knowledge important for our subject.

1. Encoding.

   How to express an algorithm or procedure in the computational environment of our choice and get it to run and produce useful results?

   Important modern computational environments include Excel (and other spreadsheet programs), MATLAB, Mathematica, Python, Ruby, PHP, JavaScript, R, SAS, SPSS, NetLogo, and much else, including core programming languages such as C, C++, and Java.

   We shall discuss coding in this book, but for the most part relegate it to later chapters, aimed at more advanced users, or at least users who would be more advanced and so wish to study programming for business analytics. Python and MATLAB will serve as our focal core programming languages, although we shall have many occasions to mention others. We will advert to Excel, GAMS, NetLogo, and OPL as needed, to the extent helpful in the context.

2. Solution design.

    Solution design is an issue that arises prior to encoding. Given a problem, how should it be represented so that it can be coded in one's computational environment of choice (Python, MATLAB, Excel, etc.)? Alternatively put: How should we design the representations—the data structures and algorithms—for solving the problem?

    Solution design, including representation, efforts occur at a level of abstraction removed from encoding efforts. The two are not entirely independent, since felicitous choice of computational environment aids and abets representation enormously. Still, good designs will ideally be accessible to, and useful for, all parties in an analytics effort, programmers and non-programmers alike.

3. Analytics: Post-solution analysis.

    Given a solution design, its encoding into a model, and successful execution of the model (or of a solver for the model), we arrive at a solution to the model. Now the real work begins! We need to undertake *post-solution anlaysis* (also known as *model analysis*) in order to validate the model, test and understand its performance, obtain information perhaps relevant to reconsidering the assumptions of the model, and so on.

    Further, how can we use data, text, and models to solve business problems (broadly conceived)? Or given a business problem, how shall we approach its solution with data, text, and models (assuming it is amenable to such)? In short, how can and how should we go about "thinking with data and models" in the context of real problems?

    Post-solution analysis (or model analysis) refers, then, to that variety of activities we undertake after a model has been designed, formulated, and solved. These activities may be directed primarily at testing and validating a model or at supporting deliberative decision making with the model, although in fact it is often pointless to maintain a distinction between the two goals. In any event, the activities characteristic of post-solution analysis are typically useful and used both for validating a model and for deliberating with it regarding what actions to take.

    The main focus of this book is on how to address these questions with specifics, to provide recommended and usable techniques, along with realistic examples. We emphasize *using* implementations that let us exercise and explore models and data for the sake of discovery, understanding, and decision making. To this end we dwell at length throughout on post-solution analysis. Its techniques and methods are not only accessible to all who would engage in thinking with models, they are also essential to actual deployment and decision making with models.

    These levels are interdependent. Coding depends on representation and representation depends on what the business problems are. To do analytics we need to specify the problem rigorously, meaning we need to find implementable representations and we need to code them up. And then, we emphasize, the real work begins: We need to use our implementations to explore and discover what our models and data can tell us. To really do analytics we need the results of representation and encoding, but that is just the beginning.

    The book is about all three levels. In the beginning, however, and for much of the book we emphasize analytics and de-emphasize representation and encoding, so that readers will

1. Recognition or detection of a problem.

   Describe the problem.

2. (Computational) solution concept.

   Develop an approach, a computational approach, for solving the problem—or at least for providing useful information about it.

3. Solution design.

   A good slogan in this context: "Algorithms + Data Structures = Programs" [158].

   (a) Data structures
   (b) Algorithms

   The result of this step is typically a representation, perhaps in *pseudocode*, that is amenable to encoding. The representation should be independent, or abstracted from, any particular programming language, although it may be developed with one in mind.

4. Encoding.

   Translate the design into a working computational entity, aka implementation, that produces results (solutions).

5. Post-solution analysis.

   Probe the encoded model in order to measure and assess its performance under a range of test conditions.

FIGURE 1.1: Computational problem solving cycle.

have much to sink their teeth into without raising the hackles of those adverse to computer programming.

Let us not forget the larger context of business analytics. We can embed our levels in the larger context of a *computational problem solving cycle*. See Figure 1.1, which we might frame as an updating or complement for the twenty-first century and new technology of Peirce's "The Fixation of Belief."[1]

The last three steps in the cycle correspond to, or encompass, our three levels. The first step—problem recognition—gets the ball rolling. Throughout, we aim to provide realistic information on representative problem situations. This will serve to motivate the analyses and to expose the reader to real world considerations.

The second step in the cycle framework is to find a solution concept for the recognized problem. By this we mean a general approach that is likely to be successful in addressing an identified aspect of the problem. Should we build a model or take a data-driven approach? If we build a model, what sort of model should we build?

Finally, to turn the list into a cycle, we emphasize that every step taken, every decision made, is provisional. Computational problem solving efforts iterate through the list and revisit steps until deliberation is—provisionally—abandoned in favor of action.

---

[1]The article [130] addresses decision making from a practical, pragmatic perspective. Indeed, it is one of the founding documents of philosophical Pragmatism. It was written for a popular audience and remains well worth reading today.

Enough abstraction for the present! We now turn to an example that touches upon and illustrates these principles.

1.2 Example: Simple Knapsack Models

Our focus in this book is very much on constrained optimization models (COModels), on solving them, and most of all on deliberating and making decisions with them. Chapter 2 surveys the field of COModels. Subsequent chapters delve into particular kinds of CO-Models. Much detail is on the way. Our purpose in this chapter is to discuss a particular kind of especially simple and clear COModel—the Simple Knapsack model—as a way of introducing concepts and terminology that we will need throughout the book. It is easier and more natural, we believe, to learn by generalizing from specific examples than it is to learn abstract generalizations and recognize particular cases when they are encountered. That, at least, is how we shall proceed to develop the core ideas of this book, starting immediately.

The Simple Knapsack model serves well as a prototypical example of a constrained optimization model (COModel). Its story is easy. DM, a decision maker, has n items. Each item has a value for DM if the item can be placed in DM's knapsack. Imagine that DM is preparing for a flight and will carry the knapsack onto the plane. An item, i, can be either in or out, and if in, its value for the DM is c_i (think: the contribution from item i, if chosen). Besides its value, each item has an amount of resource it uses up; call this w_i (think: the weight of item i). Finally, DM's knapsack has a fixed capacity, b, dimensioned the same as the w_i (think: b is the maximum weight that can be placed in the knapsack; that's how big it is). For DM, then, the weight of the item if it is placed into the knapsack is a resource that is in limited supply (as in *constrained* optimization). DM's problem is to find a right collection of items to place in the knapsack so that the capacity of the knapsack is not exceeded and so that the total value returned to DM is maximized.

Budgeting is perhaps the most widely prevalent context in which Simple Knapsack models are used. Prototypically, there will be n projects (R&D projects, advertising campaigns, capital improvement projects, production activities, and so on) that might be funded, each making an expected contribution to profit of c_i and each requiring an investment of resources (say in money) of w_i, against which a budget b is presented.

Simple Knapsack models are appropriate and used quite often. They should be considered whenever a situation arises in which

1. There are a number of distinct entities that may be chosen, each of which returns some reasonably well-known value to the DM.

2. Each entity consumes a certain amount of resource that is in limited supply.

3. The entities are independent in the sense that the value returned by choosing an entity, and the amount of resource it consumes, does not depend upon whether any other entity is chosen.

Further examples of such situations include certain types of investment portfolio problems, selecting R&D projects for funding, and even construction of multiple choice examinations. See `http://en.wikipedia.org/wiki/Knapsack_problem#Applications` (accessed 2015-06-28) for a useful list of applications of the Simple Knapsack.

Simple Knapsack models are easily represented mathematically. *Decision variables* represent the entities that may be chosen (set to 1) or not (set to 0). See Figure 1.2, where they are indicated by the x_is.

$$\max z = \sum_{i=1}^{n} c_i x_i \tag{1.1}$$

subject to the constraint

$$\sum_{i=1}^{n} w_i x_i \le b \tag{1.2}$$

with

$$x_i \in \{0,1\}, \quad i = 1, 2, \ldots, n. \tag{1.3}$$

FIGURE 1.2: Canonical form for the Simple Knapsack model represented as a mathematical program.

Points arising:

1. The Simple Knapsack model is an example of a *constrained optimization* model (aka: COModel). The x_is, $i = 1, 2, \ldots, n$ are called the *decision variables* for the model and they are what the decision maker gets to decide upon. The other mathematical entities in the model—the c_is, the w_is, and b—are *parameters*; that is, in terms of the model they are assumed to be given and fixed. A *decision* for a model is constituted by any assignment of (here, 0 or 1) values to each of the n decision variables. A *feasible decision* is one that satisfies (makes true) the constraints of the problem. Thus, decisions may or may not be feasible. In Figure 1.2, expressions (1.2) and (1.3) constitute the constraints. (Informally, it is often the case, as above, that in speaking of constraints we refer only to the equality or inequality constraints, such as expression (1.2).)

2. A note on terminology: It is perhaps more common in the literature to use *solution* where we use *decision*. The difference is that when models are solved they are said to yield solutions, which are optimal or otherwise of good quality. This is fine, but we shall often be concerned with decisions that are not optimal and maybe not good at all. So, to retain flexibility of discourse we will conduct our discussion in terms of decisions rather than solutions. That said, there will be times when the context calls for discussion of decisions of good quality produced by model solvers. In such cases we will avail ourselves of the more standard terminology. Nothing of substance turns on this; we are simply attempting to adopt terminology in service of clarity.

3. Constraints are true-or-false expressions which if true for a given decision make that decision feasible. When given as equality or as inequality expressions, as in Figure 1.2 expression (1.2), it is customary to write the constraints so that any decision variables present appear on the left-hand side of the expression. Thus, a constraint (as an equality or inequality) is said to have a *left-hand side* (LHS) and a *right-hand side* (RHS), for which the RHS is a constant (scalar or vector) and the LHS is a functional expression of the decision variables. In Figure 1.2 expression (1.2), the LHS is

$$\sum_{i=1}^{n} w_i x_i$$

and the RHS is b.

4. In Figure 1.2, expression (1.1), or rather its

$$\sum_{i=1}^{n} c_i x_i$$

part, constitutes the *objective function*, which in the present case we wish to maximize. These core elements—decision variables, objective function, and constraints—are present in all COModels expressed as mathematical programs.

5. We can use matrix notation to give an equivalent and more compact representation of the Simple Knapsack model. See Figure 1.3.

$$\max z = c^T x \tag{1.4}$$

subject to the constraints

$$w^T x \leq b \tag{1.5}$$

$$x_i \in \{0,1\}, \quad i = 1, 2, \ldots, n \tag{1.6}$$

FIGURE 1.3: Matrix form for the Simple Knapsack model.

---

x is a column vector of length n, as are c and w. (We shall follow conventions in mathematics and in MATLAB and Python with Numpy; unless otherwise indicated, vectors are columns of elements, not rows.) c^T (the transpose of c; equivalently in notation, c') is a row vector. $c^T x$ represents the dot product and resolves to a scalar. $w^T x$ is the dot product of w^T and x (also known as the *SUMPRODUCT* in Excel). It resolves to a scalar, and so b is a scalar as well.

6. The Simple Knapsack is about as simple and comprehensible a COModel as one could ask for. In consequence, it is a good basis for discussion of concepts. We shall draw upon it extensively for that purpose, and then move on to more complex problems.

7. The Simple Knapsack model serves as the basis for many important and interesting generalizations. There are already two excellent books devoted to knapsack problems: [121] and [85]. These books are also good references for applications of the Simple Knapsack model, as well as applications of more general knapsack models.

8. The Simple Knapsack model is NP-complete, which is to say that in theory it is among the problems that are challenging to solve computationally (i.e., they are "intractable" as they scale up). In practice it is often solved relatively easily and as we shall see there is an excellent heuristic for it. Other knapsack problems, and IPs in general, may be very challenging in practice.

9. If there are n decision variables in a Simple Knapsack problem, then there are 2^n possible solutions, among which we seek to find an optimal solution. Five hundred (500) decision variables is a relatively small industrial problem. $2^{500} \approx 10^{6 \times 25} = 10^{150}$. There are only about 10^{80} atomic particles in the universe. Enumerating even 10^{12} solutions is a stretch. Other methods are required if we are to find optimal or even good solutions to real-world knapsack problems.

10. Finally, another note on terminology: By definition, a Simple Knapsack model has 0-1 decision variables (they may have either the value of 0 or 1), one linear objective

function, and one linear constraint. It is often convenient to simply say "knapsack model" when Simple Knapsack model is meant, and the literature often does this. The literature, however, also recognizes many other kinds of knapsack models, distinct from the Simple Knapsack. Everything that is counted as a knapsack model is a special case of an *integer programming* (IP) model, a constrained optimization model, that is, in which *all* of the decision variables are restricted to be integers. We shall, in the interests of avoiding clutter, be parsimonious in singling out the special cases.

1.3 An Example: The Eilon Simple Knapsack Model

We now present a specific example of a Simple Knapsack model, one that is small but interesting, and has appeared in the literature [45]. This model—the *Eilon model*—has only 12 decision variables, which is quite small indeed. (See Table 4.1.) Even so, it affords a rich discussion that will apply in much larger cases. It will be useful to us in this chapter and again in Chapter 4. The associated ideas and lessons pervade the book.

Variable No.	c_i	w_i	c_i/w_i
1	4113	131	31.40
2	2890	119	24.29
3	577	37	15.59
4	1780	117	15.21
5	2096	140	14.97
6	2184	148	14.76
7	1170	93	12.58
8	780	64	12.19
9	739	78	9.47
10	147	16	9.19
11	136	22	6.18
12	211	58	3.64

TABLE 1.1: Specification of the Eilon Simple Knapsack model. $b = 620$.

The parameters for the model are fully specified in Table 4.1 and its caption. For the present we can ignore the rightmost column; we shall return to it in Chapter 4. The paper has this to say about applications of Simple Knapsack models.

The knapsack model [our Simple Knapsack model] has many applications and a useful example occurs in the field of budgeting. Suppose that a selection needs to be made from n possible projects, each having a project value of payoff Z_i [our c_i] and involving a known cost C_i [our w_i], subject to an overall budget constraint B [our b]. If a project can neither be selected or rejected (i.e. partial projects are not allowed), then this budgeting problem may be formulated as a knapsack [Simple Knapsack] model, as described above. This type of budgeting problem is commonly found in practice: selection of production activities, R&D projects, investment portfolio, computer applications, advertising campaigns, and so on. In all these cases selection is necessary because of the limitation on available budgets, and the executive's purpose is to derive the maximum value, or payoff, from the selected projects. [45, page 489]

```
 1 % File: eilon_sk.m
 2 % This is an implementation of the Eilon Simple Knapsack
 3 % model, a small and very simple model that is nonetheless
 4 % useful and appears in a thoughtful paper, Samual Eilon,
 5 % "Application of the knapsack model for budgeting," OMEGA
 6 % International Journal of Management Science, vol. 15, no.
 7 % 6, pages 489-494, 1987.
 8 % In MATLAB, type doc('intlinprog') at the command prompt to
 9 % get documentation information on the integer programming
10 % solver.
11
12 % There are only 12 decision variables.
13 % The objective coefficients are:
14 c = [4113,2890,577,1780,2096,2184,1170,780,739,147,136,211];
15 % The constraint coefficients are:
16 w = [131,119,37,117,140,148,93,64,78,16,22,58];
17 % The RHS value is:
18 b = 620;
19 % The lower bounds are all zero.
20 lb = zeros(12,1);
21 % The upper bounds are all zero.
22 ub = ones(12,1);
23 % All of our variables are integers:
24 intvars = 1:12;
25 % Call the solver with -c because we wish to maximize:
26 [x,fval] = intlinprog(-c,intvars,w,b,[],[],lb,ub);
27 % Display the decision variables at optimality:
28 disp(x')
29 % Because we are maximizing, multiply the objective value by -1:
30 fprintf('At optimality the objective value is %3.2f.\n',-fval)
31 % The left-hand-side value of the constraint is:
32 fprintf('The constraint left-hand-side value is %3.2f.\n',w*x)
```

FIGURE 1.4: Implementation in MATLAB of the Eilon Simple Knapsack Model [45]. (Line numbers added.)

Figure 1.4 presents an implementation of the Eilon model in MATLAB. It is not necessary that you understand the code in any detail. Later, in Chapter 4, we shall see how to formulate and solve the model in Excel, using Excel's Solver application. The *modeling language* approach on display in Figure 1.4 will, however, be the preferred approach to model implementation, whether in MATLAB or Python some other environment, such as GAMS, OPL, or AMPL. Best, then, to see it illustrated early on in a maximally simple example.

Most of the MATLAB code in Figure 1.4 in fact is comments that should be easily understandable. Essentially what happens in the code is that the parameter values—the c_is, the w_is, and b—are read into variables, other variables are set to reflect that this is a 0–1 integer programming problem, the MATLAB solver for linear integer programs is called, and the results are displayed, as follows:

```
Columns 1 through 6
    1.0000    1.0000    1.0000    1.0000         0    1.0000
Columns 7 through 12
         0    1.0000         0         0         0         0
At optimality the objective value is 12324.00.
The constraint left-hand-side (LHS) value is 616.00.
```

Thus we learn that at optimality decision variables 1, 2, 3, 4, 6, and 8 are 'taken', that is are put into the metaphorical knapsack, while the others are not. We also learn that the resulting value of the objective function is the 12324.0 and that the constraint LHS is 616, so there is a slack of $620 - 616 = 4$ at optimality.

1.4 Scoping Out Post-Solution Analysis

Because post-solution analysis figures so importantly in what follows we shall discuss it now in some detail, describing various facets of the concept. Later chapters, especially those in Part IV, will refer to this section and elaborate upon it.

1.4.1 Sensitivity

To begin, it is often not safe to assume that each of the model's parameters—the c_is, the w_is, and b in the case of the Simple Knapsack model—are known with a very high degree of certainty. We can estimate them, with more or less precision, and solve the model on the basis of our best estimates, but when it comes to implementation of any decision, we may find that our best estimates in fact vary from reality. This is most often the case, and if so, then a number of *sensitivity analysis* questions arise, prototypically:

> *Are there small changes in parameter values that will result in large changes in outcomes according to the model? If so, what are they and which parameters have the largest sensitivities, i.e., which parameters if changed by small amounts result in large changes in model outcomes? Which parameters have small sensitivities?*

Points arising:

1. What counts as a large change or a small change to a parameter depends, of course, entirely on the context and the ambient levels of uncertainty. Typically, we might think of changes on the order of a few percent in the level of a parameter as small. Perhaps a

better measure is the amount of uncertainty or of variation not subject to our control in the value of a parameter. In the present case, perhaps b is only comfortably certain within, say, 5% of its assumed value. If so, we have a natural range of exploration for sensitivity analysis.

2. What counts as a large change or a small change in the objective value (outcome) of the model depends in large part on the model in question and on the overall context. We will naturally be interested in whether the results of the model are feasible in this sense and, with sensitivity analysis, whether small changes in the parameter values will make feasible (constraint satisfying) results infeasible (constraint violating) and vice versa. This will be a constant theme with optimization models (our detailed discussion of these begins in Chapter 2), where we will be interested in knowing whether any small change in parameter values will make the model infeasible. That is, if we make some "small" changes (however defined) to the model's parameters and then re-solve the model, could we find that there are no longer *any* feasible solutions?

3. Less drastically, while we might be confident that small changes in the realized parameter values will not lead to infeasibility, there remains the possibility that small changes can drastically degrade the value of the optimal solution based on the estimated parameter values. In the present case, is it possible or likely that a small change to a c_i (an objective function coefficient) would greatly change the value of our present solution? Could it happen, we must ask, that the present results to the model would have disastrous consequences if implemented in the presence of very minor parameter value changes?

4. We are also interested in knowing which parameters are comparatively insensitive, so that relatively small changes to them have very little effect on outcomes upon implementing the solution to the model.

5. We will need to consider all of our sensitivity questions both in the context of changes to a single parameter and to the context of simultaneous changes to multiple parameters.

These are among the main general considerations for sensitivity analysis. The literature is extensive and the topic itself open-ended. These remarks indicate the spirit of sensitivity analysis as a means of opening, rather than closing, the discussion of its scope and extent.

Concretely, in terms of our Eilon model, we can undertake a sensitivity analysis by identifying the parameters we wish to examine (perhaps all of them) and then for each parameter identify a *consideration set* of values we wish to test. We might, because we are uncertain about c_2 and w_2, want to discover what happens as c_2 varies across the range $[2885, 2895]$ and w_2 varies in $[117, 121]$. Finally, we can *re-solve* the model for each of the new parameter settings of interest (PSoIs, sampled judiciously from the ranges of interest).

There is a broader sense of the term sensitivity analysis that is also in established use. For example,

> A possible definition of sensitivity analysis is the following: *The study of how uncertainty in the output of a model (numerical or otherwise) can be apportioned to different sources of uncertainty in the model input* [142]. A related practice is "uncertainty analysis", which focuses rather on *quantifying* uncertainty in model output. [141, page 1]

We are of course interested in all of these notions, however named. Sensitivity analysis in this broader sense might be stretched to cover most of post-solution analysis. As such it might better be termed *response analysis* or *model response analysis*, but this is non-standard. We shall avail ourselves of the term in both its narrow and broad senses. For the most part we will be using the narrow sense, and will rely on context and explicit comments to minimize confusion.

1.4.2 Policy

In a second category of post-solution analysis questions, the possibility may arise of altering decision variable levels because of considerations not reflected in the model. This possibility is often present, and if so, then a number of *policy analysis* questions arise, prototypically:

> *Are there policy reasons, not represented in the model, for deviating from either choosing a preferred (e.g., optimal) solution identified by the model or for modifying one or more parameter value? If so, what changes are indicated and what are the consequences of making them? Policy questions are about factoring into our decision making qualitative factors or aspects of the situation that are not represented in the model formulation.*

In a typical case, it may be desired for business reasons to set a decision variable to at least a minimal level. To take a hypothetical example, there may be reasons having to do with, say, maintaining supplier relationships, for putting item 5 (corresponding to decision variable 5) into our knapsack. We have seen that at optimality of the model variable 5 is not taken; it is set to 0. If for policy reasons we wish to force the taking of item 5, what are the consequences? What is the resulting optimal decision? How much of a decrease in the objective value will we experience? And so on.

The analysis question is then how best to do this and to figure out how much it will cost. If costs are too high that may well override any policy factors. In this typical mode of decision making, policy issues may be ignored and the model solved, after which an investigation is undertaken to determine whether to make changes in accordance with the policy factors.

What we are calling policy questions frequently arise as so-called *intangibles,* factors that matter but are difficult to quantify, difficult certainly to express adequately as part of a constrained optimization model (or other model) formulation. The model, however, can be most helpful for deliberations of this kind. If, to continue the example, there are policy reasons for favoring inclusion of item 5, then the model can be used to estimate our opportunity cost were we to do this. Generally speaking, having a solution to the model at hand is materially useful for evaluating the cost of taking a different decision. Which decision to take is always up to the decision maker. A model serves its purpose well when it informs the decision maker about the costs and benefits of alternatives.[2]

Concretely in terms of our Eilon model, we can reformulate the model by adding a constraint that requires x_5 to equal 1. This is easily effected in MATLAB and other modeling languages. The resulting model is no longer a Simple Knapsack model, but it helps us deliberate with one. In short, we add constraints to reflect policy initiatives, re-solve the model as many times as are needed, and deliberate with the results.

---

[2]Thanks to Frederic H. Murphy for emphasizing to us the vital importance for model-based decision making of qualitative, extra-model factors, which we call *policy questions.*

1.4.3   Outcome Reach

In our Eilon Simple Knapsack model the optimal objective function value is $z = 12{,}324$. What if instead of 12,324 we require at least 12,333? What combinations of parameter value changes will achieve this? And among them, how likely are they to be realized or achievable with action on the part of the decision maker?

This is an example of an *outcome reach* question on the improvement side. There are degradation side questions as well. For example, anything less than 12,320 might be disastrous. What combinations of parameter changes would lead to such a result, how likely are they, and what might we do to influence them? Prototypically for this third category of post-solution analysis questions we have:

> *Given an outcome predicted or prescribed by the model, and a specific desired alternative outcome, how do the assumptions of the model need to be changed in order to reach the desired outcome(s)? Is it feasible to reach the outcome(s) desired? If so, what is the cheapest or most effective way to do this? Outcome reach questions arise on the degradation side as well. By opting to accept a degraded outcome we may free up resources that can be used elsewhere.*

Concretely in terms of our Eilon model, we might guess at parameter changes that might produce the results (good or bad) of interest and then we can re-solve the model with these changes in order to discover what actually happens. Guessing, however, becomes untenable as the model becomes larger and more complex. What is needed is a systematic approach to exploring alternative versions of a model. We will introduce and develop two such systematic approaches. They are complementary.

In *parameter sweeping* (discussed in §1.5 and subsequently in the book, especially in Part IV) we characterize one or more *parameter settings of interest* or PSoI(s) and then use computational methods to sample from the set(s) and re-solve the model. This produces a corpus or plurality of solutions/decisions for the model, which we examine for the purposes of deliberation. That deliberation may pertain to outcome reach questions as well as other questions arising in post-solution analysis.

In *decision sweeping* (discussed in §1.6 and subsequently throughout the book, again especially in Part IV) we characterize one or more collections of *decisions of interest* or DoIs and then use computational methods to find sample elements—decisions for the optimization model to hand—from the DoIs. Again, this produces a corpus or plurality of solutions/decisions for the model, which we examine for the purposes of deliberation. And again, that deliberation may pertain to outcome reach questions as well as other questions arising in post-solution analysis.

1.4.4   Opportunity

We saw that at optimality of the Eilon model there was a slack of 4 for the constraint. That is, with the optimal decision in place, the LHS value of the constraint is 616, while the RHS value, b, equals 620. There are (at least) two kinds of interesting questions we can ask at this stage of post-solution analysis:

1. Relaxing: What if the value of b could be increased? How, if at all, would that improve the value of the objective function realized, what would the new optimal decision be, and how much slack would be available?

2. Tightening: What if the value of b were decreased? How, if at all, would that degrade the value of the objective function realized, what would the new optimal decision be, and how much slack would be available?

Prototypically, and more generally, for this fourth category of post-solution analysis questions we have:

> *What favorable opportunities are there to take action resulting in changes to the assumptions (e.g., parameter values) of the model leading to improved outcomes (net of costs and benefits)?*

These and related questions have also been called *candle lighting questions* in allusion to the motto of the Christopher Society, "It is better to light one candle than to curse the darkness" [86, 87, 88, 89, 102]. Given what we learn from solving the model, what might we do to change the objective conditions in the world in order to obtain a better result?

Concretely in terms of our Eilon model, we might address opportunity questions of this type by undertaking parameter sweeping on b. Of course, opportunity questions apply for all model parameters and can become complex, requiring systematic approaches, as in parameter sweeping and decision sweeping with sophisticated approaches.

1.4.5 Robustness

Something is robust if it performs well under varying conditions [92]. Wagner's characterization is representative and applies more or less to any system: "A biological system is robust if it continues to function in the face of perturbations" [150, page 1], although we will normally want to add "well" after "function". (See also [48, 104].) This general notion of robustness is apt for, is important for, and is useful in many fields, including biology, engineering, and management science. The general notion, however, must be operationalized for specific applications. We focus in this book on management science applications, especially applications to optimization.

This fifth category of post-solution analysis questions may be summarized with:

> *Which decisions or policy options of the model perform comparatively well across the full range of ambient uncertainty for the model?*

How exactly we operationalize the general concept of robustness, in the context of decision making with models, requires extensive discussion which will appear in the sequel as we discuss various examples. Here we merely note that parameter sweeping and decision sweeping will be our principal tools.

Concretely in terms of our Eilon model, elementary forms of robustness analysis can be handled by sensitivity analysis, discussed above.

1.4.6 Explanation

Models, especially optimization models, are typically used *prescriptively,* used that is to recommend courses of action. An optimal solution comes with an implicit recommendation that it be implemented. Because it is optimal the burden of proof for what to do shifts to any envisioned alternatives. In fact much of what post-solution analysis is about is deliberating whether to take a course of action other than that given by the optimal solution at hand.

It is natural in this context for any decision maker to request an explanation of *why* what is recommended is recommended. Why X rather than some favored alternative Y?

> *Why X? Given a predicted or prescribed outcome of the model, X, why does the model favor it? Why not Y? Given a possible outcome of the model, Y, why does the model not favor it over X?*

Questions of this sort are quite often encountered in practice and handled with appeal to common sense. For example, constraints might be added to the model to force outcome Y. This will result either in a degraded objective function value or outright infeasibility. In either case, the model can be helpful in explaining why Y is inferior to X, particularly by suggesting outcome reach analyses that add insight (e.g., that the price on a certain parameter would have to fall by at least a certain amount).

Although explanation questions are commonly engaged it remains true that there is neither settled doctrine or methodology on how best to undertake explanatory analysis with optimization models, nor is there bespoke software support for it. Practice remains unsystematic in this regard. The pioneering work by Harvey Greenberg in the context of linear programming— [68, 69, 70, 71]—remains the point of contemporary departure. We will not have a lot to say about explanation, important as it is, beyond drawing the reader's attention to the uses of corpora of solutions obtained by parameter sweeping and decision sweeping. These do, we believe, contribute materially to the problem, without settling it forever.

1.4.7   Resilience

Robustness and resilience are closely related concepts. Neither is precisely defined in ordinary language. More careful usage requires a degree of stipulation—"This is what we shall mean by..."—and many such stipulations have been given, if only implicitly. In short, the terms do not have established standard meanings, and the meanings that have been given are various and mutually at odds. We shall use robustness as described above: Something is robust if it works reasonably well under varying conditions. We reserve resilience for a form of robustness achieved by action in response to change. A robust defense resists attack by being strong in many ways, in the face of multiple kinds of aggressive forays. A resilient defense is robust in virtue of an effective response to an attack.

Robustness may be static or dynamic. Resilience is dynamic robustness. A resilient solution is one that affords effective decision making after now-unknown events occur. While it is generally desirable to delay a decision until more information is available, this might not be possible or it may come with a prohibitive cost. In some cases, clever design will permit delayed decision making or put the decision maker in position to adapt cheaply to a number of possible eventualities.

Summarizing, these are prototypical resilience (dynamic robustness) questions.

> *Which decisions associated with the model are the most dynamically robust? That is, which decisions best afford deferral of decision to the future when more is known and better decisions can be made?*

We will not have a lot to say about resilience, important as it is, beyond drawing the reader's attention to the uses of corpora of solutions obtained by parameter sweeping and decision sweeping. As in the case of explanation, these do, we believe, contribute materially to the problem, without settling it forever.

* * *

Figure 1.5, page 17, summarizes our seven categories of post-solution analysis questions. We emphasize that boundaries are fuzzy and urge the reader to attend less to clarifying the boundaries and more to using the framework to suggest useful questions leading to better decisions. We mean our framework to stimulate creative deliberation, rather than to encode settled knowledge. Decision making with any model occurs within a context broader than the model. That context may be simple and straightforward, it may be complex and heavily strategic, involving consideration of possible actions by many other decision makers, and it may be anywhere in between.

1. Sensitivity

   Are there small changes in parameter values that will result in large changes in outcomes according to the model? If so, what are they and which parameters have the largest sensitivities, i.e., which parameters if changed by small amounts result in large changes in model outcomes? Which parameters have small sensitivities?

2. Policy

   Are there policy reasons, not represented in the model, for deviating from either choosing a preferred (e.g., optimal) solution identified by the model or for modifying one or more parameter value? If so, what changes are indicated and what are the consequences of making them? Policy questions are about factoring into our decision making qualitative factors or aspects of the situation that are not represented in the model formulation.

3. Outcome reach

   Given an outcome predicted or prescribed by the model, and a specific desired alternative outcome, how do the assumptions of the model need to be changed in order to reach the desired outcome(s)? Is it feasible to reach the outcome(s) desired? If so, what is the cheapest or most effective way to do this? Outcome reach questions arise on the degradation side as well. By opting to accept a degraded outcome we may free up resources that can be used elsewhere.

4. Opportunity

   What favorable opportunities are there to take action resulting in changes to the assumptions (e.g., parameter values) of the model leading to improved outcomes (net of costs and benefits)?

5. Robustness

   Which solutions or policy options of the model perform comparatively well across the full range of ambient uncertainty for the model?

6. Explanation

   Why X? Given a predicted or prescribed outcome of the model, X, why does the model favor it? Why not Y? Given a possible outcome of the model, Y, why does the model not favor it over X?

7. Resilience

   Which decisions associated with the model are the most dynamically robust? That is which decisions best afford deferral of decision to the future when more is known and better decisions can be made?

FIGURE 1.5: Seven categories of representative questions addressed during post-solution analysis.

1.5   Parameter Sweeping: A Method for Post-Solution Analysis

Briefly put, a parameter sweep of a model is an experiment in which multiple model outcomes are produced by executing an algorithm numerous times that solves the model with different parameter configurations. NetLogo, with its BehaviorSpace feature, emphasizes parameter sweeping as a key part of modeling. This is from the *NetLogo User Manual*:

> BehaviorSpace is a software tool integrated with NetLogo that allows you to perform experiments with models.
>
> BehaviorSpace runs a model many times, systematically varying the model's settings and recording the results of each model run. This process is sometimes called "parameter sweeping". It lets you explore the model's "space" of possible behaviors and determine which combinations of settings cause the behaviors of interest.

Of course we can do parameter sweeping without recourse to NetLogo. To illustrate, the following code implements in MATLAB an equation known as Converse's formula [46, page 432]. (The formula has been used to evaluate the siting of retail stores.)

```
function [ Db ] = converse_formula( Dab, Pa, Pb )
%converse_formula Calculates the Converse formula.
% Dab is the distance between establishment a and
% b. Pa is the population of a, Pb is the population
% of b. Db = Dab/(1 + sqrt(Pa/Pb)). Db is the
% predicted distance from b in the direction of a
% before which customers will go to b, after which
% customers will go to a.
Db = Dab / (1 + sqrt(Pa/Pb));
end
```

It is not necessary for present purposes that you understand this code, although we imagine that its meaning is reasonably clear. The important point is that once a model is implemented and executable on a computer (as this MATLAB function is) we can write a program that iteratively supplies it with different parameter values and collects the outputs, which are the resulting values of the function.

Specifically, we can set P_a to a sequence of values, say 390,000, 395,000, 400,00, 405,000, and 410,000, and we can set P_b to a different sequence of values, say 75,000, 80,000, 85,000, 90,000, and 95,000. This yields $5 \times 5 = 25$ distinct parameter combinations for which we can obtain function values by executing the Converse function code. Table 1.2 presents the results of calculating the Converse formula for these 25 combinations of parameter values.

This simple example may serve as a prototype for much that is to come. We will confine ourselves here to just two points.

1. In two dimensions, by sweeping two parameters as in Table 1.2, it becomes natural to think of the ensemble on a geographic analogy. We see in the present case that when P_b is equal to or greater than 85,000, the formula value exceeds 31, regardless of the P_a values present. More precisely, when $P_b \geq 85,000$ or when $P_b = 80,000$ and $P_a \leq 395,000$, the function value exceeds 31; otherwise it does not. Thus, with our constraint of 31, discussed above, we see two regions or two *phases* of the parameter space revealed by the sweep.

	75000	80000	85000	90000	95000
390000	30.4845	31.1727	31.8267	32.45	33.0453
395000	30.3497	31.0362	31.6887	32.3105	32.9046
400000	30.2169	30.9017	31.5527	32.1731	32.7659
405000	30.0861	30.7692	31.4187	32.0377	32.6292
410000	29.9572	30.6387	31.2866	31.9043	32.4945

TABLE 1.2: Parameter sweep results for the Converse formula. Rows: P_a values. Columns: P_b values.

This point, that parameter sweeping may reveal regions or phases of interest, applies, of course, to ensembles of more than two parameters. And it applies to models of all kinds.

2. Generalizing further, parameter sweeping *when it is practicable* can be used as a basis for answering, or at least addressing, very many of our post-solution analysis questions. This is a subject we will develop in depth throughout the book.

1.6 Decision Sweeping

The Eilon Simple Knapsack model and other optimization models, unlike the Converse formula, have decision variables, whose values may be chosen by a decision maker. Recall that a decision for a model (with decision variables) is simply a complete setting of the decision variables for the model; each decision variable is given a value. A given decision may or may not be feasible and if feasible may or may not be any good. A main purpose of model solvers, such as `intlinprog` in MATLAB (Figure 1.4) and Solver in Excel, is to find good or even optimal decisions for a given optimization model.

Decision sweeping does for decisions what parameter sweeping does for parameters, although the techniques for undertaking the sweeps are typically very different. In parameter sweeping we

1. Identify the parameter settings of interest (PSoIs).

2. Collect model solutions for the members of the PSoIs, and

3. Use the corpus of collected solutions in deliberation and decision making.

We saw an illustration in §1.5 with Converse's formula. Decision sweeping with an optimization model consists of:

1. Identifying the decisions of interest (DoIs),

2. Systematically collecting the DoIs, and

3. Using the corpus of collected DoIs in deliberation and decision making.

DoIs for an optimization model might include feasible decisions that are non-optimal but that still have high objective function values, or infeasible decisions that are close to being feasible and have objective function values superior to what is achieved with an optimal decision.

We shall be much concerned in what follows with methods for obtaining both PSoIs and DoIs, and with how we can put them to good use in decision making.

1.7   Summary of Vocabulary and Main Points

We have introduced a number of important concepts in this chapter, concepts that will be with us throughout, either explicitly or lurking in the background. Briefly, they are as follows.

1. The computational problem solving cycle.

   This is a high-level description of what goes on in practice. It is meant as general guide or indicator, not as a methodology to be slavishly followed. Perhaps the most important point of the framework is that there are different stages of activities, with different purposes, and that they overlap and are re-engaged iteratively. The cycle repeats itself as often as necessary to get the job done, the job being to make a good decision.

2. Constrained optimization models (COModels).

   We presented the Simple Knapsack model and a particular instance, the Eilon model, as a way of introducing constrained optimization models, which are the focal decision models of this book. COModels represent situations in which we seek to maximize (or minimize) a function of one or more decision variables, subject to satisfying stated constraints on the values of those decision variables. Certain associated terminology and concepts that go with the terminology will be important throughout the book:

   (a) Constrained optimization models will have parameters and decision variables, which are normally present in mathematical expressions that constitute the models.

   (b) Parameters are constants. Their values are determined by factors outside of the model, which simply takes them as givens.

   (c) Decision variables are entities whose values may be set by a decision maker or by a model solving algorithm.

   (d) A decision is any complete setting of values for all of the decision variables in a model.

   (e) In solving a COModel we seek to find decisions that maximize (or minimize) an objective function, while at the same time satisfying the constraints of the model.

   (f) A decision is said to be optimal if (i) it is feasible, and (ii) its objective function value is equal to or superior to every other feasible decision. (A COModel may have more than one optimal decision.)

   (g) Constraints for COModels, when expressed mathematically, typically are written in a conventional form

$$\text{LHS} \ \{=, \leq\} \ \text{RHS}$$

where: (a) LHS means "left-hand side" and stands for a mathematical expression involving the decision variables and, typically, various model parameters; (b) $\{=, \leq\}$ means that the constraint is either an equality constraint, so $=$ is chosen, or an inequality constraint, so $\leq$ is chosen; and (c) RHS means "right-hand side" and stands for a model parameter given a constant value. See Figure 1.2 for illustration. The Simple Knapsack model has one constraint of this form. Expression (1.2), in Figure 1.2 on page 7, represents a single constraint in which the LHS is

$$\sum_{i=1}^{n} w_i x_i$$

the constraint type is $\leq$, and the RHS is b. Expression (1.3), in Figure 1.2 on page 7, represents $2n$ constraints that limit the values of the individual decision variables to being either 0 or 1. These constraints are not easily expressed in the LHS $\{=, \leq\}$ RHS format. The meaning of expression (1.3) is, however, clear enough.

3. Post-solution analysis.

   Post-solution analysis is a step in the computational problem solving cycle. It consists of a variety of activities we undertake after a model has been designed, formulated, and solved. These activities may be directed primarily at testing and validating a model or at supporting deliberative decision making with the model. Post-solution analysis of COModels is a main focus of this book. Figure 1.5 on page 17 lists the main types of questions addressed during post-solution analysis.

4. Parameter sweeping.

   Parameter sweeping is the systematic re-solution of a model using changing parameter values from an identified setting of parameters of interest, the PSoI. It is a principal method or technique affording post-solution analysis.

5. Decision sweeping.

   Decision sweeping is the systematic collection of decisions for a COModel from an identified collection of interest, the DoIs. It is a principal method or technique affording post-solution analysis of COModels.

A point of emphasis here and throughout the book is that analytics are not static. To do business analytics we need to exercise implementations in order to explore our models and data. Obtaining a solution—a number from a model, a plot of a body of data, and so on—is hardly enough. Business analytics is above all about *post-solution analysis and exploration*. For this purpose, implementations should be seen as providing tools for exploration. The concepts introduced in this chapter will be with us throughout as we delve into business analytics for COModels.

1.8 For Exploration

1. The Converse formula tells us that if cities A and B are 100 kilometers apart, city A has a population of 400,000, and city B has a population of 85,000, then the breakpoint

is at about 31.55 kilometers. That is, someone living on a line between A and B and living within 31.55 kilometers of B will shop at B.

An implementation of the Converse formula, say in Excel or MATLAB or any of many other appropriate environments, lets us easily and quickly calculate a solution for different values of the parameters. The implementation of the model thus supports us in exploring various *post-solution analysis and exploration* questions, such as:

(a) How large would the population of city B need to be in order for the breakpoint to be at least 37.0?

(b) How small would the population of city A need to be in order for the breakpoint to be at most 40.0?

(c) Given fixed values of the other three parameters and a D_b value that has a margin of safety of 5 kilometers for maintaining profitability, over what range of r do we remain within this margin of safety?

In addition to these post-solution analysis questions, what other interesting questions are supported by the Converse Formula NetLogo model? Discuss and explain why they are interesting or likely to be useful. What about interesting questions that are not well supported by the model?

2. Critically assess the Converse formula as a model. What are its strengths? limitations? key assumptions? Under what conditions is it likely to be useful? not very useful?

3. The following passage appears at the end of §1.1.

> ... we might frame [Figure 1.1] as an updating or complement for the twenty-first century and new technology of Peirce's "The Fixation of Belief" [130].

Really? Do you agree? or not? Discuss.

4. What is the epistemological status of Converse's formula? It is evidently not a law of nature, although it is expressed like one. Does it support counterfactuals? Is it true?

5. Reimplement the Converse formula MATLAB model in a spreadsheet program, such as Excel. Compare and contrast the two implementation environments, MATLAB and spreadsheets, for this purpose.

6. Reimplement the Converse formula MATLAB model in Python. Compare and contrast the two implementation environments, MATLAB and Python, for this purpose.

7. Assuming we have a model for which extensive parameter sweeping results can be obtained, discuss how to use parameter sweeping to answer the questions of post-solution analysis, discussed in the chapter.

8. Assuming we have a model for which extensive decision sweeping results can be obtained, discuss how to use decision sweeping to answer the questions of post-solution analysis, discussed in the chapter.

1.9   For More Information

The book's Web site has an implementation of the Converse formula in NetLogo, *Converse Formula.nlogo*. The file is also available online at the NetLogo Modeling Commons `http://www.modelingcommons.org`.

The NetLogo home page is `https://ccl.northwestern.edu/netlogo/`. There, you can download NetLogo and access a rich corpus of information about NetLogo. The MATLAB code files *converse_formula.m* and *exercise_converse.m*, which support the example in §1.5, are available on the book's Web site in association with this chapter.

There is not a lot written on parameter sweeping. It is a well-entrenched concept that flourishes and is mostly maintained in the oral tradition. "A Framework for Interactive Parameter Sweep Applications" by Wibisono et al. [154] is a welcome exception, as is: `http://www.ll.mit.edu/mission/isr/pmatlab/pMatlabv2_param_sweep.pdf` (accessed 2015-07-09):

> *Parameter sweep applications* are a class of application in which the same code is run multiple times using unique sets of input parameter values. This includes varying one parameter over a range of values or varying multiple parameters over a large multidimensional space. Examples of parameter sweep applications are Monte Carlo simulations or parameter space searches.
>
> In parameter sweep applications, each individual run is independent of all other runs.

The book's Web site is:
`http://pulsar.wharton.upenn.edu/~sok/biz_analytics_rep`.

Chapter 2

Constrained Optimization Models: Introduction and Concepts

2.1 Constrained Optimization

Beginning informally, consider a familiar kind of constrained optimization problem. You need to decide where to have lunch. You have a consideration set of several conveniently named restaurants: Burgers, Pizza, Couscous, Caminetto, Salad, Sushi, Curry, Chinese, Asian Fusion, Schnitzel, Brasserie, and Greasy Spoon. You also have information on each of these restaurants that you consider relevant to your decision: price, distance, and health value. Let us assume that for each of these you have reliable scores on a (more or less) continuous scale. Price is in the local currency. Distance is measured in time or kilometers. Health value is based on a rating scheme published by your favorite source, Whelp.

Your decision will be an easy one if these are indeed all of the relevant criteria (they may not be, but for the sake of the discussion let's say they are) *and* one restaurant is better than all of the others on each of the three attributes. It is cheapest, nearest, and serves the healthiest fare of all the options you have available. Sometimes this happens, but not often. We are concerned with the latter cases, when decision making is not trivial. In these situations we normally have tradeoffs to make. Cheaper restaurants tend to be further away, healthier food tends to cost more, and so on. In these situations you need to find a way of thinking about your problem and then you need to find a good solution, given how you are conceiving of your situation.

There are several ways to frame how to think about your decision problem regarding lunch. *One* of these ways is to view it as a constrained optimization problem. Even as a constrained optimization problem, there are many ways you might think about it. You might seek to minimize the cost of your meal, while requiring that the travel time be, say, less than 10 minutes and the Whelp quality index be at least a 6. At this point you have, or are close to having, a *constrained optimization model* (COModel).

Four features are characteristic of constrained optimization models (COModels) in general and for your model in particular:

1. One or more decision variables.

   Your model has twelve decision variables, corresponding to the twelve restaurants under consideration. In general, you seek to find attractive values for each of your decision variables, and then you decide; that is you accept certain values and act accordingly. So if you decide to choose Curry as your most preferred value, you act by having lunch at the Curry restaurant.[1]

   A *mathematical program* is a formal, mathematical representation of a COModel. We shall be much concerned with COModels expressed as mathematical programs. It is customary in a mathematical program to represent the decision variables algebraically. For the restaurant problem we might map the restaurants to integers and represent the decision variables as $x_1, x_2, \ldots x_{12}$ with x_1 corresponding to Burgers, x_2 to Pizza, and so on. The intended interpretation is $x_i = 1$ means you eat at restaurant i ($= 1$, 2, $\ldots$, 12).

   It is also useful to define index sets for the decision variables. In the present case, I is the set of restaurants under consideration:

   $$I = \{\text{Burgers, Pizza, Couscous, Caminetto}, \ldots, \text{Greasy Spoon}\} \qquad (2.1)$$

   In this representation or notation, which we usually prefer, the decision variables are written as x_{Burgers}, x_{Pizza}, and so on. The two notations—set index and numerical index—are equivalent. The reader should be prepared to recognize both.

2. One or more objective functions.

   In the restaurant problem, we have provisionally chosen to minimize the cost of the meal. This makes cost of the decision variables our objective function. Mathematically, we can represent this as follows.

   $$f(x) = c_1 \times x_1 + c_2 \times x_2 + \ldots + c_{12} \times x_{12} \qquad (2.2)$$

   Equivalently, as is customary in mathematics, we drop the multiplication sign $\times$ when doing so does not lead to confusion:

   $$f(x) = c_1 x_1 + c_2 x_2 + \ldots + c_{12} x_{12} \qquad (2.3)$$

   Expression (2.3) is more compactly—and preferably—written as:

   $$f(x) = \sum_{i \in I} c_i x_i \qquad (2.4)$$

   or as

   $$f(x) = \sum_{i=1}^{n} c_i x_i \qquad (2.5)$$

---

[1] Alternatively, you might model your problem as having one decision variable taking on one of twelve possible values. There are pluses and minuses of each representation.

when there are n elements in I. Equation (2.4) uses the *set indexing* representation (the values of i take on the elements of I), while equation (2.5) uses what we have called the *numerical indexing* representation (the values of i are integers that ultimately are mapped to the elements of I). The two are equivalent once we match set elements to numerical indices. Oftentimes one will be more convenient than the other, so we shall be using both representations in the sequel.

In either of the two equivalent cases, c_i is the cost of restaurant i and x_i is its decision variable. The c_i's are examples of *parameters* in the model. They are determined antecedently and do not change during the solution seeking process. Unlike the decision *variables*, parameters are *constants*. What to count as a variable or as a parameter is a modeling decision. Very often, for example, we fix variables, treat them as parameters in a model, then revisit them once we have solved the model. We shall have very much to say about this *post-solution analysis* process throughout this book; it is a major theme.

3. Objective goal: maximize or minimize.

In the present case we seek to minimize the objective function, i.e., the cost of the meal. We write:

$$\text{Minimize} \quad z = \sum_{i \in I} c_i x_i \tag{2.6}$$

Note that whether we minimize or maximize is a matter of convenience. Multiplying the objective function by -1 converts the goal from one to the other. In our present case,

$$\text{Maximize} \quad z = -1 \times \sum_{i \in I} c_i x_i = \sum_{i \in I} -c_i x_i \tag{2.7}$$

will yield the same answer.

4. Constraints.

We will have four kinds of constraints in our running example: a constraint on distance traveled, on quality, on the number of restaurants we may choose, and on the possible values for the decision variables.

Proceeding in reverse order, the x_i (the decision variables) should be limited to having values of 0 or 1, with the intended meaning that if the value of the variable is 1 we eat at the corresponding restaurant and if it is 0 we do not. We write:

$$x_i = \{0, 1\}, \quad \forall i \in I \tag{2.8}$$

(The expression $\forall i$ conventionally means "for every i" or "for all i." $\forall$ is called the *universal quantifier*. It is used notationally in logic and throughout mathematics.)

Next we need to have a constraint limiting us to just one restaurant for lunch. We can express this as follows:

$$\sum_{i \in I} x_i = 1 \tag{2.9}$$

With every x_i having a value of either 0 or 1, this constraint guarantees that exactly one of the x_i's will get the value of 1.

On the matter of quality, we will need parameters q_i for the quantity score of restaurant i. These are analogous to the cost parameters in the objective function. The constraint on quality can be written as follows.

$$\sum_{i \in I} q_i x_i \geq 6 \tag{2.10}$$

This serves to guarantee that whatever restaurant is chosen, its quality score will be at least 6.

Finally, distance works like quality. The d_i parameters represent the distance to restaurant i. The constraint is written as follows.

$$\sum_{i \in I} d_i x_i \leq 10 \tag{2.11}$$

Pulling this all together, we represent our example COModel standardly as a mathematical program as follows:

$$\text{Minimize} \quad z = \sum_{i \in I} c_i x_i \tag{2.12}$$

Subject to:

$$x_i = \{0, 1\}, \forall i \in I \tag{2.13}$$

$$\sum_{i \in I} x_i = 1 \tag{2.14}$$

$$\sum_{i \in I} q_i x_i \geq b_1 \tag{2.15}$$

$$\sum_{i \in I} d_i x_i \leq b_2 \tag{2.16}$$

Notice that we have replaced the numbers with parameters b_1 and b_2 on the right-hand sides of the quality and distance constraints. The resulting expression of the model is said to be the *model class* or *model type* or *model template* or *model pattern* or *model schema* or even just *model form*. Because none of the parameters have been given values, the *model class* is in a sense a template or pattern or schema for a specific model. Once we declare values for I, the c_i's, the q_i's, the d_i's, and b_1 and b_2 we will have a *model instance* or *model token* that we could actually solve.

In discussing particular models we will normally present them in *model type* format (our preferred term), with at least some of the parameters left uninstantiated. Solving a model will of course require that a full instance or token of the model be specified. This is normally done by mapping the model's parameters to specific values.

A model type (or schema) is an abstraction—and a useful one—of many particular model tokens (instances). Further abstractions are possible and useful. Figure 2.1 presents a very general way of expressing mathematical programs.

In the figure, x is a set of one or more decision variables, f is a function on x returning a scalar value, $\mathcal{C}$ is the set of permitted solutions, and $x \in \mathcal{C}$ stands for the constraints of the model. A *decision*[2] for the model is *any* complete assignment of values to the elements of x. This is perhaps a counter-intuitive terminology but it is standard and we shall adhere to the standard. There will in general be very many decisions for a COModel. A *feasible decision* is one that also satisfies the constraints of the model. In the Figure 2.1 representation $\mathcal{C}$ is

---

[2]Or *solution*, but we prefer the term decision for reasons presented in Chapter 1.

$$\text{Maximize} \quad z = f(x) \qquad (2.17)$$

$$\text{Subject to:}$$

$$x \in \mathcal{C} \qquad (2.18)$$

FIGURE 2.1: Abstract mathematical programming formalization for constrained optimization models.

a set of decisions. Any member of this set is said to be feasible for the model. That is, if x' is any decision for the model and if $x' \in \mathcal{C}$ (i.e., if x' is a member of the set of feasible solutions), then we say that x' is a *feasible decision* or simply that x' is *feasible*.

Figure 2.1 shows the *constrained optimization abstract model formalization* and from it we can easily read off the four components of a COModel described above: decision variables, objective function, optimization goal, and constraints.

	Objective and all constraints linear	Objective function or some constraints nonlinear
All decision variables real valued	(1) linear program (LP)	(4) nonlinear program (NLP)
All decision variables integer valued	(2) integer linear program (ILP)	(5) nonlinear integer program (NLIP)
Some decision variables real valued, and some integer valued	(3) mixed integer linear program (MILP)	(6) mixed integer nonlinear program (MINLP)

FIGURE 2.2: A classification of mathematical programs.

2.2   Classification of Models

COModels, represented as mathematical programs, may usefully be classified with a 3×2 framework. See Figure 2.2. We will discuss each cell separately.

2.2.1   (1) Linear Program (LP)

Figure 2.3 presents a canonical form (there are others) for the linear programming model type (or schema).

$$\text{Maximize} \quad z = \sum_{j \in J} c_j x_j \tag{2.19}$$

$$\text{Subject to:}$$

$$\sum_{j \in J} a_{ij} x_j \leq b_i, \quad \forall i \in I \tag{2.20}$$

$$x_j \geq 0, \quad \forall j \in J \tag{2.21}$$

FIGURE 2.3: A canonical schema for linear programming models.

In Figure 2.4 expression (2.21) represents what are called the *variable constraints*. These are constraints on the values of individual variables; here all are constrained to be non-negative. A conventional assumption made in formalizing mathematical programs is that, unless otherwise stated, the decision variables have values that are real numbers, in distinction to being limited to integers. By definition, decision variables in an LP are real valued (and continuous).

Figure 2.4 presents a conforming LP token (instance) of the pattern in Figure 2.3, with three decision variables.

$$\text{Maximize} \quad z = 12.5x_1 + 11.0x_2 + 9.7x_3 \tag{2.22}$$

$$\text{Subject to:}$$

$$23.4x_1 + 19.0x_2 + 7.9x_3 \leq 34.6 \tag{2.23}$$

$$11.1x_1 - 17.4x_2 - 7.7x_3 \leq 12.0 \tag{2.24}$$

$$x_j \geq 0, \quad \forall j \in \{1, 2, 3\} \tag{2.25}$$

FIGURE 2.4: A token of an LP type in the canonical form of Figure 2.3, having three decision variables.

What makes a mathematical program (a COModel represented mathematically) an LP are three conditions:

1. All of the decision variables are real valued (and continuous). None are restricted to be integers or to be discontinuous in any way.

2. The objective function is a linear combination of the decision variables. The function sums the products of each of the decision variables and their coefficients; no decision variable is a function of another decision variable.

3. The constraints are similarly all linear combinations of the decision variables.

LPs may of course have constraints that are equalities or that are $\geq$'s. This is a detailed discussion which we defer to the next chapter.

2.2.2 (2) Integer Linear Program (ILP)

Integer linear programs (ILPs) are defined exactly as LPs are defined, except that the decision variables are restricted to be integers. Very often, the decision variables in an ILP are permitted the values of 0 or 1 only. Such cases are common in practice and are called *binary integer (linear) programs* and *0–1 integer (linear) programs*. Figure 2.5 presents in a canonical schema the so-called *Simple Knapsack model*, an example of a widely applied binary ILP model. We shall discuss it at length in Chapter 4. Note that the integrality requirements for the decision variables are stipulated in constraints (2.28).

$$\text{Maximize} \quad z = \sum_{i \in I} c_i x_i \tag{2.26}$$

Subject to:

$$\sum_{i \in I} w_i x_i \le b \tag{2.27}$$

$$x_i \in \{0, 1\}, \quad i \in I \tag{2.28}$$

FIGURE 2.5: A canonical schema for the Simple Knapsack model.

Recalling our restaurant problem, it should be clear that it, or rather its mathematical programming formulation, is an ILP with binary (0–1) decision variables.

2.2.3 (3) Mixed Integer Linear Program (MILP)

Mixed integer linear programs (MILPs) are defined exactly as LPs are defined, except that some of the decision variables are restricted to be integers and some are restricted to be continuous (and real) numbers. Figure 2.7 illustrates a commonly used technique. Suppose we have an LP (or for that matter any other mathematical programming schema). Let the *LP relaxation* of the model be as in Figure 2.5, that is, the model without the requirement that the decision variables be integers (shown in Figure 2.6).

$$\text{Maximize} \quad z = \sum_{i \in I} c_i x_i \tag{2.29}$$

Subject to:

$$\sum_{i \in I} w_i x_i \le b \tag{2.30}$$

FIGURE 2.6: LP relaxation of the Simple Knapsack model in Figure 2.5.

Suppose now that we want to add a disjunctive constraint. For example, we want to require that either

$$3x_1 + 4x_2 \le 18 \tag{2.31}$$

or

$$5x_1 + 3x_2 \le 15 \tag{2.32}$$

To express this more complex constraint we use the so-called "big M" method. We introduce a new parameter, M, and choose a value for it that is very large given the specific instance of the problem. We also introduce a binary integer variable, call it y. We then add the two constraints (2.35) and (2.36) of Figure 2.7.

$$\text{Maximize} \quad z = \sum_{i \in I} c_i x_i \tag{2.33}$$

Subject to:

$$\sum_{i \in I} w_i x_i \leq b \tag{2.34}$$

$$3x_1 + 4x_2 \leq 18 + yM \tag{2.35}$$

$$5x_1 + 3x_2 \leq 15 + (1 - y)M \tag{2.36}$$

$$y \in \{0, 1\} \tag{2.37}$$

FIGURE 2.7: A schema for a mixed integer linear program (MILP).

To see how this works we can reason as follows. The decision variable y must be set either to 0 or to 1. If it is set to 0 then constraint (2.35) amounts to expression (2.31) and constraint (2.36) will be satisfied because it amounts to

$$5x_1 + 3x_2 \leq 15 + M \tag{2.38}$$

and M is a very large number. On the other hand, if y is set to 1, then constraint (2.35) amounts to

$$3x_1 + 4x_2 \leq 18 + M \tag{2.39}$$

and is satisfied because M is so large. Further, constraint (2.36) now amounts to expression (2.32). In sum, it will be the case that there is a requirement for either (2.31) or (2.32), just as we had intended. By adding constraints (2.35) and (2.36) we have effectively added the requirements specified by (2.31) or (2.32), without adding any other constraint.

2.2.4   (4) Nonlinear Program (NLP)

Nonlinear programs (NLPs) are defined exactly as LPs are defined, except that either the *left-hand side* of at least one constraint or the objective function is nonlinear. Recall from Chapter 1 that by the left-hand side of a constraint we mean this. Constraints are expressed as inequalities ($\leq$, $=$, $\geq$) in which the right-hand side as written (see the examples above) is a scalar number and the left-hand side is a function of one or more of the decision variables. In a linear programming model, the left-hand side of every constraint is a linear function of the decision variables. That is, it can be expressed by multiplying each variable by a coefficient and summing the products. If the left-hand-side function or the objective function cannot be expressed this way, then we have a nonlinear programming problem (assuming that the decision variables are all continuous).

Multiplying decision variables together and applying functions to them individually are two principal ways in which nonlinearities are created. We give a simple example in the next section, on nonlinear integer programming.

2.2.5 (5) Nonlinear Integer Program (NLIP)

Nonlinear integer programs (NLIPs) are defined exactly as nonlinear programs (NLPs), except that all the decision variables are constrained to be integers. The so-called *maximum diversity problem*, having many applications in practice, will serve as an example. It is quadratic in its objective function, making it nonlinear, and it has 0–1 (binary) decision variables. It is also perhaps the simplest possible example for the case in point.

The problem may be described as follows [57, 118]: Given a set N of n elements (typically these are small geographic areas), find K a subset of N with k elements such that the sum of the pairwise distances among the elements of K are maximized. Roughly put, find k elements of N that are maximally far apart from one another. The problem may be concisely stated as a zero-one integer programming problem with a quadratic objective function:

$$\text{Maximize } z = \sum_{i=1}^{n} \sum_{j=1}^{n} d_{ij} x_i x_j \qquad (2.40)$$

Subject to:

$$\sum_{i=1}^{n} x_i = k \qquad (2.41)$$

$$x_i \in \{0, 1\}, \quad i = 1, \ldots, n \qquad (2.42)$$

Here, d_{ij} is the distance (under a suitably chosen metric, such as Euclidean distance) between elements i and j of N.

2.2.6 (6) Mixed Integer Nonlinear Program (MINLP)

Mixed integer nonlinear programs (MINLPs) are defined exactly as other nonlinear programs, except that they have decision variables that are continuous and some decision variables that are integers. We repeat the point that if the objective function is nonlinear in the decision variables or if any of the constraints are nonlinear in the decision variables, then the program is nonlinear.

2.3 Solution Concepts

Given a mathematical programming model for a real world constrained optimization decision problem, we standardly define the set of optimal decisions for the model and seek a member of the set. Finding such a decision constitutes what is called *solving* the optimization problem. This standard terminology may invite confusion. Recall that a decision (aka: solution) to an optimization *model* (including any mathematical programming model) is simply any complete assignment of values to its decision variables. A decision for an optimization *problem* posed by an optimization model is a best feasible solution to the model. (This is distinct from solving an associated real world, and hence not formalized, optimization problem. It is important to distinguish real world problems or situations from models of them. Our focus at present is on the models. We shall not, however, forget that the purpose of modeling is to be useful for real world problems.)

It will serve the cause of clarity to formalize a bit this notion of solving an optimization problem. Let our given constrained optimization model have the form shown in Figure 2.1, repeated here:

$$\text{Maximize} \quad z = f(x) \tag{2.43}$$

$$\text{Subject to:}$$

$$x \in \mathcal{C} \tag{2.44}$$

Recall that in this representation x is a set of one or more decision variables, f is a function on x returning a scalar value, and $\mathcal{C}$ is the set of feasible decisions for the model. We then standardly define the set of globally optimal decisions for the model, $\mathcal{O}^G$, as

$$\mathcal{O}^G \stackrel{\text{def}}{=} \{x | x \in \mathcal{C} \wedge \forall y (y \in \mathcal{C} \to f(x) \geq f(y))\} \tag{2.45}$$

Expressed verbally, expression (2.45) may be read "$\mathcal{O}^G$ is the set of decisions, such that each individual decision, x, is feasible ($x \in \mathcal{C}$) and for any decision, y, if y is feasible, then the objective function value for x is at least as large as that for y ($\forall y (y \in \mathcal{C} \to f(x) \geq f(y))$)." Put otherwise, expression (2.45) defines the set of globally optimal decisions for a model, $\mathcal{O}^G$, as the set consisting of every decision for the model that (i) is feasible and (ii) has an objective function value at least as large as every other feasible solution. This definition, of course, assumes that our optimization goal is to maximize.

A globally optimal decision is at least as good as every other decision. A locally optimal decision, $\mathcal{O}^L$, is a decision that is at least as good as every other decision in its suitably defined neighborhood. Formally,

$$\mathcal{O}^L \stackrel{\text{def}}{=} \{x | (x \in \mathcal{C} \wedge \forall y (y \in \mathcal{C} \wedge y \in \mathcal{N}(x)) \to f(x) \geq f(y))\} \tag{2.46}$$

$\mathcal{N}(x)$ is the set of decisions that are in the neighborhood of decision x. The definition of $\mathcal{N}$ will generally be model or problem specific. We discuss this and related matters in the sequel. Note, however, the potential for complexity. A set of locally optimal decisions, $\mathcal{O}^L$, need not be unique for a problem. Potentially, $\mathcal{O}^L_i \neq \mathcal{O}^L_j$ for distinct indexes i and j. If this is possible, and often it is, we must speak of <u>a</u> set of locally optimal decisions, rather than <u>the</u> set.

A set of optimal decisions, $\mathcal{O}$ (whether global or local), associated with any specific model may be empty, may have exactly one member, or may have any larger number of members.[3] The *optimization problem* associated with an optimization model is twofold: (i) determine whether $\mathcal{O}$ is non-empty, and (ii) if it is, then find one member of it. This is what is standardly meant by solving an optimization model: Find and produce an optimal decision if there is at least one, or show that there are no optimal decisions for the model. Of course, it is generally preferred to find a global, rather than merely local, optimum. Whether this can be done depends on characteristics of the problem. In general, if there are multiple local optima in the problem at hand, then there can be no guarantee of finding a globally optimal solution, although that may be our goal. For many kinds of problems, however, it can be shown that any local optima are also globally optimal. This is true in particular for linear programs.

In summary, the optimization methods we have (more on this below) can only find, at best, local optima or approximations thereof. If and only if the underlying problem is known to have only global optima then we can be assured that any local optimum our solver discovers is also globally optimal. With this understanding we will proceed by using $\mathcal{O}$ to mean $\mathcal{O}^L$, for that is the most we ever get from our solvers.

So formulated, a generalization of the optimization problem immediately presents itself: (i) determine whether $\mathcal{O}$ is non-empty, and (ii) if it is, then find all members of it. This generalization is an instance of what we call a *solution pluralism (SP)* problem. Such problems

---

[3]It may even be uncountably infinite.

are characterized by a request for a collection, a plurality, of solutions. When $\mathcal{O}$ is large it will be impractical to find all of its members. In this case we might request, for example, a sample of at least a certain size from $\mathcal{O}$. This, too, is a solution pluralism problem because we are asking for a plurality (if it exists) of solutions.

We are concerned in this book with a further generalization. Given an optimization model we may define $\mathcal{P}$ (in distinction to $\mathcal{O}$) as a set of *decisions of interest (DoIs)* that may, and normally would, include non-optimal solutions. A template or schema for such a definition may be stated as follows:

$$\mathcal{P} \overset{\text{def}}{=} \{x | (x \in \mathcal{C} \wedge \phi) \vee (x \notin \mathcal{C} \wedge \psi)\} \tag{2.47}$$

Expressed verbally, expression (2.47) defines the body (corpus) of decisions of interest for a model, $\mathcal{P}$, as the set consisting of every solution to the model that (i) is feasible and (ii) has the property ϕ or (iii) is infeasible and (iv) has the property ψ. This template definition, of course, has not specified what the two properties ϕ and ψ are. That is a matter for users and analysts to determine, as we shall discuss. We make matters specific and concrete in Chapter 13.

We call $\{x | (x \in \mathcal{C} \wedge \phi)\}$ the set of *feasible decisions of interest* or FoIs and $\{x | (x \notin \mathcal{C} \wedge \psi)\}$ the set of *infeasible decisions of interest* or IoIs.

Finally, $\mathcal{P}$ will in many cases be quite large or even infinite, presenting practical obstacles to its use. We introduce in consequence the concept of a *sampling function* on $\mathcal{P}$ that can return a suitable subset of $\mathcal{P}$. We call such a function S. It yields a subset of $\mathcal{P}$. Once more, we emphasize that our construct, here $S(\mathcal{P})$, is a schema or template. It may be instantiated, tokenized, in many different ways, and serves mainly to organize the discussion of such ways, discussion that continues throughout the remainder of this book.

2.4   Computational Complexity and Solution Methods

Computational complexity is a highly technical and challenging subject with a large associated literature. As relevant and as interesting as it is, we can safely ignore much of the topic for present purposes. Approximate accuracy is sufficient for much of what we need to cover. This should be kept in mind while reading what follows.

1. Optimization problems fall into two theoretically fundamental and distinct categories: tractable and intractable.

2. Tractable optimization problems are often reliably solved (an optimal solution produced if there is one) for very large problems, having say millions of decision variables and millions of constraints, using commercial off the shelf (COTS) computing resources.

3. Intractable optimization problems, said to have high computational complexity, in theory cannot be solved reliably unless their scale is small. For some common industrial applications, problems having as few as 30 variables and 20 constraints are beyond the reach of existing COTS computing resources. Larger problems are beyond the reach of foreseeable high performance research computing systems. Intractability, however, is assessed for the "worst case." Specific problems may in fact be solved rather easily in large scale. Identifying such problems is an empirical matter.

4. Considering the six categories in Figure 2.2 of optimization models expressed as mathematical programming problems, only linear programming (1) is entirely tractable. Some of the problems in (4) NLP—those that are *convex optimization* problems[4]— are tractable, although generally more difficult than LPs. All others are, in their worst or general cases, intractable.

5. Two kinds of solution methods, implemented as *solvers,* are available for optimization problems: *exact methods* and *heuristics.*

6. Exact methods will recognize an optimal solution (i.e., a locally optimal solution) if they find one, and will report it as such.

   Linear programming optimization problems are routinely solved in very large scale with either of two (families of) exact solution methods: the *simplex algorithm* and *interior point methods.*

   Exact methods also exist for the intractable problems in the other five categories of optimization models in Figure 2.2. These include, prominently, methods based on *branch-and-bound* search procedures or on *dynamic programming* approaches. Solvers using branch-and-bound methods are generally available for very many specialized classes of mathematical programs in categories 2–6 of Figure 2.2. See for example the GAMS library of solvers, `http://www.gams.com`, which includes solvers for such specialized cases as integer programs with quadratic constraints (but linear objective functions). Discussion of these methods is beyond the scope of this book.

7. It is important to note that exact methods applied to intractable problems may well fail to find optimal decisions because of scale and problem complexity. This is in fact very often an issue in industrial applications: The available exact methods simply fail to provide optimal solutions in acceptable time and the decisions they do find are unsatisfactory. Mitigating this circumstance is that branch-and-bound solvers will report a "duality gap," that is, an upper and lower bound on the optimal solution, when they fail to find an optimal solution. Nevertheless, in many cases of practical importance exact solvers are unsatisfactory because of problem scale.

8. Heuristic methods are not exact methods in that while they may find optimal, even globally optimal, decisions they have no way of recognizing them as such. There are a limited number of results, applying to a small number of heuristics, that prove their worst case performance to be within a small tolerance of optimality. For nearly all real world applications, however, we are left to trusting our heuristics based on experiments and general experience.

9. Exact methods fall short for very many real world applications, leaving use of heuristics to solve optimization problems as the best option. They are the best we can do and so we study them here.

10. Our discussion so far in this section has focused on optimization problems. Solution pluralism problems normally include optimal solutions in their defined solutions of interests. When this occurs, they can be no more tractable than their associated optimization problems. In consequence, whenever heuristics are mandated for optimization problems, they will often be useful if not necessary for any associated solution pluralism problems.

---

[4]In convex optimization or programming, the objective function is convex and the set of feasible solutions is also convex. The technical details of this important topic are beyond the scope of this book, but quadratic functions are convex and quadratic constraints lead to convex feasible sets. Good textbooks are available, among them [19].

11. Even when exact methods are available for solving an optimization problem, they are normally of limited value for addressing associated solution pluralism problems. (Linear programming is an exception.) Heuristics, as we shall see, will often be most useful in these cases as well.

2.5 Metaheuristics

There are heuristics specific for particular classes of COModel. Some are very effective and we shall see examples in this book. A principal limitation is just that they are specific, not general. Fortunately we do have general heuristics. These are abstractions of or schema for heuristics that are created by tailoring a general heuristic to a specific problem, much in the way that a mathematical programming template or pattern may be instantiated to a full specific problem with its parameters assigned actual values.

General heuristic patterns go by the name of *metaheuristics*. They may fairly be classified as falling into two broad categories: local search metaheuristics and population based metaheuristics. We shall have much to say about both kinds in subsequent chapters. They are in fact so important that we give a very basic introduction here in this section. The concepts, like the other concepts in this chapter, will be immediately useful and used.

2.5.1 Greedy Hill Climbing

Optimization heuristics search the space of solutions, looking for good ones. What makes a search procedure a heuristic is that it is not guaranteed to find a good solution even if there is one. In saying something is a heuristic we presuppose that it has been tested as such and can be warranted as worthy of attention.

The *exploration-exploitation* distinction is something to keep in mind whenever heuristics are discussed. When we search heuristically we need both. We need to explore for new information and we need to exploit the information we have. The challenge is how to allocate our efforts between the two distinct purposes. This is another optimization problem (or schema for such) and there is no general answer for it. Different heuristics will use different allocations and in the end we can only go by experience in deciding how best to search.

Purely random search lies at the exploration extreme of the exploration-exploitation range of approaches to heuristic search. It consists simply of examining at random a number of solutions and choosing the best encountered. This is a terrible approach in almost all cases of practical importance, unlikely in many cases to find even a feasible solution, let alone a good or optimal one.

The place to begin our discussion, or to go next after dismissing random search, is at the other extreme, extreme exploitation. The name for this is *greedy hill climbing*. There is—at least as yet—no agreed-upon measure to place an algorithm along a line running between pure exploration and pure exploitation. Other dimensions may be relevant, too. Still, it is useful to talk this way, provided we keep in mind our inability to cash in the talk in any very precise way. The fact is that we can agree that random search is extremely explorative, at the expense of exploitation of any findings, and that greedy hill climbing is extremely exploitative with regards to its finding and does very little exploration. So let us now see how it works.

In saying "greedy hill climbing" we of course speak metaphorically. The metaphor is nonetheless a live one and we can exploit it for gaining understanding. Imagine you are situated in some place, say a wilderness. You need to get to the highest point in it, but you do not have access to visual information. You cannot see distant peaks. In fact, you are blindfolded or operating in a fog on a moonless night. What would you do? (Screaming for help doesn't count.) Very likely you would do something like what Figure 2.8 describes.

---

1. Obtain a starting position.

   You have to start somewhere so you start where you find yourself.

2. Define the neighborhood within which you will explore.

   Because vision is of little avail, you explore one step away from where you are, in multiple directions.

3. Set your exploration budget.

   This is the level of effort you are willing to expend on a given exploration step.

   Let us say eight steps evenly placed around you, in effect: N, NE, E, SE, S, SW, W and NW.

4. Explore.

   Probe the neighborhood until your exploration budget is exhausted, noting what you find.

5. Exploit.

   If you have found at least one point that improves your current position go to the best such point, resolving ties randomly, then continue at step 4; otherwise (no improvement is found so you) quit. You are at a local optimum.

FIGURE 2.8: Greedy hill climbing procedure.

A few comments. First, many variations on this basic hill climbing idea are possible. For example, neighborhood definitions and exploration budgets may be dynamic and change during execution of the procedure. We'll stick to the KISS principle (keep it simple, stupid) for now.

Second, the hill climbing procedure has an exploration-exploitation *loop* (steps 4 and 5). It is something, in various different forms, that we see in many other search procedures. This raises all sorts of policy and design issues for search procedures. How much exploration should be done before making an exploitation move? Having made an exploitation move, how committed to it should the procedure be? And so on. These will be themes that get addressed as we proceed.

Third, in converting this metaphor into a computer algorithm we will need to implement a procedure for searching the neighborhood. For example, if our solutions have n decision variables, $x_1, x_2, \ldots, x_n$, then we might explore a neighborhood of $2n$ variations on a given solution, n in which each of the n variables is increased slightly one at a time and n in which they are decreased. There will be cases for which defining the neighborhood of a solution is problematic. We will see examples in subsequent chapters.

Fourth, the usual practice in implementing a greedy hill climbing algorithm is to start with a random solution at step (1). There is usually no better option realistically available. Random starts, however, climb only the hill that happens to be chosen randomly. When multiple local optima are present (the usual case), a single ascent of a greedy hill climbing algorithm is unlikely to find a high quality local optimum. For this reason the normal practice is to repeat the algorithm hundreds or thousands of times, retaining the best solution found on each run. Over many runs one might hope to find a high quality solution.

Greedy hill climbing is conceptually foundational for heuristic search, yet it performs poorly on many problems of practical interest and in fact is rarely implemented and tested in practice. Variants very close to it are widely used and used successfully. We shall have much to say about them, especially in Chapter 11. In the next section we introduce one of these important alternatives.

2.5.2 Local Search Metaheuristics: Simulated Annealing

Greedy hill climbing is a form, about the simplest possible, of a *local search* metaheuristic. Search is conducted by examining a neighborhood of solutions for an incumbent solution, then possibly replacing the incumbent with a neighbor, depending on what is found.

What makes greedy hill climbing greedy is that it always accepts—replaces the incumbent solution with—the best observed solution in the neighborhood, if it is better than the incumbent. Why not? The name is a hint. "Greedy" suggests disapproval, deficiency, and was chosen for that reason. "Myopic" would have worked just as well. The fundamental flaw with greedy hill climbing is that it cannot explore a hill other than the one it lands on at the outset. Moreover, it has no way of escaping a flat region, even if it is locally minimal rather than maximal.

What to do? The general answer is to be more explorative. Simulated annealing is a simple variant in this direction (although not the simplest as we shall see in Chapter 11). Simulated annealing works exactly like greedy hill climbing with one exception. If the best discovered solution in the neighborhood is worse than the incumbent solution, simulated annealing will with a certain probability accept that solution and replace the incumbent. The probability of doing so declines with each passage through the explore-exploit loop in Figure 2.8 (steps (4–5)). Remarkably, with some tuning of the acceptance probability and its rate of decline, simulated annealing works very well on very many difficult, theoretically intractable, problems of practical import. It is very widely used and is absolutely a key member of the business analytics armamentarium. We discuss simulated annealing in some detail in Chapter 11.

2.5.3 Population Based Metaheuristics: Evolutionary Algorithms

Population based metaheuristics characteristically work by first randomly creating a plurality of solutions (called a *population* by analogy with biological evolution), and then modifying the population's solutions in light of information received during the search process. What makes the population aspect essential is that modifications to the population depend in part on the constituents of the population at the time of modification. The solutions in the population in effect communicate with each other during the search process. The heuristic cannot be duplicated by multiple independent trials with individual solutions.

Evolutionary computation, said to be "inspired" by biological evolution, is a prototypical case of a population based metaheuristic. In a typical form of evolutionary computation, a population is created at initialization consisting of a certain number (say 100) of randomly created solutions, then a loop is entered for a certain number of *generations* (say 2000; the name is by analogy with biological evolution). During each pass through the loop, every solution in the current population is evaluated and receives a score based on how well it performs on the problem at hand. This is called the solution's *fitness* by analogy with fitness in biological evolution. Then a subset of the population, biased in favor of those solutions with better fitness scores, is chosen to serve as "parents" for the next generation. This is called *selection* by analogy with selection in biological evolution. A new population is created from these parents both by copying them exactly and by systematically perturbing (changing) them with what are called *genetic operators*. These may be of various kinds, prominently *mutation* and *recombination* by analogy with biological evolution. Once the new population is fully created, it replaces the existing population and a single passage through the loop (of one discrete "generation") is completed. This continues until the requisite number of passages through the loop (the number of generations) has been achieved. A running record of the best solutions encountered is consulted to report the results of the run.

As in the case of simulated annealing, evolutionary algorithms of various kinds are very widely used in practice and routinely produce excellent results. Knowledge of them is also absolutely a key member of the business analytics armamentarium. Chapter 12 focuses specially on evolutionary algorithms, although we will encounter them often in other chapters.

2.6   Discussion

This chapter has presented a litany of concepts and conceptual tools that we will draw upon for the remainder of the book. With this essential background in place, we turn now to exploring, both with exact methods and heuristics, a series of mathematical programming models that have been found to be very useful in practice and are widely applied.

2.7   For Exploration

1. Binary decision variables are often used to represent logical relationships. For example, if x and y are binary variables, with 1 for true and 0 for false, then

$$x \leq y$$

   has the interpretation or meaning "If x, then y." Explain.

2. Using binary variables, state one or more equations ($=$, $\leq$, or $\geq$) that mean "x if and only if not y."

3. Using binary variables, state one or more equations ($=$, $\leq$, or $\geq$) that mean "x or y," with "or" taken inclusively to mean "one or the other or both."

4. Using binary variables, state one or more equations ($=$, $\leq$, or $\geq$) that mean "x or y," with "or" taken exclusively to mean "one or the other but not both."

5. Returning to the searching in the wilderness analogy we relied upon in Figure 2.8 and our discussion of greedy hill climbing, in what sort of landscape would greedy hill climbing work well? work poorly?

6. Find or formulate examples of mathematical programming models for each of our six categories. In each case, explain the model and discuss the application that motivated it.

2.8 For More Information

How to Solve It: Modern Heuristics by Michalewicz and Fogel [123] has a nice—clear and insightful—discussion of hill climbing, simulated annealing and very much else in the bailiwick of... modern heuristics. The *Handbook of Metaheuristics* [55] accessibly discusses in depth the main kinds of metaheuristics now in use.

The classic work on computational complexity is [54]. It remains an excellent and essential point of departure for the subject. A standard textbook on convex optimization or programming is [19].

Chapter 11 is devoted to dicussion of local search metaheuristics, including simulated annealing and threshold accepting algorithms, which are even simpler variants on greedy hill climbing than simulated annealing. Chapter 12, as noted above, focuses especially on evolutionary algorithms, although we will encounter them often in other chapters.

Chapter 3

Linear Programming

3.1   Introduction

The presence of a real-world constrained optimization problem occasions, at a high level, three tasks:

1. Formulate

   Conceive and represent the problem, translate it into a useful formalism—*model* it in short—so that it can be implemented and solved.

2. Implement and solve

   Convert the problem as modeled to a form in which it can be solved, and then solve it.

3. Conduct post-solution analysis and make a decision

   Undertake post-solution investigations with and based upon the solved model, and then decide what to do with regard to the original real-world problem.

These tasks as stated are certainly idealizations of actual modeling processes. There is inevitably much iteration among them and boundaries are often fuzzy. Even so, they constitute a useful *modeling process framework*. (See Chapter 1 for a somewhat more elaborated version of the framework.)

Our focus in this chapter is on optimization problems that may be formulated as linear programs (LPs). Further, and in regard to the framework, our primary concern is with post-solution analysis of linear programming models. We introduce the subject in the next section by working in the context of a specific example.

3.2   Wagner Diet Problem

Suppose we are given the formulated LP model, shown in Figure 3.1. The model seeks to optimize the mixture of three different grains to be fed to livestock. It minimizes the cost of purchase, subject to constraints on delivered nutrition. Buying no grain is cheaper than buying some grain, but then the livestock will starve. The problem is to find the cheapest way to feed the livestock in a satisfactory fashion.

$$\text{Minimize} \quad z = 41x_1 + 35x_2 + 96x_3 \tag{3.1}$$

Subject to:

$$2x_1 + 3x_2 + 7x_3 \geq 1250 \tag{3.2}$$

$$1x_1 + 1x_2 + 0x_3 \geq 250 \tag{3.3}$$

$$5x_1 + 3x_2 + 0x_3 \geq 900 \tag{3.4}$$

$$0.6x_1 + 0.25x_2 + 1x_3 \geq 232.5 \tag{3.5}$$

$$x_1 \geq 0, x_2 \geq 0, x_3 \geq 0 \tag{3.6}$$

FIGURE 3.1: Wagner Diet linear programming model [151].

Called the Wagner Diet model after [151, page 79] (intended for livestock), it is indeed representative of a frequently employed use of linear programming, which is to optimize a mixture subject to minimality and/or maximality constraints. In this particular case, we have the basic form:

$$\text{Minimize} \quad z = c^T x \tag{3.7}$$

Subject to:

$$Ax \geq b \tag{3.8}$$

$$x \geq 0 \tag{3.9}$$

In either case, the interpretation is as follows:

1. Decision variables

    (a) x_1, units of grain 1 to be used in the mixture
    (b) x_2, units of grain 2 to be used in the mixture
    (c) x_3, units of grain 3 to be used in the mixture

    Alternatively, x_j, units of grain j to be used in the mixture, $j \in \{1, 2, 3\}$.

2. Objective function coefficients

    (a) c_1, cost per unit of grain 1
    (b) c_2, cost per unit of grain 2
    (c) c_3, cost per unit of grain 3

    Alternatively, c_j, cost per unit of grain j, $j \in \{1, 2, 3\}$.

3. Right-hand-side (RHS) values

    (a) b_1, net amount of nutrition ingredient A
    (b) b_2, net amount of nutrition ingredient B

(c) b_3, net amount of nutrition ingredient C

(d) b_4, net amount of nutrition ingredient D

Alternatively, b_i, net amount of nutrition ingredient i, $i \in \{A, B, C, D\}$.

4. Constraint matrix coefficients

a_{ij}, amount of nutrition ingredient i per unit of grain j, $i \in \{A, B, C, D\}$, $j \in \{1, 2, 3\}$.

Upon solving the model we find that at optimality $z = 19550$, $x_1 = 200$, $x_2 = 50$, and $x_3 = 100$. It is, in consequence, a settled matter: If we purchase these amounts of each of the three grains, mix them, and feed them to our livestock, then we will (according to the model) have fed the livestock a minimal cost nutritionally sound diet (for the duration of the feed).

At this point there may be occasions in which we would simply implement this optimal solution by obtaining and mixing the identified optimal amounts of grains, and then proceed to feed the livestock. If experience has shown that this and similar models are reliably high performers and if we continue to view the model's assumptions as known with great accuracy, then we may well be warranted in proceeding to implement the (model's) optimal solution.

Very often, however, we are not warranted in proceeding without deliberation to implement an optimal solution discovered for the model. These situations require *post-solution analysis* of some sort. We give a comprehensive, although high-level, discussion of post-solution analysis in the next section.

3.3 Solving an LP

Figure 3.2 shows how to implement the Wagner Diet LP model in Excel. Rows 15 through 29 provide documentation on the named ranges in the model. For example, in row 18 we learn that the name `dvars` refers to the range A3:C3 on the worksheet, that these cells hold the values of the decision variables for the model.

The displayed values of the decision variables, `x_1`, `x_2`, and `x_3`, are those of an optimal solution found by solving the model. Initially, they were set to 2s. The value of 19550 shown in cell A7 is the resulting objective function value at optimality. The underlying formula for cell A7 is `=SUMPRODUCT(dvars,objcoeffs)`. The LHS values are computed in a similar fashion. The formula underlying cell F10 is `=SUMPRODUCT(A10:C10,dvars)` and the formulas for cells F11:F13 are analogous. And that is pretty much all there is to say regarding the implementation in Excel of the Wagner Diet model.

Figure 3.3 on page 47 shows the Excel Solver settings for solving the Wagner Diet model as implemented in Figure 3.2. Notice how the use of named ranges, documented in Figure 3.2, leads to this simple and elegant filled-in form of directives to the solver.

If we now go ahead and click the Solve button, the Excel Solver will solve the LP and alert us when finished with the execution report as in Figure 3.4. The figure shows the report after we have elected to keep the Solver solution and selected the three reports available, Answer, Sensitivity, and Limits. Clicking the OK button causes Excel to generate the three reports as new worksheets in the workbook and to set the optimal values it has found on the model formulation, as shown in Figure 3.2.

	A	B	C	D	E	F	G	H
1								
2	x_1	x_2	x_3					
3	200	50	100					
4	c_1	c_2	c_3					
5	41	35	96					
6	min z =							
7	19550							
8								
9	subject to:					LHS values		RHS values
10	2	3	7	=	Ingredient_A	1250	>=	1250
11	1	1	0	=	Ingredient_B	250	>=	250
12	5	3	0	=	Ingredient_C	1150	>=	900
13	0.6	0.25	1	=	Ingredient_D	232.5	>=	232.5
14								
15	c_1	='Wagner Diet'!$A$5			objective function coefficient			
16	c_2	='Wagner Diet'!$B$5			objective function coefficient			
17	c_3	='Wagner Diet'!$C$5			objective function coefficient			
18	dvars	='Wagner Diet'!$A$3:$C$3			decision variables			
19	Ingredient_A	='Wagner Diet'!$F$10			LHS value, ingredient A			
20	Ingredient_B	='Wagner Diet'!$F$11			LHS value, ingredient B			
21	Ingredient_C	='Wagner Diet'!$F$12			LHS value, ingredient C			
22	Ingredient_D	='Wagner Diet'!$F$13			LHS value, ingredient D			
23	LHS	='Wagner Diet'!$F$10:$F$13			LHS values			
24	objcoeffs	='Wagner Diet'!$A$5:$C$5			objective function coefficients			
25	objective	='Wagner Diet'!$A$7			objective value			
26	RHS	='Wagner Diet'!$H$10:$H$13			RHS values			
27	x_1	='Wagner Diet'!$A$3			decision variable			
28	x_2	='Wagner Diet'!$B$3			decision variable			
29	x_3	='Wagner Diet'!$C$3			decision variable			

FIGURE 3.2: Excel formulation of the Wagner Diet problem.

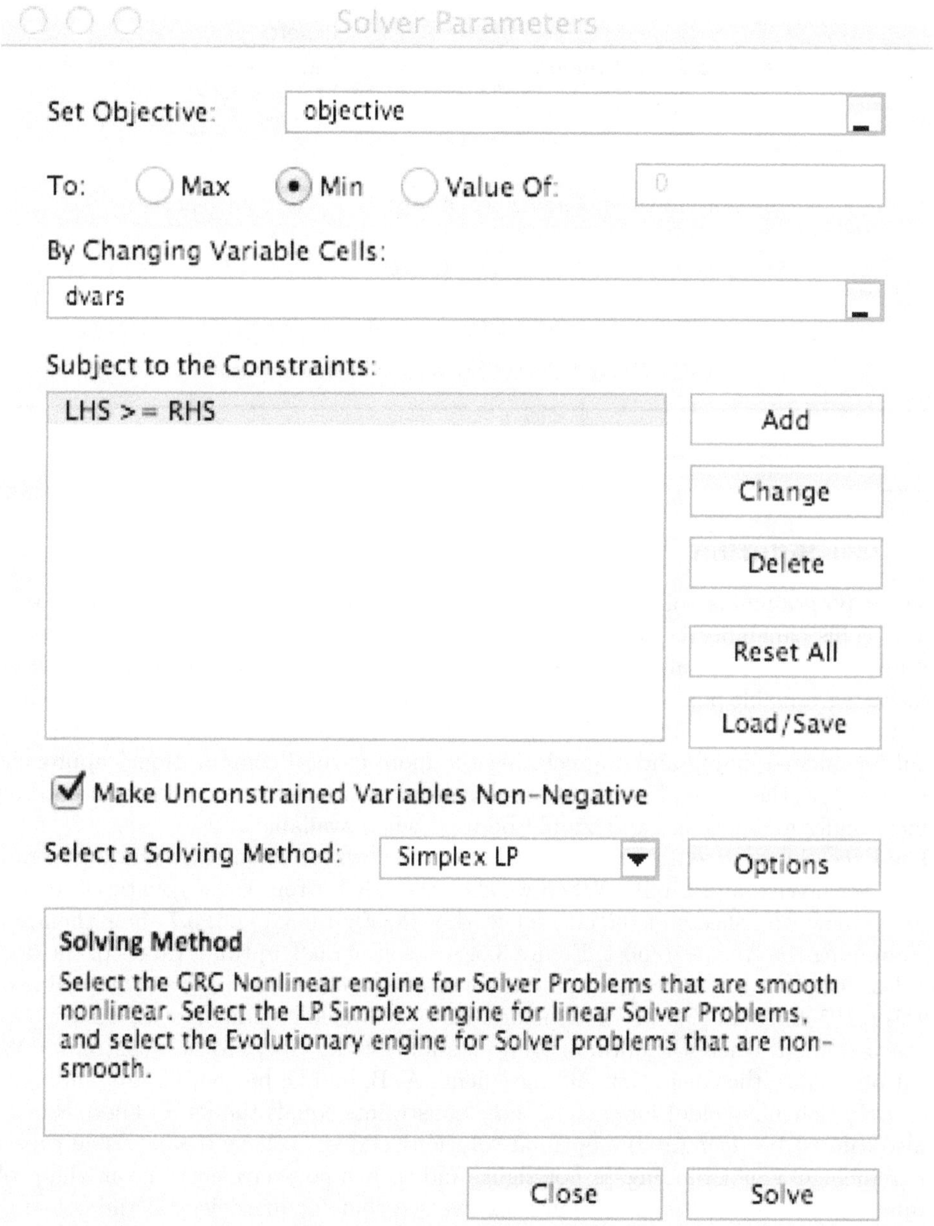

FIGURE 3.3: Excel Solver setup for Wagner Diet problem.

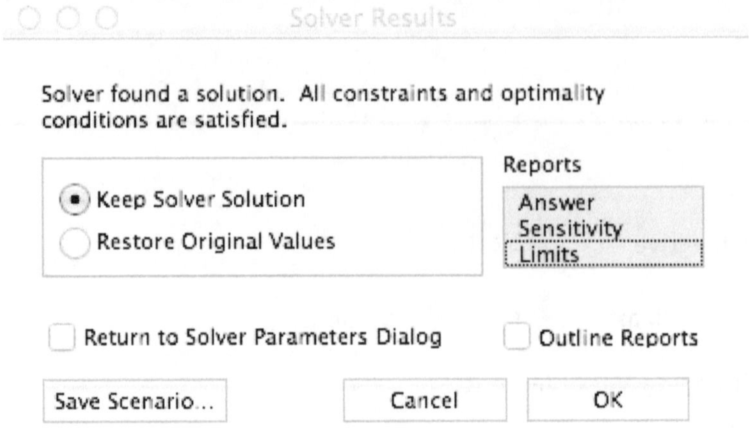

FIGURE 3.4: Excel Solver execution report.

3.4   Post-Solution Analysis of LPs

Linear programming solvers are able to generate valuable information for post-solution analysis. This capability is unique to LPs (and its solvers) in virtue of the mathematical structure of linear programs. We shall, in the sequel, explore general computational means of generating valuable post-solution analysis information. Here, in this chapter, we will focus on post-solution analysis information that is specific to LPs and their solvers. This will be helpful for understanding and appreciating the more general computational approaches to come (based on the principle of solution pluralism). Moreover, the LP-specific information is conceptually foundational and should be used when available.

The Excel Solver execution report in Figure 3.4 shows selection of three solution reports, Answer, Sensitivity, and Limits. When we click the OK button, Excel generates these three reports as new worksheets in the current workbook. Figure 3.5, 3.6, 3.7 show them.

Examining the Answer report, Figure 3.5, we see the final, optimal values of the decision variables, 200 for x_1, 50 for x_2, and 100 for x_3, as well as the final, optimal value of the objective, 19550. All of this we knew before. Rows 26–31 do contain new information for us, information about where the optimal solution stands with respect to the constraints. We see that at optimality the constraints on ingredients A, B, and D, but not C, are binding. That is, the LHS (left-hand side) for each of these constraints equals the RHS. These constraints are also said to be *tight* (at the optimal solution) and so to have 0 *slack*. The constraint on ingredient C is not binding, is not tight, and so has non-zero slack. From the problem formulation, see Figure 3.1 and Figure 3.2, we see that for ingredient C the constraint is that LHS $\geq$ 900. In Figure 3.5 we see at D30 in the figure that the LHS of the constraint has a value of 1150 at optimality. The slack, at cell G30, is 250 = 1150 - 900.

This new information is quite relevant to post-solution analysis. Recall that we are minimizing and that our constraints are all $\geq$ constraints. We have learned that there is considerable slack on the constraint for ingredient C and in consequence the constraint could be tightened considerably (in the case increasing the RHS value for the $\geq$ constraint) without affecting our optimal solution or its value to us. Conversely, it would not pay us to relax that constraint (decreasing its RHS value) at all. The other three constraints are all binding, so tightening them (here, increasing the RHS values) will likely result in a degraded (increased) objective function value and loosening them (here, decreasing the RHS values) might well result in an improved (lower, because we are minimizing) objective function value.

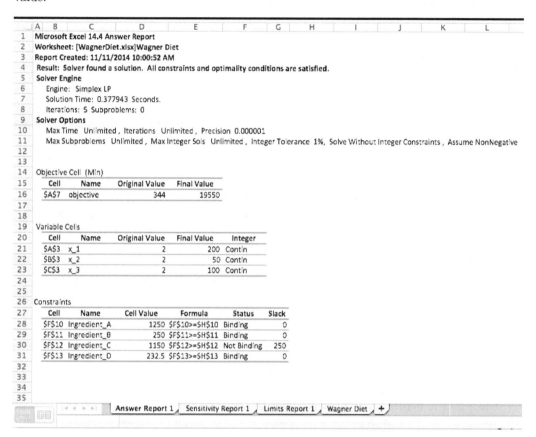

FIGURE 3.5: Excel Answer report for the Wagner Diet model.

These reflections can serve to direct exploration of the model by intelligently re-solving it repeatedly. For example, if ingredients A, B, C, and D can be obtained as dietary supplements (in small amounts), it may pay us to relax one or more of the constraints and use the savings to purchase some supplements. Focusing for the sake of an example on the constraint for ingredient A, suppose it were possible to relax the RHS value by 100 units, decreasing it from 1250 to 1150. What would be the improvement in the objective value? Would this be enough to justify purchase of 100 units of A supplement? We can answer the first question by changing the RHS value and re-solving the model. We can answer the second question by using the information from the first question and knowledge of the market price of 100 units of A supplement.

The process generalizes: With repeated re-solving of the model we can answer any number of interesting post-solution questions. The issue this raises is the amount of effort needed to do this when the model is large enough to require substantial time and computational resources to re-solve. Linear programming has a happy response to this dilemma, as we shall now see from examining the Sensitivity report.

	A B	C	D	E	F	G	H
1	Microsoft Excel 14.4 Sensitivity Report						
2	Worksheet: [WagnerDiet.xlsx]Wagner Diet						
3	Report Created: 11/11/2014 10:00:52 AM						
4							
5							
6	Variable Cells						
7			Final	Reduced	Objective	Allowable	Allowable
8	Cell	Name	Value	Cost	Coefficient	Increase	Decrease
9	$A$3	x_1	200	0	41	27.6	2.76
10	$B$3	x_2	50	0	35	19.71428571	1.568181818
11	$C$3	x_3	100	0	96	2.653846154	78.85714286
12							
13	Constraints						
14			Final	Shadow	Constraint	Allowable	Allowable
15	Cell	Name	Value	Price	R.H. Side	Increase	Decrease
16	$F$10	Ingredient_A	1250	8	1250	431.25	172.5
17	$F$11	Ingredient_B	250	1	250	265.3846154	67.12062257
18	$F$12	Ingredient_C	1150	0	900	250	1E+30
19	$F$13	Ingredient_D	232.5	40	232.5	24.64285714	61.60714286
20							

FIGURE 3.6: Excel Sensitivity report for the Wagner Diet model.

Figure 3.6 contains the Excel Solver Sensitivity report for the Wagner Diet model. All other LP solvers have similar reports. Excel's is merely convenient and representative.

The report in Figure 3.6 actually contains two reports. Rows 6–11 report on the decision variables and rows 13–19 report on the constraints. We begin with the decision variables. Looking at row 9, we see that the decision variable x_1 has a final value of 200, exactly as indicated in the Answer report, Figure 3.5. Next we see that the reduced cost for x_1 is zero. The reduced cost can be interpreted as the amount that the objective function *coefficient* needs to be improved for the associated decision variable to have non-zero value in the optimal solution. Because x_1 is at 200, its reduced cost is, and must be zero. The same reasoning applies to the other decision variables. When it is the case, as it often is, that some variables are at zero in the optimal solution, then the reduced cost information is valuable for post-solution analysis. The flip side of indicating how much a coefficient would have to be improved (increased in the case of maximization, decreased in the case of minimization) is that the reduced cost tells us the unit cost (degradation) of setting a variable that is 0 at optimality to some larger value.

Next in row 9 of Figure 3.6 we see recorded the objective function coefficient of x_1, which is 41 (see also Figure 3.1 and Figure 3.2). The allowable increase—27.6 in the present case—is the amount we can increase the coefficient—41—and still have the present decision variable values optimal, keeping everything else constant. Of course if we do this we will also degrade (increase) the value of the optimal solution by 200 × the amount of the increase.

Remarkably, this is all computed from the Sensitivity report, without having to re-solve the model. The reader should check that indeed if the coefficient on x_1 is set to 41+27.6= 68.6 and the model is re-solved, then the optimal values of the decision variables do not change, the objective value increases by $200 \times 27.6 = 5520$, and the new allowable increase for x_1 is 0.

Finally in row 9 of Figure 3.6, the allowable decrease for the objective function coefficient on x_1 is 2.76. The reasoning is quite analogous to that for the allowable increase. We may decrease the coefficient by up to 2.76 without changing the optimal settings of the decision variables, keeping everything else unchanged. Because we are minimizing, a decrease in the coefficient results in an improvement (decrease) in the objective value. The amount of improvement is 200 (the value of the decision variable) × the amount of the decrease. And all of this applies equally well to the other decision variables, rows 10 and 11 in the present instance.

We turn now to the constraints report, rows 13–19 in Figure 3.6. Focusing on row 16 we see that the final value of the Ingredient A constraint is 1250. This is the left-hand-side (LHS) value of this constraint at optimality, at the optimal setting of the decision variables. The RHS value is restated two columns over. It too is equal to 1250. Recall from the Answer report (Figure 3.5 line 28) that the status of the constraint is binding (or tight) and the slack is $0 = \text{LHS} - \text{RHS}$ (for a $\geq$ constraint) = 1250 - 1250 in the present case.

The Ingredient A constraint has a shadow price of 8 (cell E8 in the figure). This introduces a new concept for our discussion, one of great importance for post-solution analysis. It may be characterized as follows. The shadow price is the per unit change that is realized upon re-solving the problem to optimality for *increases* in the RHS value of the associated constraint. This applies only for changes within the allowable range, as specified by the allowable increase and allowable decrease for (the RHS value of) the constraint. In the present case, if we increase the RHS value of 1250 by 3, to 1253, this is permitted because it does not exceed the allowable increase of 431.25 and upon re-solving the problem we would obtain a new objective value of $19550 + (8 \times 3) = 19574$. Similarly, if we decrease the RHS value by 4, that is increase it by -4, to 1246, this is permitted because it does not exceed the allowable decrease of 172.5 and upon re-solving the problem we would obtain a new objective value of $19550 + (8 \times -4) = 19518$. Of course, we need to actually re-solve the problem to get the new optimal values of the decision variables.

As usual, the reasoning applies to the other constraints in the model and to LPs in general. Notice that if we increase a RHS value in a $\geq$ constraint and, as we do in the present example, we are minimizing, then we can expect that the objective value will deteriorate, that is will increase. This is because increasing the RHS of a $\geq$ constraint can only reduce the size of the feasible region and so our objective value cannot improve (cannot decrease when minimizing, cannot increase when maximizing) and would typically degrade (typically increase when minimizing, typically decrease when maximizing). If the constraint is $\leq$, then increasing its value cannot degrade the optimal value of the objective. With equality constraints, $=$, things might go either way.

The shadow price information on the constraints of solved LPs is tailor made for post-solution analysis of opportunity questions, for it allows us to explore and answer many such questions without having to re-solve the model.

The Excel Solver Limits report, shown in Figure 3.7, is of little value in the present case, is generally the least important of the three reports, and is typically not provided by other solvers. We include it here for completeness. The Excel Solver Limits report is created by re-running the Solver model with each of the decision variables, one at a time, set as the objective (both maximizing and minimizing), with all other variables held fixed at their optimal values for the original problem.

Put otherwise, for each decision variable the original problem is solved twice with the following twists. First, all of the other decision variables are set to their optimal values for the original problem, so that there is only one decision variable left to adjust. Second, the resulting problem is solved twice, once as a maximization and once as a minimization. This is repeated in turn for each of the original decision variables. The resulting report tells us the results. The lower limit for each variable is the smallest value that the variable can take while satisfying the constraints of the original problem and holding all of the other variables constant. The upper limit is the largest value the variable can take under these circumstances. We see from Figure 3.7 that none of the variables can be decreased from their optimal values to improve the objective function, while remaining feasible. We also see that increasing any one of the variables produces an infeasible solution.

Microsoft Excel 14.4 Limits Report
Worksheet: [WagnerDiet.xlsx]Wagner Diet
Report Created: 11/11/2014 10:00:57 AM

	Objective	
Cell	Name	Value
$A$7	objective	19550

	Variable		Lower	Objective	Upper	Objective
Cell	Name	Value	Limit	Result	Limit	Result
$A$3	x_1	200	200	19550	#N/A	#N/A
$B$3	x_2	50	50	19550	#N/A	#N/A
$C$3	x_3	100	100	19550	#N/A	#N/A

FIGURE 3.7: Excel Limits report for the Wagner Diet model.

Stepping back and taking an overview, we will generally need to deliberate whether we wish to choose a decision or solution that is suboptimal in terms of the model, but that affords us a degree of cushion in the face of uncertainty. For example, the constraints on the set of ingredients in our present case, $\{A, B, C, D\}$, may be only approximate. Further, when the grain is actually delivered to the livestock chance events may intercede and leave some animals undernourished even when there is on average sufficient food on hand. So we may want to plan with a degree of slack in hand. Any number of other considerations may be relevant and subject to deliberation. Examples include:

1. Cost uncertainties. The costs as given, the c_j's, will typically be estimates and so will differ from the actual prices available at time of purpose. If the optimal solution depends critically on a cost level that is known to be relatively variable, then we might want to find another solution that is not sensitive to a volatile quantity, providing the loss from doing so is acceptably small.

2. RHS side uncertainties. The b_i's, may also be subject to uncertainty. What happens if some animals by chance do not quite receive the necessary level of nutrition? Is a small miss consequential? How much of a safety buffer should be planned for and at what cost? Additionally, some grains may have multiple uses outside of the model, perhaps for other forms of livestock, or other locations. Does it make sense to shade the decision so as to have more of a reserve in grains that might be used for other purposes?

3. Uncertainties in the constraint matrix coefficients. The a_{ij}'s, representing the amount of nutrition ingredient per unit of grain, may as well be subject to uncertainty. Values may change with different sourcing of grains, due to damage during storage and

transportation, and so on. The possibilities are endless. Depending on circumstances, the a_{ij}'s may be standard values that underestimate, or low ball, the level of nutrition effectively delivered. If so, then taking decisions that shade these values slightly may be warranted. Whether this is the case—the values could as easily be biased in the other direction, or different values could have different biases—is of course entirely dependent upon the actual situation. The information is outside the model.

Figure 3.8 presents a heavily commented MATLAB script for solving the Wagner Diet problem. Lines beginning with % are comments and not code that is executed. When the script is executed the following output results:

```
>> wagner_diet
Optimization terminated.
  200.0000   50.0000  100.0000
At optimality the objective value is 19550.00.
Shadow prices:
    8.0000
    1.0000
    0.0000
   40.0000
```

Points arising:

1. Even without knowledge of MATLAB the reader should have little difficulty understanding the basics of the script and the output, especially with the Excel solution present.

2. The MATLAB script is an example of a modeling language approach to implementing and solving a COModel. This contrasts with the graphical or spreadsheet approach used by Excel and other products. The two approaches each have their own pluses and minuses. Two significant advantages of the modeling language approach are (i) it affords model manipulation under program control, and (ii) the core code works even with very large scale models, thus affording ease of inspection and debugging.

3. As it happens, the Excel solver provides a much richer information set for post-solution analysis of LPs than does the MATLAB solver.

This section has indicated how LP solvers produce special information that can assist us in dealing with cost and RHS uncertainties, but not with constraint matrix coefficient uncertainties. All of this is for one parameter at a time. Before concluding with a discussion of the general situation, we turn to a discussion of how the methods of the present section may be extended to more than one parameter at a time.

3.5   More than One at a Time: The 100% Rule

We now broaden our discussion from investigating changes in one parameter at a time to several. Throughout, out discussion is focused on the special case of LP and using the information for post-solution analysis that it uniquely provides, which applies to RHS values and objective function coefficients.

```
% File: wagner_diet.m
% This is an implementation of the Wagner Diet
% model, a small and very simple model that is nonetheless
% useful for understanding linear programming.
% In MATLAB, type doc('linprog') at the command prompt to
% get documentation information on the integer programming
% solver.

% There are only 3 decision variables.
% The objective coefficients are:
c = [41, 35, 96];
% There are four constraints, plus nonnegativity conditions
% on the decision variables. The four constraints have the
% >= form, so we multiply by -1 to get <= as required by
% MATLAB.
% The LHS constraint coefficients are:
A = [-2, -3, -7;
     -1, -1, 0;
     -5, -3, 0;
     -0.6, -0.25, -1];
% The RHS constants are (again multiplying by -1):
b = [-1250; -250; -900; -232.5];
% The lower bounds are all zero.
lb = zeros(3,1);
% Call the solver, which natively minimizes:
[x,fval,exitflag,output,lambda] = linprog(c,A,b,[],[],lb);
% Display the decision variables at optimality:
disp(x')
fprintf('At optimality the objective value is %3.2f.\n',fval)
fprintf('Shadow prices:\n')
disp(lambda.ineqlin)
```

FIGURE 3.8: MATLAB script to solve the Wagner Diet problem.

The Excel Sensitivity report, as in Figure 3.6, states allowable increases and decrease for the individual objective coefficient values and for the individual constraint RHS values. (We use the Excel report merely for convenience; other LP solvers will provide this information.) In the previous section we emphasized that the analyses described were only valid if the parameter in question was changed within its allowable range, that is, if it is increased then the amount of increase must be less than or equal to the allowable increase and if it is decreased then the amount of decrease much be less than or equal to the allowable decrease. These allowable amounts are parameter-specific for any given solution to the model. Of course, if we are interested in examining a parameter change beyond its allowable range, we can always set it to a desired level and re-solve the problem. The present discussion is about what we can learn *without* re-solving.

Abstracting just a little the requirement that the contemplated change in a parameter, p (either objective coefficient or RHS value) must be within its allowable range, we can define a function that tells us how much of the permitted change has occurred with any new value p^n of the parameter in question. Let:

$$f(p^n) = \begin{cases} 0 & \text{if } p^n = p \\ \frac{p^n - p}{p^h - p} & \text{if } p^n > p \wedge p^n \leq p^h \\ \frac{p - p^n}{p - p^l} & \text{if } p^n < b \wedge p^n \geq p^l \\ \text{undefined} & \text{otherwise} \end{cases} \tag{3.10}$$

The first clause says that if p is not changed (if $p^n = p$), then $f(p^n) = 0$; that is none of the permitted change has been used up. The second clause addresses increases in p, the cases in which the new value is greater than p and less than or equal to p^h, the "high" value of p, that is p plus its allowable increase. When these conditions obtain, then $f(p^n) =$

$$\frac{p^n - p}{p^h - p}$$

That is, $f(p^n)$ is the fraction of the allowable increase consumed by p^n. Conversely, the third clause covers the case when p^n is a decrease within the allowable range, with p^l the "low" value of p, that is p minus its allowable decrease. Finally, the last clause tells us that $f(p^n)$ is undefined for values outside its permitted range, $[p^l, p^h]$.

Notice that if $f(p^n)$ is defined, it takes on a value between 0 and 1 inclusive, and may be interpreted as the fraction of allowable change being consumed by p^n. Remarkably, we can use $f(p^n)$ to determine whether multiple parameter changes are on balance within the permitted amount of change overall and if they are, we can determine the new value of the objective function at optimality. There are two formulas, one for objective function coefficients and one for RHS values:

$$\Sigma_b = \sum_i f(b_i^n) \tag{3.11}$$

$$\Sigma_c = \sum_j f(c_j^n) \tag{3.12}$$

If $\Sigma_{b/c}$ (either b or c) is defined, that is if none of the $f(p^n)$ values is undefined, and if $\Sigma_{b/c} \leq 1$, then the individual changes are collectively valid for the analysis. This is called the *100% rule* for obvious reasons. (There are really two rules, one for bs and one for cs.) A little notation to help us see how this works: Let:

1. b_i = the RHS value for constraint i in the current model

2. b_i^h = the high value on the allowable range of b_i in the current model

3. b_i^l = the low value on the allowable range of b_i in the current model

4. b_i^n = a new value for b_i

5. $\Delta b_i = b_i^n - b_i$ (may be positive or negative or zero)

6. c_j = the objective coefficient for decision variable x_j in the current model

7. c_j^h = the high value on the allowable range of c_j in the current model

8. c_j^l = the low value on the allowable range of c_j in the current model

9. $\Delta c_j = c_j^n - c_j$ (may be positive or negative or zero)

From the previous section we know that if we are changing a single objective function coefficient c within its allowable range then the new value of the objective function is given by expression (3.13) in Figure 3.9.

$$z' = z^* + \Delta c_j \times x_j^*$$ (3.13)

FIGURE 3.9: Objective coefficient formula. z^* = current optimal value of objective function. Δc_j amount of increase in c_j (objective function coefficient for decision variable x_j). z' = value of objective function with the change to c_j. Valid for Δc_j within the allowable range. The optimal decision variable values remain unchanged

Also from the previous section we know that if we are changing a single RHS value b within its allowable range then the new value of the objective function is given by expression (3.14) in Figure 3.10.

$$z' = z^* + \Delta b_i \times \lambda_i$$ (3.14)

FIGURE 3.10: Shadow price formula. z^* = current optimal value of objective function. Δb_i amount of increase in b_i (right-hand-side value of the constraint). λ_i = shadow price of the constraint. z' = value of objective function with an optimal solution to the revised problem. Valid for Δb_i within the allowable range.

The 100% rule tells us that we may combine these expressions provided we do not exceed 100% of the permitted changes. Figure 3.11 and Figure 3.12 state it explicitly for constraint RHS values and for objective function coefficients.

If

$$\Sigma_b = \sum_i f(b_i^n)$$ (3.15)

is defined and ≤ 1, then

$$z' = z^* + \sum_i \Delta b_i \times \lambda_i$$ (3.16)

FIGURE 3.11: 100% rule for constraint RHS values. z^* = current optimal value of objective function. Δb_i amount of increase (positive or negative) in b_i (right-hand-side value of the constraint). λ_i = shadow price of the constraint (the change in the objective function value per unit increase in b_i). z' = value of objective function with an optimal solution to the revised problem. Valid for Δb_is within their allowable ranges.

If

$$\Sigma_c = \sum_j f(c_j^n) \tag{3.17}$$

is defined and ≤ 1, then

$$z' = z^* + \sum_j \Delta c_j \times x_j^* \tag{3.18}$$

FIGURE 3.12: 100% rule for objective function coefficients. z^* = current optimal value of objective function. Δc_j amount of increase (positive or negative) in c_j (objective function coefficient on decision variable x_j). x_j^*s = optimal values of the problem's decision variables, x_js. z' = value of objective function to the revised problem (for which the x_j^*s are unchanged). Valid for Δc_js within their allowable ranges.

Maximize $z = 5x_1 + 4.5x_2 + 6x_3$ $\tag{3.19}$

Subject to:

$$6x_1 + 5x_2 + 8x_3 \le 60 \tag{3.20}$$

$$10x_1 + 20x_2 + 10x_3 \le 150 \tag{3.21}$$

$$1x_1 + 0x_2 + 0x_3 \le 8 \tag{3.22}$$

$$x_1 \ge 0, x_2 \ge 0, x_3 \ge 0 \tag{3.23}$$

FIGURE 3.13: Custom-molder linear programming model.

3.6   For Exploration

1. Use Excel Solver or any other LP solver to implement and solve the LP in Figure 3.13. What is the optimal solution you find? What is the corresponding optimal value of the objective function? Which constraints are binding? Which are not and how much slack do they have? What are the shadow prices on the constraints?

2. Discuss the seven categories of questions for post-solution analysis, summarized in Figure 1.5 and in each case whether the Excel Answer, Sensitivity, and Limits reports (and similar reports from other LP solvers) can handle the questions and if so how. Give examples of credibly interesting post-solution analysis questions for an LP that can be answered with these reports, and give examples of questions that cannot be answered with them.

3.7   For More Information

Linear programming is covered in very many books, at very many levels of depth. Winston [157] is a good, standard, introductory yet thorough example.

The 100% rule originated in [20, chapter 3] and was extended in [26].

As noted above, the pioneering work on post-solution analysis by Harvey Greenberg in the context of linear programming— [68, 69, 70, 71]—remains an important point of contemporary departure.

Part II

Optimization Modeling

Chapter 4

Simple Knapsack Problems

4.1 Introduction

We introduced the Simple Knapsack model and an example of it, the Eilon model, in Chapter 1. Our discussion here presumes familiarity with that material. We begin where the discussion in Chapter 1, in particular the discussion in §§1.2–1.3, left off.

4.2 Solving a Simple Knapsack in Excel

Recall that the Eilon model [45] has only 12 decision variables, which is quite small indeed. Even so, it affords a rich discussion that will apply in much larger cases. The

Variable No.	c_i	w_i	c_i/w_i
1	4113	131	31.40
2	2890	119	24.29
3	577	37	15.59
4	1780	117	15.21
5	2096	140	14.97
6	2184	148	14.76
7	1170	93	12.58
8	780	64	12.19
9	739	78	9.47
10	147	16	9.19
11	136	22	6.18
12	211	58	3.64

TABLE 4.1: Specification of the Eilon Simple Knapsack model. $b = 620$.

model itself is fully specified in Table 4.1 (even ignoring the rightmost column to which we return soon) and implemented in Excel, as shown in Figure 4.1. The Excel file, *Eilon_Omega_1987.xlsx*, is available from the book's Web site. The reader may easily reconstruct the implementation from Figure 4.1 with a small aliquot of information: (1) Cell C15 contains the formula =SUMPRODUCT(dvars,objectivecoeffs), which computes the objective function value for the current solution, and (2) Cell D16 contains the formula =SUMPRODUCT(dvars,constraintcoeffs), which computes the LHS value for the current solution. With this setup, the Excel's Solver tool can be used as indicated to solve the model to optimality. The pattern generalizes to larger models, subject to the limitation of Excel's solver.

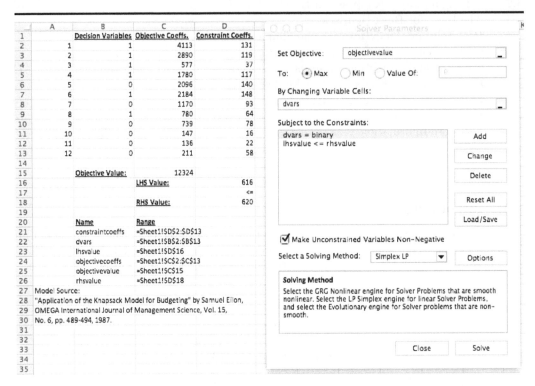

FIGURE 4.1: Solving the Eilon Simple Knapsack model in Excel.

4.3 The Bang-for-Buck Heuristic

We saw, in the previous section, how the Eilon Simple Knapsack model is easily solved exactly—to guaranteed optimality—in Excel. Now we will explore how the model can be solved heuristically, that is, how we can reliably find a good—although usually not optimal—solution to the model. Why bother with a heuristic solution (method) since an exact solution (method) is available? There are at least two good reasons.

1. Exact solution methods are not always available or available at acceptable cost (either in time or computing resources). The Simple Knapsack model is among the most easily

solved models with exact methods and even it will overwhelm these methods when it gets large enough. When this happens we have no choice but to rely on heuristic methods for solving the problem.

2. We will be using and exploring heuristics throughout this book. They are valuable and pervasive throughout business analytics. It is helpful to introduce and explore heuristics in contexts for which exact methods are available. And so we shall.

Microsoft Excel Solver, as we encountered it in the previous section, is limited to 200 decision variables http://support.microsoft.com/kb/75714 (accessed 2015-07-20); see also http://www.solver.com/standard-excel-solver-dealing-problem-size-limits (accessed 2015-07-20)). This is small by current day problem standards. Other commercial solvers (including MATLAB's, which we saw at work in Chapter 1) will work with much larger problems, but it will be handy to have a "good enough" (heuristic) solver of our own.

The bang-for-buck heuristic for Simple Knapsack problems is a reliable and excellent heuristic in practice and in theory [121]. It is intuitive, very fast, and easy to implement. The underlying idea is simple and intuitively attractive: Order the decision variables in descending order of profit divided by weight, c_i/w_i, then proceeding in descending order, if the item fits into the knapsack put it in. Here, a bit more formally, is how it works:

i. Initialize the current solution to all zeros—begin with no items inserted into the knapsack.

   (This is guaranteed to be a feasible solution, although it is a disappointing one.)

ii. For each decision variable, i, obtain the ratio c_i/w_i.

   (This is the bang-for-buck of the variable.)

iii. Sort the bang-for-buck ratios in descending order.

iv. For each decision variable, x_i, starting with the variable with the largest bang-for-buck ratio and continuing in descending order:

   (a) Set the value of x_i in the solution to 1.
   (b) Check the resulting solution for feasibility.
   (c) If the resulting solution is feasible, continue at Step iv with the next decision variable; otherwise (the resulting solution is infeasible, so) set the value of x_i in the solution to 0 and continue at step iv with the next decision variable.

In a nutshell, the bang-for-buck of an item is the ratio of its contribution to its cost (weight or resource usage), and we greedily add the best available item on the bang-for-buck measure if it will fit. Points arising:

1. The bang-for-buck heuristic generally finds a very good solution. Often it does not find an optimal solution, however.

2. The bang-for-buck heuristic is very fast; it requires but one pass through the decision variables. In consequence it can handle very large models and it can solve models multiple times, without undue consumption of resources. The latter property is one we shall make much use of in the sequel, when we come to analyze Simple Knapsack models.

3. In Figure 4.1, page 61, we have the bang-for-buck ratios in the rightmost column. As it happens, the variables are already in descending order of bang-for-buck.

Figure 4.2 and 4.3 illustrate solving the Eilon model "by hand" using the bang-for-buck heuristic, with $b = 645$. Recall that the decision variables are already presented in descending order by c_i/w_i. So, all we need to do is zero out the dvars range (B2:B13), then starting at the top convert 0s to 1s. If a conversion to a 1 leads to an infeasibility (D16 > D18), we covert back to 0, and proceed to the next variable. Figure 4.2 shows the first infeasible solution we find during this process. Since the last added item (last variable set to 1) is the sixth, we reset it to 0 and proceed. Figure 4.3 shows where we end up. We have a feasible solution, with an objective value of 12,626 and a slack of 8 = RHS − LHS.

Figure 4.4 shows the results of actually finding the optimal solution using Solver in Excel. We have a feasible solution, with an objective value of 12,714 and no slack at all. While bang-for-buck gives us a good solution, it is too greedy and the solution it finds is not optimal.

4.4 Post-Solution Analytics with the Simple Knapsack

There is a saying among management scientists that in modeling studies the real work only begins once the data have been collected, the model has been built and tested, and the model has been solved. Put otherwise, Figure 4.1 and Figure 4.4 represent merely the beginning, not the end, of the important work with the Eilon model. The point is apt and in fact the great bulk of model analytics work is done after a model has been solved.

Stated more carefully, this *post-solution* analysis (aka: *post-evaluation* analysis) refers to investigations for the purposes of deliberation and decision making that happen *after* a model has been formulated and an optimizing or heuristic evaluator has been applied. A presumably good, or even optimal, solution *to this model* is at hand. Call it $\vec{x}^+$ with objective function value z^+. Before actual decisions are taken, however, it is normally prudent to ask various kinds of *deliberation questions.* Two main types concern us here (see Chapter 1 for a more extensive list):

1. *Sensitivity analysis* questions: Do small changes in the model's parameters have relatively small effects on either (a) the value realized for the objective function, or (b) the accepted solution?

2. *Opportunity* (or *candle lighting analysis*) questions: Are there advantageous opportunities to change the assumptions of the model? For example, would loosening one or more constraints yield a significantly improved objective value? If so, does the cost of doing so net out to a profit? Conversely, are there good solutions available upon tightening one or more constraints? If so, can the extra resource(s) be profitably sold or used for some other purpose?

Both sorts of questions are quite important in practice. They constitute key aspects of deliberating with models for decision making. We shall revisit them throughout in what follows. Here, we shall merely touch upon them, by way of introduction.

4.4.1 Sensitivity Analysis

In sensitivity analysis we examine the consequences of relatively small changes in model parameters. For example, $c_6 = 2184$ is a parameter of the Eilon model. It is the value of the objective function coefficient on decision variable 6, x_6. With $b = 645$, $x_6 = 1$ at optimality (see Figure 4.4) but $x_6 = 0$ in the bang-for-buck solution (see Figure 4.3).

	A	B	C	D
1		**Decision Variables**	**Objective Coeffs.**	**Constraint Coeffs.**
2	1	1	4113	131
3	2	1	2890	119
4	3	1	577	37
5	4	1	1780	117
6	5	1	2096	140
7	6	1	2184	148
8	7	0	1170	93
9	8	0	780	64
10	9	0	739	78
11	10	0	147	16
12	11	0	136	22
13	12	0	211	58
14				
15		**Objective Value:**	13640	
16			**LHS Value:**	692
17				<=
18			**RHS Value:**	645
19				
20		**Name**	**Range**	
21		constraintcoeffs	=Sheet1!$D$2:$D$13	
22		dvars	=Sheet1!$B$2:$B$13	
23		lhsvalue	=Sheet1!$D$16	
24		objectivecoeffs	=Sheet1!$C$2:$C$13	
25		objectivevalue	=Sheet1!$C$15	
26		rhsvalue	=Sheet1!$D$18	
27	Model Source:			
28	"Application of the Knapsack Model for Budgeting" by Samuel Eilon,			
29	OMEGA International Journal of Management Science, Vol. 15,			
30	No. 6, pp. 489-494, 1987.			

FIGURE 4.2: Solving the Eilon Simple Knapsack model with bang-for-buck, first infeasibility with $b = 645$.

	A	B	C	D
1		**Decision Variables**	**Objective Coeffs.**	**Constraint Coeffs.**
2	1	1	4113	131
3	2	1	2890	119
4	3	1	577	37
5	4	1	1780	117
6	5	1	2096	140
7	6	0	2184	148
8	7	1	1170	93
9	8	0	780	64
10	9	0	739	78
11	10	0	147	16
12	11	0	136	22
13	12	0	211	58
14				
15		**Objective Value:**	12626	
16			**LHS Value:**	637
17				<=
18			**RHS Value:**	645
19				
20		**Name**	**Range**	
21		constraintcoeffs	=Sheet1!$D$2:$D$13	
22		dvars	=Sheet1!$B$2:$B$13	
23		lhsvalue	=Sheet1!$D$16	
24		objectivecoeffs	=Sheet1!$C$2:$C$13	
25		objectivevalue	=Sheet1!$C$15	
26		rhsvalue	=Sheet1!$D$18	
27	Model Source:			
28	"Application of the Knapsack Model for Budgeting" by Samuel Eilon,			
29	OMEGA International Journal of Management Science, Vol. 15,			
30	No. 6, pp. 489-494, 1987.			

FIGURE 4.3: Solving the Eilon Simple Knapsack model with bang-for-buck, solution with $b = 645$.

	A	B	C	D
1		**Decision Variables**	**Objective Coeffs.**	**Constraint Coeffs.**
2	1	1	4113	131
3	2	1	2890	119
4	3	1	577	37
5	4	1	1780	117
6	5	0	2096	140
7	6	1	2184	148
8	7	1	1170	93
9	8	0	780	64
10	9	0	739	78
11	10	0	147	16
12	11	0	136	22
13	12	0	211	58
14				
15		**Objective Value:**	12714	
16			**LHS Value:**	645
17				<=
18			**RHS Value:**	645
19				
20		**Name**	**Range**	
21		constraintcoeffs	=Sheet1!$D$2:$D$13	
22		dvars	=Sheet1!$B$2:$B$13	
23		lhsvalue	=Sheet1!$D$16	
24		objectivecoeffs	=Sheet1!$C$2:$C$13	
25		objectivevalue	=Sheet1!$C$15	
26		rhsvalue	=Sheet1!$D$18	
27	Model Source:			
28	"Application of the Knapsack Model for Budgeting" by Samuel Eilon,			
29	OMEGA International Journal of Management Science, Vol. 15,			
30	No. 6, pp. 489-494, 1987.			
31				

FIGURE 4.4: Solving the Eilon Simple Knapsack model to optimality, with $b = 645$.

Would a small change in the value of c_6, keeping everything else constant, result in x_6 being included in the bang-for-buck solution? If not a small change, how large a change would be needed for bang-for-buck to include x_6? As part of post-solution deliberation and analysis, we will be interested in both questions. The former is conventionally classed as a sensitivity analysis question, the latter is more naturally seen as a candle lighting question. Very little turns on the terminology we use; what matters is that these distinct questions are asked and answered.

In order to begin to answer these questions for the case at hand, we will vary the value of c_6, resolve the model for each variation, and report the results. Figure 4.5 constitutes that report. What we have done is to set c_6 to 2180, 2181, ..., 2220 and resolved for each of these 41 cases. In each of the three panels of Figure 4.5 the abscissa (x-axis) represents these varying values of c_6. In the top panel, the ordinate (y-axis) indicates whether x_6 is in ($=1$) or is not in ($=0$) the corresponding bang-for-buck solution. We see that so long as $c_6 \leq 2215$, x_6 is not in the (heuristically optimal) bang-for-buck solution.

Panel 2, in the middle of Figure 4.5, plots the objective function values as a function of c_6. Varying c_6, so long as x_6 is not in the knapsack, does not influence z. Once x_6 is in, the value of z jumps by more than 100 points, and increases modestly from there. Together, these two panels tell us that modest improvements in the contribution of x_6 will not affect our decision *not* to have it in the solution. However, if we can improve c_6 to at least 2216, we can effect a significant improvement in z.

Finally, panel 3, at the bottom of Figure 4.5, plots the slacks (RHS - LHS) as a function of c_6. It tells us that if we do improve c_6 to at least 2216, we will also lose 8 units of slack. This may or may not be deemed a good thing, depending upon circumstances.

As a second example of sensitivity analysis we examine a constraint coefficient, again for the Eilon model with $b = 645$. Both the optimal and bang-for-buck solutions put x_2, with the second highest c_i/w_i ratio, into the knapsack. As $w_2 = 119$ increases the value of the c_2/w_2 ratio will decrease. We would like to understand the dynamics here. If w_2 increases sufficiently x_2 will be dropped from the solution. We would like to know when that will happen, among other things.

As in the previous case, in order to begin to answer these questions we will vary the value of a parameter, now w_2, resolve the model for each variation, and report the results. Figure 4.6 constitutes that report. What we have done is to set w_2 to 116, 117, ..., 200 and resolved for each of these 85 cases. In each of the three panels of Figure 4.6 the abscissa (x-axis) represents these varying values of w_2. In the top panel, the ordinate (y-axis) indicates whether x_2 is in ($=1$) or is not in ($=0$) the corresponding bang-for-buck solution. We see that so long as $w_2 \leq 195$, x_2 is in the (heuristically optimal) bang-for-buck solution.

Panel 2, in the middle of Figure 4.6, plots the objective function values as a function of w_2. Now the picture is more complex. Varying w_2, with it still in the knapsack, does influence z at intervals. The direct influence is on the slack, shown in panel 3 of Figure 4.6. Increasing w_2 either (and usually) decreases the slack without affecting the objective function value *or* it increases the slack and reduces the value of z, because it forces some other variable out of the solution. The pattern holds until w_2 increases to 196, at which point it itself is forced out of the solution (panel 1) and further increases in w_2 become irrelevant.

Points arising:

1. *Uncertainty* is fundamentally why sensitivity analysis and post-solution analysis needs to be undertaken. The solutions we get are solutions to models, not to problems. The models merely represent—with varying degrees of accuracy—the real world problems that interest us.

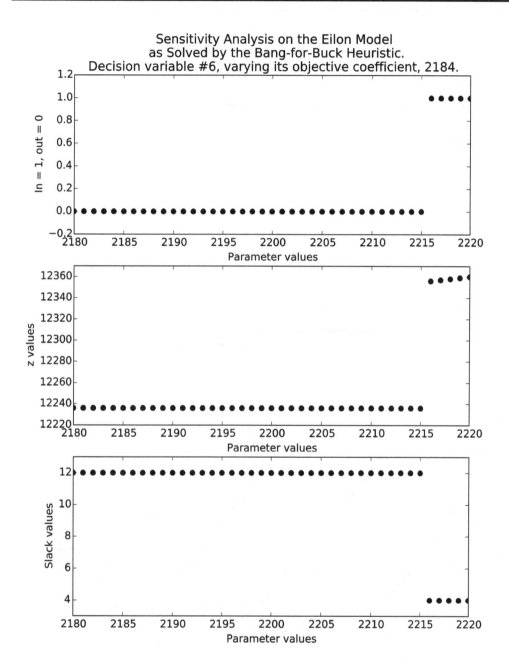

FIGURE 4.5: Sensitivity report on the Eilon Simple Knapsack model, with $b = 645$ and varying c_6, as solved multiple times (41 in all) by the bang-for-buck heuristic.

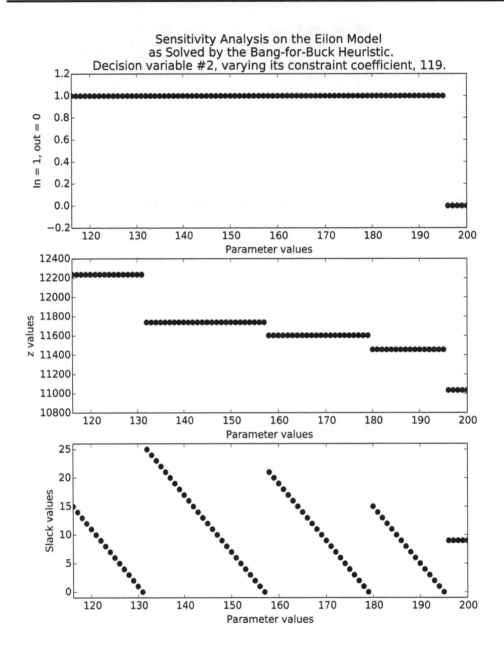

FIGURE 4.6: Sensitivity report on the Eilon Simple Knapsack model, with $b = 645$ and varying w_2, as solved multiple times (85 in all) by the bang-for-buck heuristic.

2. *Solution pluralism* is the principle or philosophy we invoke for undertaking post-solution analytics, both sensitivity analysis and candle lighting analysis. In doing so, we define a set of *decisions of interest* (DoIs), we solve the model multiple times in order to obtain the DoIs (or a good sample of them), and we deliberate—base our actual decisions in part—on the discovered DoIs.

3. Even if a model can be solved to optimality in acceptable time, resolving it tens, hundreds, even thousands of times (or more!) to optimality is seldom practicable. In consequence, fast heuristics, such as bang-for-buck, are most welcome even when the original COModel can be solved to optimality.

4.4.2 Candle Lighting Analysis

Sensitivity analysis focuses on the consequences—especially for stability and robustness—of small changes to the model. Candle lighting analysis asks a different question: *Are there opportunities proactively to change the assumptions of the model in order to obtain a globally superior result?* As in the case of sensitivity analysis, we shall be much concerned with candle lighting analysis throughout this book and we seek merely to introduce and illustrate it in this section. To that now.

The Simple Knapsack model has one constraint. The RHS value of that constraint in the case of the Eilon model is, we shall say, $b = 620$. The value of b represents a quantity of resource (say, budget dollars) that is in limited supply. If we had more of it, presumably we could improve on z. If we had less of it we might not reduce z at all (if the reduction is within the available slack) or we might reduce it by a small, acceptable amount and deploy the saved resource elsewhere where it is more valuable. Candle lighting analysis is about investigating these possibilities. Its results will depend upon both what we find with the model and what conditions external to the model are.

Figure 4.7 plots the objective value (as found by bang-for-buck) against the RHS value, above and below its default of 620. Predictably, z increases step-wise as b (the RHS value) increases. Immediately we see a possible opportunity. At $b \in [620, 623]$, $z = 12236.0$ and at $b = 624$, $z = 12383.0$. If we can increase b by 4 units, we can increase z by 47 units. Is it worth it? The model cannot tell us. We need to know the price and availability of more b. What the analysis has done is to alert us to an interesting possibility, one that merits at least some attention. Notice that the analysis also teaches us something important: Any increase of b to a value below 624 will yield no improvement in z (according to the bang-for-buck solver). So we need at least 4 units of b, but a little more than 4 is no more valuable than just 4.

The analysis works on the downside too. At $b = 607$, $z = 11739.0$. Further, from Figure 4.8 when $b = 620$ we have 12 units of slack available; our requirements for the solution at $b = 620$ are for 608 units, leaving 12 units of slack, which we might deploy elsewhere for better purposes. At $b = 607$, we have 25 units of slack available for redeployment. Now the question is: Is the loss of 497.0 units of z worth the gain of 13 units of b (or the availability of a full 25 units of b)? As on the upside, the model cannot decide for us, but the analysis has identified an opportunity worth checking out.

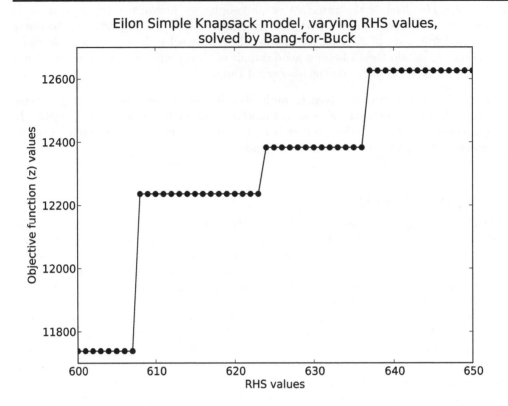

FIGURE 4.7: Candle lighting analysis: The Eilon model objective value as a function of its constraint RHS value.

4.5   Creating Simple Knapsack Test Models

Martello and Toth [121, pages 67f] present standard generator routines for Simple Knapsack models. We have implemented the "weakly correlated" method (which they claim produces problems typical of practice) in the Python module *genweakcorknapsack.py*, which is available from the book's Web site. If you run it as illustrated below, it will generate a new Simple Knapsack problem, printing the key parameter values (objective function coefficients, constraint coefficients, and right-hand-side value) to the screen. Moreover, it produces two files on output: *aknapsack.txt* and *aknapsackexcel.txt*. Note a trivial notation change: a_i replaces w_i in our previous formulations.

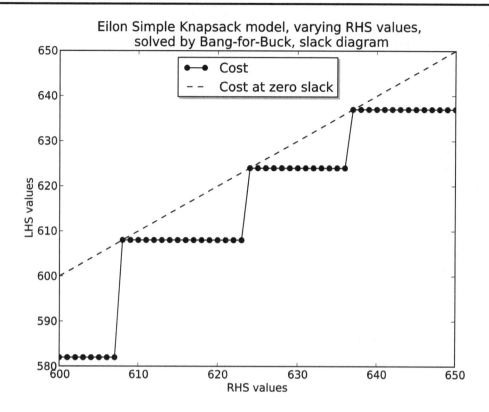

FIGURE 4.8: Candle lighting analysis: Available slack with different values of b in the Eilon model.

```
Number of decision variables? (e.g., 20) 10

v for [5,v] uniform range on constraint coefficients, a_i?
(e.g., 15.6 (> 5)) 15.6

r for [a_i -r, a_i + r] uniform range on objective coefficients, c_i?
(e.g., 3.7 (< 5)) 3.7

RHS deflator for deflator * sum(a_i)? (e.g., 0.5) 0.5

Constraint coefficients, a_i:

14.603918758
7.15158945418
14.270862245
11.9389616363
15.5246945281
12.718997089
6.64374020162
10.1536803755
6.28336889707
```

11.9140352696

Objective coefficients, c_i:

18.2302540759
8.97807309728
13.3684628354
9.88298869709
17.0518516553
11.4023195199
3.81478383404
7.61088109316
5.07592314463
14.6473647135

Right-hand-side (RHS) value, b:

55.6019242272

Let us call this problem the aknapsack1 problem. The *aknapsack.txt* and *aknapsackexcel.txt* files, produced by this run of the *genweakcorknapsack.py* Python program, have been re-named *aknapsack1.txt* and *aknapsackexcel1.txt*. They are available on the book's Web site.

If you launch Excel and import aknapsackexcel1.txt you will get a well formulated Excel knapsack problem, as shown in Figure 4.9.

It is then straightforward to direct Excel's Solver to solve the problem. Note that "Simplex LP" is the solving method, even though we have indicated binary (integer) decision variables. The nomenclature is non-standard, but Solver gets it right and gives us integer solutions. See Figure 4.10 for the Solver setup parameter values.

When you ask Solver to solve the problem (with this setup) it does so, producing the output report shown in Figure 4.11.

4.6 Discussion

The Simple Knapsack model is a useful and widely used model. Its importance for us goes beyond this fact, for the model serves as an especially clear example of a constrained optimization model and it affords some discussion of the key issues of model analytics, including solvers, heuristics, model inaccuracies, and post-solution analytics. With more complex models these challenges become . . . more complex. It will help to have the lessons of this chapter kept in mind as a conceptual foundation.

4.7 For Exploration

1. Excel's Solver supports three solution methods. Generate some larger problems and try them out with the other solution methods. What do you find? Compare with the bang-for-buck solutions.

	1	2	3	4	5	6	7	8	9	10	11	12
1												
2	Decision variables:	1	0	1	0	1	0	1	0	1	0	
3	Objective Coefficients:	18.23025408	8.978073097	13.36846284	9.882988697	17.05185166	11.40231952	3.814783834	7.610881093	5.075923145	14.64736471	
4	LHS Coefficients:	14.60391876	7.151589454	14.27086225	11.93896164	15.52469453	12.71899709	6.643740202	10.15368038	6.283368897	11.91403527	
5	Maximize z =											
6	57.54127555											
7	Subject to:											
8	LHS =	57.32658463	<=		55.60192423	'= RHS						
9												

FIGURE 4.9: The file aknapsackexcel1.txt imported as data into Excel.

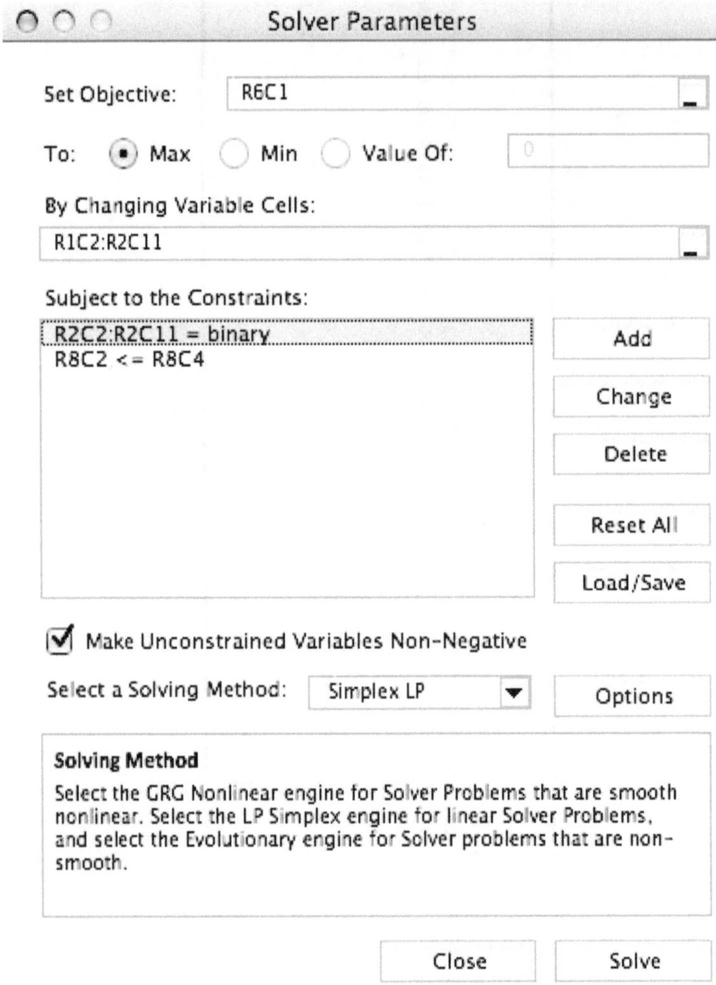

FIGURE 4.10: Excel Solver setup for aknapsack1.

1	2	3	4	5	6	7	8
1	Microsoft Excel 14.3 Answer Report						
2	Worksheet: [aknapsack1.xlsx]Sheet1						
3	Report Created: 9/2/2013 5:48:14 PM						
4	Result: Solver found a solution. All constraints and optimality conditions are satisfied.						
5	Solver Engine						
6	Engine: Simplex LP						
7	Solution Time: 0.709564 Seconds.						
8	Iterations: 2 Subproblems: 10						
9	Solver Options						
10	Max Time Unlimited, Iterations Unlimited, Precision 1e-06, Use Automatic Scaling						
11	Max Subproblems Unlimited, Max Integer Sols Unlimited, Integer Tolerance 1%, Assume NonNegative						
12							
13							
14	Objective Cell (Max)						
15		Cell	Name	Original Value	Final Value		
16	R6C1		Maximize z =	57.54127555	63.98346669		
17							
18							
19	Variable Cells						
20		Cell	Name	Original Value	Final Value	Integer	
21	R1C2			0	0	Contin	
22	R1C3			0	0	Contin	
23	R1C4			0	0	Contin	
24	R1C5			0	0	Contin	
25	R1C6			0	0	Contin	
26	R1C7			0	0	Contin	
27	R1C8			0	0	Contin	
28	R1C9			0	0	Contin	
29	R1C10			0	0	Contin	
30	R1C11			0	0	Contin	
31	R2C2		Decision variables:	1	1	Binary	
32	R2C3		Decision variables:	0	1	Binary	
33	R2C4		Decision variables:	1	0	Binary	
34	R2C5		Decision variables:	0	0	Binary	
35	R2C6		Decision variables:	1	1	Binary	
36	R2C7		Decision variables:	0	0	Binary	
37	R2C8		Decision variables:	1	0	Binary	
38	R2C9		Decision variables:	0	0	Binary	
39	R2C10		Decision variables:	1	1	Binary	
40	R2C11		Decision variables:	0	1	Binary	
41							
42							
43	Constraints						
44		Cell	Name	Cell Value	Formula	Status	Slack
45	R8C2		LHS =	55.47760691	R8C2<=R8C4	Not Binding	0.12431732
46	R2C2:R2C11= Binary						
47							

FIGURE 4.11: Excel Solver solution output for aknapsack1.

2. The following passage occurs on page 68.

> Finally, panel 3, at the bottom of Figure 4.5, plots the slacks (RHS − LHS) as a function of c_6. It tells us that if we do improve c_6 to at least 2216, we will also lose 8 units of slack. This may or may not be deemed a good thing, depending upon circumstances.

Discuss the concluding sentence. Why might loss of slack be seen as a good thing? As a bad thing?

3. The following passage occurs on page 68.

> *Uncertainty* is fundamentally why sensitivity analysis and post-solution analysis needs to be undertaken. The solutions we get are solutions to models, not to problems. The models merely represent—with varying degrees of accuracy—the real world problems that interest us.

Discuss why if uncertainty is the underlying problem we do not simply build and solve models that are highly accurate. Is this possible? When? When not?

4. Our exploration via sensitivity analysis and candle lighting analysis of the Eilon model has served the purpose of illustration. It is not a full and systematic investigation. What would be? Discuss how you would conduct a thoroughgoing, comprehensive—rather than complete and exhaustive—post-solution analysis of a knapsack model. It will be useful to specify a (likely imagined) context of decision making as you do this.

5. Imagine you are undertaking a candle lighting analysis on a Simple Knapsack model. You are able to discover a comprehensive price schedule for b, covering the entire range of interest for your problem. Describe how you would use this information to set an optimal value for b.

6. Imagine you are undertaking a candle lighting analysis on a Simple Knapsack model. The real value of b is not in fact known with certainty and may change somewhat after you need to make the decision of which projects to fund. Unfortunately, if b comes in low and your solution is infeasible, you will experience a loss because at least one project will have been started that will have to be abandoned. Discuss how you would go about deciding what to do, specifically, whether to fund a low slack "optimal" solution or a higher slack "sub-optimal" solution. Hint: Revisit Figure 4.8.

4.8   For More Information

See [13] for a gentle introduction to Simple Knapsack models and an independent endorsement of the bang-for-buck's usefulness.

As noted above, there are two excellent books devoted to knapsack problems: [121] and [85]. The Wikipedia entries `http://en.wikipedia.org/wiki/Knapsack_problem` and `http://en.wikipedia.org/wiki/List_of_knapsack_problems` (accessed 2013-09-02) are useful as well.

There is a solid literature on sensitivity analysis, post-solution analysis, and relata. See [127] for a thoughtful and very useful treatment in the context of policy analysis. Saltelli and colleagues have produced volumes on variance-based approach to sensitivity analysis [140, 141, 142]. The work is well worth looking at.

See [22, 86, 87, 88, 89, 95, 102] for elaboration of the candle lighting concept.

The code-based examples and computations for this chapter were based on three items: the Excel workbook file *Eilon_Omega_1987.xlsx*, and the Python 2.7X files *genweakcorknapsack.py* and *simpleknapsackutilities.py*. All three are available from the book's Web site: `http://pulsar.wharton.upenn.edu/~sok/biz_analytics_rep`.

Chapter 5

Assignment Problems

5.1   Introduction

In the *simple assignment problem*,[1] we seek to pair up elements from two disjoint sets of equal size. We might call these sets A and J to suggest agents (who will do work) and jobs (work for agents to do). Essential for the simple assignment problem is that the number of jobs equals the number of agents ($|A| = |J| = n$), each agent can do each job, but at a cost; jobs must be done by exactly one agent; and agents can only do exactly one job. We seek to assign all of the jobs to agents (or equivalently, all of the agents to jobs) in such a way as to minimize cost.

Assignment problems more generally—in which we seek to pair up elements from two, disjoint, but not necessarily equal-sized sets, all in the presence of constraints of various sorts—occur very often in practice. The SAP itself applies in many real world cases, but it is especially interesting as a starter example of this rich class of problems (and models to go with them).

Figure 5.1 presents a formulation of the SAP as an integer program. The x_{ij}s are the decision variables, set to 1 if agent i is assigned job j, and 0 otherwise. c_{ij} is the cost of assigning job j to agent i and we would like to minimize it. A decision for the problem is any assignment of either 1 or 0 to each of the decision variables. The objective function, expression (5.1), totals the cost of a decision. In the model formulation there are two groups of constraints plus the constraints on the decision variables to be 0 or 1 (5.4). These latter are called *integrality constraints* because they constrain the variables to be integers, here limited to 0 and 1. Expression (5.2) expresses $|A|$ constraints, one for each job j. Collectively they state that for every job, there is exactly one agent that is assigned to it. Similarly, the constraints covered by expression (5.3) state that for every agent i there is exactly one job that it does (or is assigned to).

---

[1]Or SAP, usually just called the assignment problem in the literature.

$$\min z \;=\; \sum_{i=1}^{n}\sum_{j=1}^{n} c_{ij}x_{ij} \tag{5.1}$$

subject to

$$\sum_{i=1}^{n} x_{ij} \;=\; 1 \quad (j = 1, 2, \ldots, n) \tag{5.2}$$

$$\sum_{j=1}^{n} x_{ij} \;=\; 1 \quad (i = 1, 2, \ldots, n) \tag{5.3}$$

$$x_{i,j} \;\in\; \{0, 1\} \tag{5.4}$$

FIGURE 5.1: The simple assignment problem formulated as an integer program.

A remarkable fact is that if we drop the integrality requirement in the expression (5.4) constraints and simply require that decision variables range from 0 to 1—

$$0 \le x_{ij} \le 1 \tag{5.5}$$

—and then we solve the resulting model with a linear programming solver, we are guaranteed to get a decision in which all the decision variables are integers, either 0 or 1 (assuming the problem is feasible).

Figure 5.2 displays an implementation in Excel of a very small SAP, $n = 5$, solved by linear programming. It is important to understand that while LP will work for SAPs, if you modify the problem in any material way, say by adding an additional constraint, then it is unlikely that the mathematical requirements will be met for LP to applicable. If these requirements are not met and you solve a modified SAP with LP, then you are likely to get fractional—not 0 or 1—decision variable values at optimality.

5.2 The Generalized Assignment Problem

The generalized assignment problem (GAP) is a widely studied combinatorial optimization problem with many practical applications [85], including assigning people to jobs in a services firm (consulting, law), and machines to jobs in a manufacturing setting. The problem may be summarized as follows. Given m agents (or processors) and n tasks (or jobs), the GAP aims at finding a maximum-profit assignment of each task to exactly one agent, subject to the bounded capacity of each agent. The GAP can be formulated as follows. Let $I = \{1, ..., m\}$ index the set of agents and $J = \{1, ..., n\}$ index the set of jobs. A standard integer programming formulation for GAP is given in expressions (5.6–5.9) where c_{ij} is the contribution to profit from assigning job j to agent i, a_{ij} the resource required for processing job j by agent i, and b_i is the capacity of agent i. The decision variables x_{ij} are set to 1 if job j is assigned to agent i, 0 otherwise. See Figure 5.3. The constraints, including the integrality condition on the variables, state that each job is assigned to exactly one agent, and that the bounded capacities of the agents are not exceeded [85, 121].

Assignment Problems 83

C	D	E	F	G	H
Cost Matrix (agents by jobs):	1	2	3	4	5
1	13	4	7	6	12
2	1	11	5	4	9
3	6	7	2	8	4
4	1	3	5	9	13
5	7	8	3	10	11
Decision variables:	1	2	3	4	5
1	0	1	0	0	0
2	0	0	0	1	0
3	0	0	0	0	1
4	1	0	0	0	0
5	0	0	1	0	0
minimize z =	16				
subject to:					
	1	2	3	4	5
columns (jobs)	1	1	1	1	1
=	=	=	=	=	=
1	1	1	1	1	1
rows (agents)					
1	1	=	1		
2	1	=	1		
3	1	=	1		
4	1	=	1		
5	1	=	1		

FIGURE 5.2: A simple assignment problem solved in Excel.

$$\max \; z = \sum_{i \in I} \sum_{j \in J} c_{ij} x_{ij} \qquad (5.6)$$

subject to:

$$\sum_{i \in I} x_{ij} = 1 \quad \forall j \in J \qquad (5.7)$$

$$\sum_{j \in J} a_{ij} x_{ij} \leq b_i \quad \forall i \in I \qquad (5.8)$$

$$x_{ij} \in \{0,1\} \quad \forall i \in I \quad \forall j \in J \qquad (5.9)$$

FIGURE 5.3: The generalized assignment problem formulated as an integer program.

Unlike simple assignment problems, solving a GAP can be challenging. There are no known simple heuristics for GAP and solving linear relaxations of them will normally not yield integer decisions. Let's look at a tiny example, from [121, page 194]. It may be specified as follows.

$$C = \begin{bmatrix} 6 & 9 & 4 & 2 & 10 & 3 & 6 \\ 4 & 8 & 9 & 1 & 7 & 5 & 4 \end{bmatrix} \qquad (5.10)$$

$$A = \begin{bmatrix} 4 & 1 & 2 & 1 & 4 & 3 & 8 \\ 9 & 9 & 8 & 1 & 3 & 8 & 7 \end{bmatrix} \qquad (5.11)$$

$$B = \begin{bmatrix} 11 \\ 22 \end{bmatrix} \qquad (5.12)$$

$$X = \begin{bmatrix} 1 & 1 & 0 & 1 & 0 & 1 & 0 \\ 0 & 0 & 1 & 0 & 1 & 0 & 1 \end{bmatrix} \qquad (5.13)$$

$$X^{ix0} = \begin{bmatrix} 0 & 0 & 1 & 0 & 1 & 0 & 1 \end{bmatrix} \qquad (5.14)$$

With reference to Figure 5.3, c_{ij} is the value at the intersection of row i and column j of the array C. a_{ij} is the analog for array A, x_{ij} is the analog for array X, and b_i is the ith row of array B.

In addition, X^{ix0} is the *0-indexed* version of X. Here, we assume that rows and columns in our arrays are numbered (indexed) starting with 0. Thus, $a_{0,0} = 4$, $a_{1,2} = 8$ and so on. The row indices of A, B, and C go from 0 to 1. The column indices of A and C go from 0 to 6. Returning to X^{ix0}, it is a row vector whose elements indicate the row of X in which the job associated with the jth entry of X^{ix0} is equal to 1. So, for example, $x_2^{ix0} = 1$ means that job 2 in X is assigned to agent (row) 1 in X. And so it is. This 0-indexing scheme is useful because it is very compact and easy to read, and it generalizes easily when there are more than 2 agents, in which case the entries in X^{ix0} will include integers larger than 1.

5.3 Case Example: GAP 1-c5-15-1

Let's now look at a larger problem, GAP 1-c5-15-1 from Beasley's OR Library http://people.brunel.ac.uk/~mastjjb/jeb/info.html. In this problem there are 5 agents (numbered 0 to 4) and 15 jobs (numbered 0 to 14). The data for the model's elements are given in Figures 5.4, 5.5, and 5.6.

$$C = \begin{bmatrix} 17 & 21 & 22 & 18 & 24 & 15 & 20 & 18 & 19 & 18 & 16 & 22 & 24 & 24 & 16 \\ 23 & 16 & 21 & 16 & 17 & 16 & 19 & 25 & 18 & 21 & 17 & 15 & 25 & 17 & 24 \\ 16 & 20 & 16 & 25 & 24 & 16 & 17 & 19 & 19 & 18 & 20 & 16 & 17 & 21 & 24 \\ 19 & 19 & 22 & 22 & 20 & 16 & 19 & 17 & 21 & 19 & 25 & 23 & 25 & 25 & 25 \\ 18 & 19 & 15 & 15 & 21 & 25 & 16 & 16 & 23 & 15 & 22 & 17 & 19 & 22 & 24 \end{bmatrix} \quad (5.15)$$

FIGURE 5.4: Objective function coefficients for GAP 1-c5-15-1. Rows: 5 agents. Columns: 15 jobs.

$$A = \begin{bmatrix} 8 & 15 & 14 & 23 & 8 & 16 & 8 & 25 & 9 & 17 & 25 & 15 & 10 & 8 & 24 \\ 15 & 7 & 23 & 22 & 11 & 11 & 12 & 10 & 17 & 16 & 7 & 16 & 10 & 18 & 22 \\ 21 & 20 & 6 & 22 & 24 & 10 & 24 & 9 & 21 & 14 & 11 & 14 & 11 & 19 & 16 \\ 20 & 11 & 8 & 14 & 9 & 5 & 6 & 19 & 19 & 7 & 6 & 6 & 13 & 9 & 18 \\ 8 & 13 & 13 & 13 & 10 & 20 & 25 & 16 & 16 & 17 & 10 & 10 & 5 & 12 & 23 \end{bmatrix} \quad (5.16)$$

FIGURE 5.5: Constraint coefficients for GAP 1-c5-15-1. Rows: 5 agents. Columns: 15 jobs.

$$B = \begin{bmatrix} 36 \\ 34 \\ 38 \\ 27 \\ 33 \end{bmatrix} \quad (5.17)$$

FIGURE 5.6: Constraint right-hand-side values for GAP 1-c5-15-1.

By contemporary standards this is a small problem. It is readily solved by Excel's Solver, even though the basic search space is $5^{15} = 30,517,578,125$. Figure 5.7 shows a setup for the problem in Excel, as well as an optimal decision. Figure 5.8 shows the Solver settings for doing the optimization. We see from Figure 5.7 that at optimality for this decision, jobs 0 and 1 are assigned to agent/server 1, and so on.

We also see that at optimality the value of the objective function is 336. Although we get an optimal decision and its objective function value from running the Excel Solver on our problem, we do not get much else. In this regard, the Excel Solver is typical of standard mathematical programming solvers, such as CPLEX. Because this problem cannot be properly formulated as an LP, we do not get sensitivity analysis information or other forms of post-solution analysis information. How might we be able to obtain such information?

This takes us to a main theme of this book: using metaheuristics to provide decision makers with information for deliberation that is not otherwise readily available. We begin this discussion incrementally, in the next section.

Normally in practice (as distinguished from teaching), GAPs in particular and constrained optimization problems in general are solved by implementing them in commercial solvers, such as CPLEX. After being represented in a front-end *algebraic modeling language*, a language in which a model can be represented and then solved by automatically translating the representation to a format understandable to the solver. IBM's OPL language is an example of such an algebraic modeling language. Figure 5.9 shows our GAP problem represented in OPL, within the IBM ILOG CPLEX Optimization Studio, a commercial product from IBM that serves as an integrated development environment (IDE) for optimization modeling. Once a model is properly represented in OPL, the user can click a button in the IDE and have the model solved by CPLEX (or another solver). The overall effect is like what we have seen in Excel, but now models are represented in an algebraic modeling language—which is a much superior representation form for large models—and "industrial strength" solvers are available.

MATLAB is an example of a prominent language and supporting system that can be used as an algebraic modeling language for optimization. Figure 5.10 shows a MATLAB implementation for GAP models and Figure 5.11 presents a test script in MATLAB for exercising the function in Figure 5.10. MATLAB and OPL are alternative modeling languages, along with AMPL and GAMS, for doing constrained optimization.

5.4   Using Decisions from Evolutionary Computation

In EC—*evolutionary computation*—we seek to solve problems, such as GAP and other constrained optimization problems, in a manner that reflects evolution by natural selection. To do this we use *evolutionary algorithms*, or EAs for short, which implement one form or another of EC. Of course selection is hardly natural in our implementations! We select decisions according to how well they do at solving the problem we have set. By analogy with biological evolution, we say that better decisions are more *fit* than the ones that do less well. EC is an example of a metaheuristic. It is a heuristic because, while we have evidence that it will generally perform well in solving our problems, there is no guarantee in any one instance that it will not perform badly. Nor is there ever any guarantee that it will likely find an optimal decision for any particular instance of a problem.

The essential idea of EC is simple. At a high level—as a metaheuristic—it goes as in Figure 5.12. In essence, we begin with a randomly created population of decisions, then we undertake an evaluate-select-perturb loop. In every pass through the loop we evaluate each of the decisions in the current population; we create a new population by drawing on the current population in a manner that is biased towards decisions with better evaluation scores (said to have higher "fitnesses"); we perturb the new population by making changes to at least some of its members; and we replace the current population with this new

	A	B	C	D	E	F	G	H	I	J	K	L	M	N	O	P
3	Decision	0	0	0	0	1	0	1	0	1	0	0	0	1	0	0
4	Variables:	1	1	0	0	0	0	0	1	0	0	0	0	0	0	0
5		0	0	0	1	0	0	0	0	0	0	0	0	0	0	1
6		0	0	1	0	0	0	0	0	0	1	1	1	0	0	0
7		0	0	0	0	0	1	0	0	0	0	0	0	0	1	0
8																
9	Objective	17	21	22	18	24	15	20	18	19	18	16	22	24	24	16
10	Coefficients:	23	16	21	16	17	16	19	25	18	21	17	15	25	17	24
11		16	20	16	25	24	16	17	19	19	18	20	16	17	21	24
12		19	19	22	22	20	16	19	17	21	19	25	23	25	25	25
13		18	19	15	15	21	25	16	16	23	15	22	17	19	22	24
14																
15	Constraint	8	15	14	23	8	16	8	25	9	17	25	15	10	8	24
16	Coefficients:	15	7	23	22	11	11	12	10	17	16	7	16	10	18	22
17		21	20	6	22	24	10	24	9	21	14	11	14	11	19	16
18		20	11	8	14	9	5	6	19	19	7	6	6	13	9	18
19		8	13	13	13	10	20	25	16	16	17	10	10	5	12	23
20																
21	Column Sum:	1	1	1	1	1	1	1	1	1	1	1	1	1	1	1
22		=	=	=	=	=	=	=	=	=	=	=	=	=	=	=
23		1	1	1	1	1	1	1	1	1	1	1	1	1	1	1

Objective: 336

	LHS Values:		RHS Values:
	35	<=	36
	32	<=	34
	38	<=	38
	27	<=	27
	32	<=	33

25	columnsums	=GAP1 c5l5-1 336!$B$21:$P$21
26	constraintcoefficients	=GAP1 c5l5-1 336!$B$15:$P$19
27	dvars	=GAP1 c5l5-1 336!$B$3:$P$7
28	lhsvalues	=GAP1 c5l5-1 336!$I$26:$I$30
29	objectivecoefficients	=GAP1 c5l5-1 336!$B$9:$P$13
30	requreunique	=GAP1 c5l5-1 336!$B$23:$P$23
31	rhsvalues	=GAP1 c5l5-1 336!$K$26:$K$30

FIGURE 5.7: GAP 1-c5-15-1: A small GAP solved in Excel.

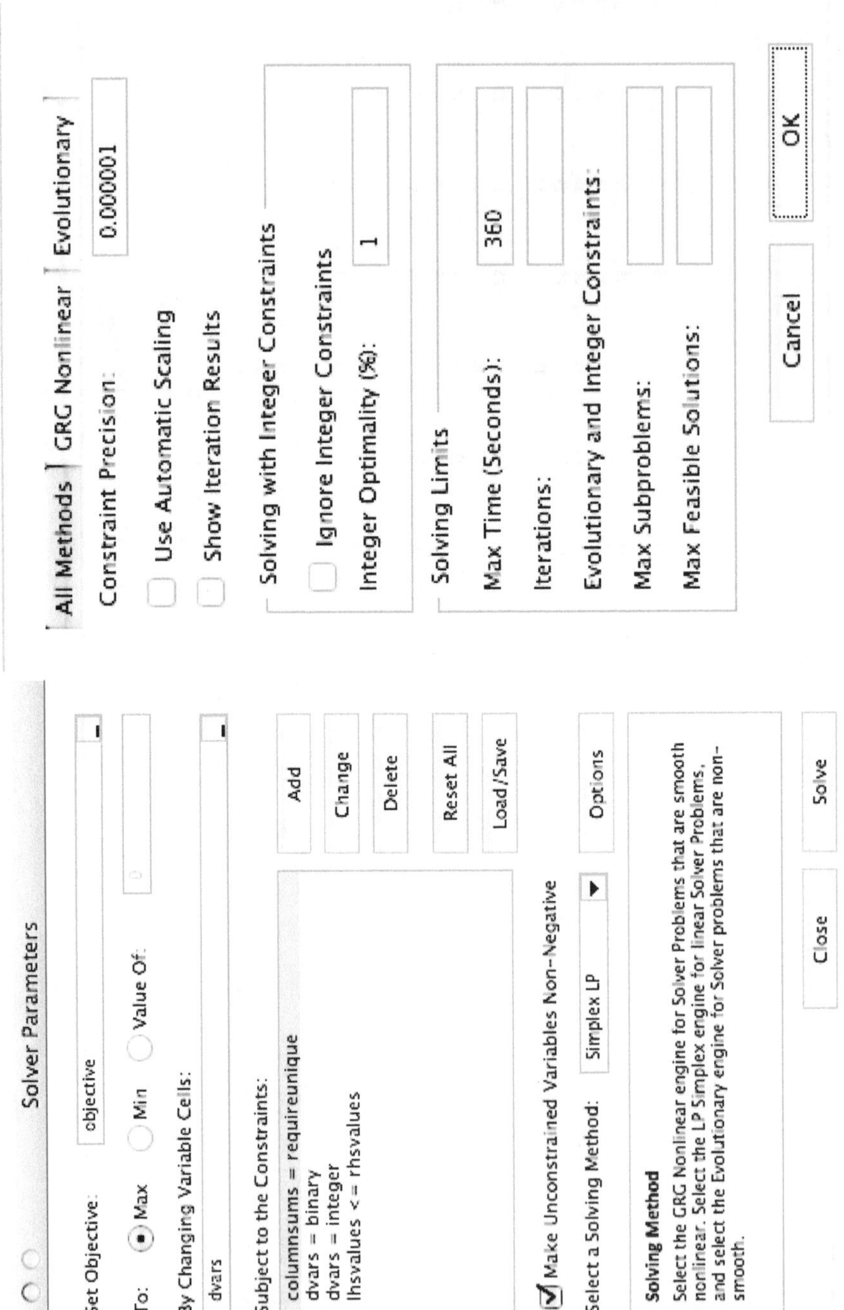

FIGURE 5.8: Excel Solver settings for GAP 1-c5-15-1.

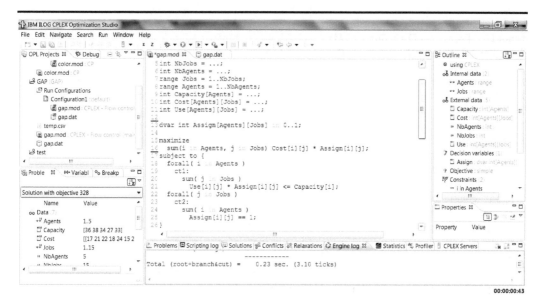

FIGURE 5.9: A generalized assignment problem model implemented in OPL.

one. Finally, we set a stopping condition—such as the number of times we go through the evaluate-select-perturb loop (called the number of generations by analogy with biological processes)—and we continue looping until the stopping condition is met. Then we see what we have.

EC is a *meta*heuristic because it really describes an indefinitely large number (or family) of particular implementations. These particular implementations are called *evolutionary algorithms* (EAs). In fact evolutionary algorithms, such as genetic algorithms, evolutionary programming, genetic programming, and so on, are themselves quite general and may be instantiated into many distinct algorithms, so it is appropriate to speak of these evolutionary algorithms as metaheutritstics as well. There are very, very many useful ways to implement particular algorithms that follow or even just resemble strongly the general pattern outlined in Figure 5.12. There is no official definition and indeed new particular heuristics are continually being invented and developed. For present purposes, such details are distractions.[2]

Consider now the table in Table 5.1. It contains the 21 best feasible decisions to the GAP 1-c5-15-1 problem that were discovered in a run of an EA. The decisions are sorted in descending order by objective value, labeled obj in column 1 of the table. We see that the first decision, row 0 of the table, has an objective value of 336, so from the Excel decision, we know it is optimal. Further, we see that this decision is identical to the optimal decision found by Excel and shown in Figure 5.7. In Table 5.1 we use 0-indexing to represent the decisions, so there is one decision per row. The column labeled x0 holds the 0-indexed assignments for job 0 for each of the decisions in the table, and similarly for x1, x2, ... x14. Thus, we can verify that in fact the optimal decision found by the EA is identical with the optimal decision found by Excel Solver.

The columns labeled b0, b1, b2, b3, and b4 in Table 5.1 hold the *constraint slack* values for the corresponding decision (row). Because $b_1 = 34$ in our problem, the b_1 value of 2.0 for

---

[2]We go into considerable detail later in the book, for example in Chapter 12. The EC data discussed in the present section was produced by a simple evolutionary program (or EP); see §12.2 for relevant details.

```
function [x,z,exitflag,output,Aineq,choices] = gap(C, A, b )
%gap Formulates a GAP IP from C, A, and b and calls intlinprog  to solve it.
% Returns the full intlinprog output (x,fval, exitflag,output), except that fval
% is replaced by z = -fval because GAPs are (assumed to be) maximizing and
% MATLAB's intlinprog minimizes. Type doc intlinprog in the command window to see
% MATLAB's  documentation of intlinprog.  C is an array constituting the
% objective function coefficients in the standard GAP formulation.
% It has dimensions number of agents by number of jobs. A is an array
% constituting the agent constraint coefficients in the
% standard GAP formulation. It has the same dimensions as C.
% b is an array constituting the right-hand-side values of the
% agent constraints in the standard GAP formulation. It has
% dimensions number of agents by 1.
%    On the output side, x will be a number of variables by 1 array,
% where the number of variables is the product of the number of
% agents and the number of jobs. This is necessitated by MATLAB's
% formulation requirements. All related matters are handled by the
% function, where xx = reshape(x,jobs,agents); xx = xx'; produces
% xx as a number of agents by number of jobs 0-1 array constituting
% a valid (and optimal!) decision to the original problem as
% standardly formulated.
%    Also on the output side we have Aineq and choices. Aineq*x <= b.
% So, the slacks at optimality are (b - Aineq*x).
% Aineq is the reformulated version of A in the standard GAP. choices
% is a number of jobs by 1 vector whose contents are the IDs of the
% assigned agent for the job at the given index. If choices(8) has the
% value of 3, then job 8 has been assigned to agent 3.
[agents,jobs] = size(C); %agents/jobs=number of agents/jobs.
numvars = agents * jobs;
C = C *  -1; % becuase intlinprog minimizes and we want to maximize
% Convert C to f, the vector form required by intlinprog.
C = C'; f = C(:);
% Set up Aineq, the analog of A in the standard formulation.
Aineq = zeros(agents,numvars);
for ii=1:agents
    start = ((ii - 1)*jobs) + 1;
    stop = start + jobs - 1;
    Aineq(ii,start:stop) = A(ii,:);
end
% Set up the constraints that each job is assigned exactly once.
Aeq = repmat(eye(jobs),1,agents); beq = ones(jobs,1);
% Make this a 0-1 integer program.
lb = zeros(numvars,1); ub = ones(numvars,1);
% Call the intlinprog solver.
[x,fval,exitflag,output] = ...
    intlinprog(f,(1:numvars),Aineq,b,Aeq,beq,lb,ub);
% Finish up by setting various output factors.
z = -1*fval;
xx = reshape(x,jobs,agents); xx = xx';
out=find(xx(:)>0); choices = mod(out,agents)';
choices = choices + ((choices == 0)*agents);
end
```

FIGURE 5.10: GAP solver function in MATLAB.

```
% MATLAB script to test the gap solver function on
% GAP c5-15-1 from Beasley's OR-Library.
% http://people.brunel.ac.uk/~mastjjb/jeb/info.html
C = [ 17 21 22 18 24 15 20 18 19 18 16 22 24 24 16;
 23 16 21 16 17 16 19 25 18 21 17 15 25 17 24;
 16 20 16 25 24 16 17 19 19 18 20 16 17 21 24;
 19 19 22 22 20 16 19 17 21 19 25 23 25 25 25;
 18 19 15 15 21 25 16 16 23 15 22 17 19 22 24];
A = [8 15 14 23 8 16 8 25 9 17 25 15 10 8 24;
 15 7 23 22 11 11 12 10 17 16 7 16 10 18 22;
 21 20 6 22 24 10 24 9 21 14 11 14 11 19 16;
 20 11 8 14 9 5 6 19 19 7 6 6 13 9 18;
 8 13 13 13 10 20 25 16 16 17 10 10 5 12 23];
b = [36; 34; 38; 27; 33];
[x,z,exitflag,output, Aineq,choices] = gap(C,A,b);
```

FIGURE 5.11: MATLAB script to test the *gap* solver function.

decision (row) 0 tells us that for this decision the amount of resource consumed for agent 1 is $34 - 2 = 32$. The constraint for agent 1 is said to have a slack of 2. In consequence, it would be possible to reallocate up to 2 units of b_1 for some other purpose, without making the decision infeasible and of course without changing its objective value. The constraint slacks for agents 2 and 3, however, are 0.0. These constraints are thus said to be *tight*. If we reduce their values—b_2 or b_3—decision 0 will no longer be feasible; the decision requires all of the resources associated with agents 2 and 3.

The table in Table 5.1 has two other column headings. The third column is labeled sslacks for "sum of the slacks." The sslacks value is just that, the sum of the corresponding values for b0, b1, b2, b3 and b4. It is a convenient statistic, for in general more slack is better. Finally, the second column is labeled svs for "sum of the constraint violations." A decision that is infeasible will have at least one constraint that uses more than its associated b_i value. Because the b_i values in the table represent (b_i—the amount of resource i consumed by the decision), b_i (in the table) will be negative (< 0) for any such constraint. svs is the sum of the b_i values that are negative for the decision. We know that all of the decisions in Table 5.1 are feasible because svs is uniformly equal to 0.0.

Having explained the table's meaning and learned a little from its contents, what else can we learn from Table 5.1 that might be useful in a real decision making situation? Very much, it turns out. Here are just a few observations.

1. With the exception of decisions 8 and 12, all of the decisions in Table 5.1 come with more slack than the optimal decision (decision 0). Some have considerably more, such as decision 3 which has an objective value of 334, off just 2 from decision 0, and a total slack of 15, instead of 4. Moreover, the slacks for decision 3 are all either equal—in the case of b2 and b3—to those for decision 0, or are much larger. This may well present an opportunity to implement decision 3 and divert the extra slack resources to another purpose. Similar opportunities are evident for other decisions as well.

2. There are interesting similarities and differences in the uniformities of the assignments. Every one of these 21 decisions assigns agent 2 to job 14. Job 11 is assigned to agent 3 in all cases but one, decision 11, where it is assigned to agent 4. Compare the b_i values

1. We create a number of decisions to the problem we wish to solve, thereby creating what we call a *population* of decisions by analogy with biological processes.

2. We evaluate each member of the population and obtain its *fitness score*.

3. We choose which individuals will be represented in the next generation.

   In doing this we sample randomly, but with bias towards individuals with higher fitnesses. Keeping the population size constant, some individuals will be represented by more than one ("daughter") decision in the next generation.

4. We alter the decisions in the next generation.

   In doing so we apply "genetic operators" such as computer program analogs of mutation and recombination in biology. Alterations are performed in a randomized fashion, resulting in a next generation population that resembles but is different from the current generation population.

5. If our stopping condition (such as number of generations) has not been met, we return to step (2), using the next generation population as our new current generation.

FIGURE 5.12: Outline of evolutionary computation procedures (EC).

for decisions 10 and 11. Both decisions have an objective value of 330.0, but their slacks are different. Every b_i value in decision 10 is equal to or larger than the corresponding b_i value in decision 11. Thus, decision 11 is weakly dominated by decision 10. It is difficult to imagine a circumstance in which one would prefer decision 11 to decision 10.

3. The constraint on agent 2 is tight in every one of these 21 decisions. Perhaps agent 2 is comparatively more efficient, and so comparatively more in demand. Perhaps agent 2 is under-resourced. The table on its own cannot tell us why this is happening, but it can flag unusual patterns for our attention.

4. The constraint on agent 3 is usually tight with a few exceptions. Decision 13 is notable with its value of 7.0. Depending on circumstances, this may represent an opportunity to redeploy some of this usually tight resource.

We could go on. In the exercises we ask you to go on. The larger point is that Table 5.1 and by extension the method that produced it offer valuable information for deliberation and reasoning with models.

Table 5.2 is the infeasible decisions counterpart of Table 5.1. Its interpretation is identical, except that the decisions it contains are the 37 best infeasible decisions discovered by the EA, sorted first by svs and then by obj. In the present case, the EA only found one infeasible decision with and objective value greater than the optimal value, here decision 0 at 342.0. Notice that the constraint violations are limited to one unit of agent 1. If that unit were available at the right price, another 6 units of objective value could be obtained. The EA has again found an opportunity worth considering.

	obj	svs	sslacks	b0	b1	b2	b3	b4	x0	x1	x2	x3	x4	x5	x6	x7	x8	x9	x10	x11	x12	x13	x14
0	336.0	0.0	4.0	1.0	2.0	0.0	0.0	1.0	1	1	3	2	0	4	0	1	0	3	3	3	0	4	2
1	335.0	0.0	6.0	1.0	2.0	0.0	0.0	3.0	1	1	3	2	4	4	0	1	0	3	3	3	0	0	2
2	335.0	0.0	5.0	3.0	2.0	0.0	0.0	0.0	0	4	3	2	0	4	1	1	0	3	3	3	1	0	2
3	334.0	0.0	15.0	3.0	7.0	0.0	0.0	5.0	4	1	3	2	0	4	0	1	0	3	3	3	1	0	2
4	334.0	0.0	8.0	1.0	1.0	0.0	0.0	5.0	4	1	3	2	0	4	3	1	0	1	3	3	0	0	2
5	333.0	0.0	13.0	3.0	2.0	0.0	1.0	8.0	1	1	3	2	0	4	0	1	0	3	3	3	4	0	2
6	332.0	0.0	11.0	1.0	5.0	0.0	0.0	5.0	4	1	3	2	0	4	1	1	0	3	3	3	0	0	2
7	332.0	0.0	7.0	2.0	0.0	0.0	0.0	5.0	4	1	3	2	0	4	0	1	1	3	3	3	0	0	2
8	332.0	0.0	3.0	1.0	2.0	0.0	0.0	0.0	1	4	3	2	0	4	3	1	0	3	1	3	0	0	2
9	331.0	0.0	11.0	3.0	7.0	0.0	0.0	1.0	0	1	3	2	0	4	0	1	4	3	3	3	1	4	2
10	331.0	0.0	11.0	4.0	3.0	0.0	0.0	4.0	0	4	3	2	1	1	0	1	0	3	3	3	1	0	2
11	331.0	0.0	6.0	1.0	2.0	0.0	0.0	3.0	1	1	3	2	0	4	3	1	0	3	3	4	0	4	2
12	331.0	0.0	4.0	1.0	1.0	0.0	1.0	1.0	0	1	3	2	0	4	3	1	0	1	3	3	4	0	2
13	330.0	0.0	11.0	3.0	1.0	0.0	7.0	0.0	4	1	3	2	0	4	0	1	0	1	3	3	0	0	2
14	330.0	0.0	13.0	3.0	7.0	0.0	0.0	3.0	0	1	3	2	4	4	0	1	0	3	3	3	4	3	2
15	330.0	0.0	6.0	5.0	1.0	0.0	0.0	0.0	4	1	0	2	0	4	3	1	0	1	3	3	1	0	2
16	330.0	0.0	14.0	2.0	1.0	0.0	2.0	9.0	4	1	3	2	0	4	0	1	4	1	3	3	4	3	2
17	330.0	0.0	6.0	1.0	1.0	0.0	1.0	3.0	0	1	3	2	4	4	3	1	0	1	3	3	0	0	2
18	330.0	0.0	12.0	2.0	2.0	0.0	1.0	7.0	1	1	3	2	0	3	0	1	4	3	4	3	0	0	2
19	330.0	0.0	5.0	1.0	2.0	0.0	2.0	0.0	1	1	4	2	0	4	3	1	0	3	3	3	0	0	2
20	329.0	0.0	13.0	3.0	7.0	0.0	0.0	3.0	0	1	3	2	0	4	3	1	0	3	4	3	1	0	2
21	329.0	0.0	13.0	11.0	1.0	0.0	1.0	0.0	4	3	3	2	0	4	3	1	0	1	3	3	4	0	2

TABLE 5.1: FoIs: Interesting feasible decisions found by EC for the GAP 1-c5-15-1 problem. Decision 0 is optimal.

	obj	svs	sslacks	b0	b1	b2	b3	b4	x0	x1	x2	x3	x4	x5	x6	x7	x8	x9	x10	x11	x12	x13	x14
0	342.0	-1.0	3.0	3.0	-1.0	0.0	0.0	0.0	1	4	3	2	0	4	0	1	0	3	3	3	1	0	2
1	332.0	-1.0	6.0	3.0	-1.0	2.0	1.0	0.0	1	4	2	3	0	4	0	1	0	3	3	3	1	0	2
2	332.0	-1.0	3.0	1.0	2.0	0.0	-1.0	0.0	1	0	4	2	4	1	0	1	0	2	3	3	0	3	2
3	331.0	-1.0	8.0	5.0	3.0	0.0	0.0	-1.0	4	1	3	2	0	3	0	1	4	3	3	3	1	0	2
4	330.0	-1.0	14.0	3.0	-1.0	0.0	1.0	10.0	1	4	3	2	0	1	0	1	0	3	4	3	1	0	2
5	330.0	-1.0	8.0	5.0	3.0	0.0	0.0	-1.0	4	0	3	2	0	4	3	1	4	3	4	3	1	0	2
6	328.0	-1.0	6.0	6.0	0.0	0.0	-1.0	0.0	4	1	0	2	0	4	0	1	1	3	3	3	4	3	2
7	328.0	-1.0	10.0	5.0	5.0	0.0	-1.0	0.0	4	1	0	2	4	3	1	1	0	3	3	3	4	3	2
8	327.0	-1.0	14.0	3.0	-1.0	0.0	1.0	10.0	1	4	3	2	0	3	0	1	0	3	4	4	1	0	2
9	327.0	-1.0	12.0	11.0	-1.0	0.0	1.0	0.0	1	4	3	2	0	4	0	1	0	3	3	3	1	3	2
10	327.0	-1.0	10.0	3.0	7.0	0.0	-1.0	-1.0	0	1	4	2	0	3	0	1	0	3	3	3	1	0	2
11	327.0	-1.0	11.0	2.0	1.0	0.0	8.0	-1.0	4	1	3	2	4	3	0	1	4	1	4	3	0	0	2
12	327.0	-1.0	14.0	4.0	8.0	0.0	2.0	-1.0	0	4	3	2	0	3	0	1	4	1	3	3	4	0	2
13	327.0	-1.0	13.0	10.0	1.0	0.0	2.0	2.0	4	1	3	2	0	3	0	1	4	1	3	3	0	3	2
14	326.0	-1.0	11.0	1.0	8.0	0.0	-1.0	-1.0	4	4	3	2	0	3	0	1	0	1	4	3	0	0	4
15	326.0	-1.0	20.0	12.0	7.0	0.0	1.0	0.0	4	4	3	2	0	4	0	1	4	3	4	3	1	3	2
16	326.0	-1.0	5.0	1.0	2.0	2.0	-1.0	5.0	1	1	3	2	0	4	0	1	0	2	3	3	0	0	2
17	326.0	-1.0	6.0	1.0	-1.0	0.0	0.0	5.0	1	1	3	2	0	4	0	1	0	3	4	3	0	3	2
18	326.0	-1.0	10.0	5.0	0.0	0.0	-1.0	2.0	4	4	0	2	4	1	3	1	4	3	3	3	1	1	2
19	326.0	-1.0	10.0	5.0	3.0	0.0	-1.0	-1.0	4	1	3	2	0	3	0	1	4	3	4	3	0	3	2
20	326.0	-1.0	13.0	10.0	1.0	0.0	2.0	-1.0	4	4	3	2	0	3	3	1	4	1	4	3	1	3	2
21	325.0	-1.0	20.0	2.0	17.0	0.0	1.0	-1.0	4	1	3	2	0	1	0	1	4	3	3	3	0	0	2
22	325.0	-1.0	14.0	2.0	6.0	0.0	6.0	0.0	4	1	3	2	4	1	0	1	4	3	3	3	0	0	2
23	325.0	-1.0	16.0	10.0	6.0	0.0	0.0	0.0	4	1	3	2	0	1	0	1	0	3	3	3	0	0	2
24	325.0	-1.0	17.0	4.0	13.0	0.0	0.0	-1.0	0	4	3	2	0	4	0	1	4	3	4	4	4	0	2
25	325.0	-1.0	6.0	1.0	5.0	0.0	-1.0	0.0	0	1	4	2	0	1	1	1	0	3	3	3	0	3	2
26	325.0	-1.0	10.0	5.0	3.0	0.0	-1.0	2.0	4	4	0	2	4	3	3	1	0	3	3	3	1	3	2
27	325.0	-1.0	17.0	1.0	1.0	0.0	-1.0	15.0	4	1	3	2	0	3	0	1	4	1	4	3	0	3	2
28	324.0	-1.0	11.0	2.0	1.0	0.0	8.0	-1.0	4	4	3	2	4	1	0	1	0	1	3	3	0	0	2
29	324.0	-1.0	6.0	3.0	1.0	0.0	-1.0	2.0	4	1	0	2	0	1	1	1	4	3	3	4	0	3	2
30	324.0	-1.0	16.0	10.0	6.0	0.0	0.0	-1.0	4	4	3	2	0	3	3	1	4	3	4	3	0	0	2
31	324.0	-1.0	6.0	2.0	2.0	0.0	-1.0	2.0	4	4	3	2	0	3	1	1	0	0	4	3	1	3	2
32	324.0	-1.0	11.0	1.0	8.0	0.0	-1.0	2.0	4	4	3	2	0	1	0	1	0	3	3	4	0	3	2
33	324.0	-1.0	13.0	12.0	1.0	0.0	0.0	-1.0	0	4	4	4	0	2	1	1	4	3	3	3	4	0	2
34	323.0	-1.0	22.0	3.0	-1.0	12.0	0.0	7.0	1	1	3	4	0	4	0	1	0	3	3	3	1	3	2
35	323.0	-1.0	19.0	1.0	2.0	16.0	-1.0	0.0	1	1	2	2	0	3	0	1	4	3	3	3	0	0	2
36	323.0	-1.0	20.0	12.0	7.0	0.0	1.0	-1.0	4	1	3	2	0	3	0	1	4	3	3	3	1	0	2

TABLE 5.2: IoIs: Interesting infeasible decisions found by EC for the GAP 1-c5-15-1 problem.

5.5 Discussion

Points arising:

1. The GAP is a paradigmatic constrained optimization problem of the linear integer programming type. As such it is a computationally challenging problem for exact solvers, although much progress has been made in recent years.

2. In real applications, it will often be necessary to deviate from a pure GAP formulation and to add constraints of various kinds. For example, some jobs may have to be processed by a subset of the available agents, or certain combinations of jobs may not be processable by a single agent. The possible additional complexity is extensive and quite real. The additional complexity will often prove difficult to handle by exact solvers.

3. Heuristic solvers, including those based on evolutionary computation, have proved effective at solving difficult GAPs along with extensions of them.

4. An important by-product of EC-based heuristic solvers is that they can produce, as a by-product of solving a problem, a plurality of decisions, both FoIs and IOIs, which have an important role in post-solution analysis.

5.6 For Exploration

A number of questions in this section, questions 1–4 below, refer to the data in this file, which are partially on display in Table 5.1.

The file *feas1500.csv* contains the data included in Table 5.1, and in fact 100 feasible decisions of interest in all. This file may be found on the book's Web site in the *Data/* directory. Use it for your answers.

1. In reference to the 100 FoIs in *feas1500.csv*, some of which are shown in Table 5.1, which decision provides the largest total slack and what is its objective function value? Answer this question (a) by inspecting the file as loaded into Excel or another spreadsheet program, and (b) by using the data filtering features in Excel (or another spreadsheet program) to produce the answer, e.g., by sorting the data and having the answer(s) appear at the top of the displayed data.

2. In reference to the 100 FoIs in *feas1500.csv*, some of which are shown in Table 5.1, which decision provides the largest objective function value if job 14 (counting from 0) is not assigned to agent 2? What agent is the job assigned to and what is the resulting objective function value? Answer this question (a) by inspecting the file as loaded into Excel or another spreadsheet program, and (b) by using the data filtering features in Excel (or another spreadsheet program) to produce the answer, e.g., by sorting the data and having the answer(s) appear at the top of the displayed data.

3. In reference to the 100 FoIs in *feas1500.csv*, some of which are shown in Table 5.1, which decision provides the largest objective function value if job 14 (counting from 0) is not assigned to agent 2? What agent is the job assigned to and what is the resulting

objective function value? Answer this question (a) by inspecting the file as loaded into Excel or another spreadsheet program, and (b) by using the data filtering features in Excel (or another spreadsheet program) to produce the answer, e.g., by sorting the data and having the answer(s) appear at the top of the displayed data.

4. In reference to the 100 FoIs in *feas1500.csv*, some of which are shown in Table 5.1, what is the opportunity cost of assigning job 1 to agent 4 (counting from 0)? If this is done, what other assignments change at optimality? Answer this question (a) by inspecting the file as loaded into Excel or another spreadsheet program, and (b) by using the data filtering features in Excel (or another spreadsheet program) to produce the answer, e.g., by sorting the data and having the answer(s) appear at the top of the displayed data.

5. Formulate and solve in Excel the tiny GAP discussed in §5.2. Repeat in OPL. Repeat in MATLAB.

6. Use the formulas from [159] to generate new GAPs, then formulate and solve them.

5.7　For More Information

The books [85] and [121] contain excellent general overviews of assignment problems. The monograph [25] is also excellent, but addressed to the specialist. [159], besides being a useful article proposing an interesting heuristic for GAPs, has a clear presentation of the standard algorithms or formulas for generating various classes of GAP test problems. [161] discusses another modern heuristic for GAPs. [160] is a short and very useful characterizing of GAPs and their generalizations.

Chapter 6

The Traveling Salesman Problem

6.1 Introduction

A sales person is planning a business trip that takes her to certain cities in which she has customers (or jobs to perform) and then returns her back home to the city in which she started. Between some of the pairs of cities she has to visit, there is direct air service; between others there is not. Can she plan the trip so that she (a) begins and ends in the same city while visiting every other city only once, and (b) pays the lowest price in airfare possible (assuming she can only buy direct air tickets)? The goal is not just finding a solution—a tour that visits every city on the agenda once—but an optimal solution, one with the lowest airfare.

The Traveling Salesman Problem (TSP) arises and can be expressed in many different ways in real world applications. We will discuss two.

- **Example 1: Theme Park Routing Problem:** A visitor in a theme park (say Disneyland or Universal Studios) likes to visit a number of attractions within the park in the shortest possible time. In what order shall she visit the attractions?

- **Example 2: Automated Teller Machine Problem:** A bank has many ATM machines. Each day, a courier goes from machine to machine to make collections and service the machines. In what order should the machines be visited so that the courier's route is the shortest possible?

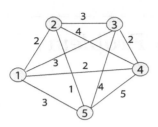   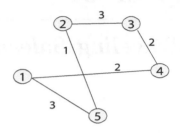

(a) Example TSP problem.   (b) And its solution.

FIGURE 6.1: TSP example with solution.

6.2   Problem Definition

We now offer a more formal definition of the TSP. Given a graph (an ensemble of nodes and arcs that connect them) with n nodes and costs on the arcs, the TSP involves finding a tour (i.e., a cyclical sequence of visits to all of the nodes, called a *Hamiltonian cycle*) with the minimum total traveling cost. Note that in a valid tour: (i) each node must be visited exactly once, and (ii) after departing from its starting node, the tour much return to it after visiting each of the other nodes. Figure 6.1 provides an example of a TSP instance and its optimal solution.

Finding an optimal TSP tour from a given graph is known to be a computationally hard problem. To illustrate, one might count the total number possible of unique solutions, which is $(n-1)!/2$ (assuming the costs are symmetric); so if there are 50 nodes to visit, the total number of solutions is $49!/2 = 3.04 \times 10^{62}$. Formally, Richard M. Karp showed in 1972 that deciding whether a Hamiltonian cycle exists in a given graph was so-called *NP-complete*, which provides a scientific explanation for the apparent computational complexity of finding optimal tours. Nonetheless, great progress was made in the late 1970s and 1980, when Grötschel, Padberg, Rinaldi and others managed to find exact solutions to TSP instances with up to 2392 cities, using cutting planes and branch-and-bound methods. For details on historical milestones in the solution of TSP instances, refer to the Traveling Salesman Problem page maintained by the University of Waterloo at http://www.math.uwaterloo.ca/tsp/history/milestone.html.

In the 1990s, Applegate, Bixby, Chvátal, and Cook developed the Concorde package that is freely available and widely used to produce optimal solutions for problems with thousands of nodes [3]. Gerhard Reinelt published the TSPLIB in 1991, a collection of benchmark instances of varying difficulty, which has been used extensively in research for comparing results. The TSPLib website is found in http://comopt.ifi.uni-heidelberg.de/software/TSPLIB95/index.html. In 2006, Cook and others computed an optimal tour through an 89,500 city instance given by a microchip layout problem, the largest solved TSPLIB instance to date (see http://www.math.uwaterloo.ca/tsp/pla85900/index.html). For many other instances with millions of cities, solutions can be found using heuristics that are provably within 2-3% of an optimal tour [134].

6.3 Solution Approaches

There are many methods to solve the Traveling Salesman Problem. In this section, let us look at two broad classes of approaches: exact algorithms and basic heuristics. Exact algorithms will not always be able to produce an optimal solution (i.e., one with the least cost) because of problem difficulty, but if they find an optimal solution they can recognize it and report it. Heuristics, as always, are only able to report the best solution they have found, without being able to determine whether or not it is optimal.

6.3.1 Exact Algorithms

Finding exact solutions or decisions for large TSPs is something of an international competitive sport. An exact solution/decision for 15,112 German towns from TSPLIB was found in 2001 using the cutting-plane method proposed by George Dantzig, Ray Fulkerson, and Selmer Johnson in 1954 [36], using a linear programming based algorithm. The total computation time was equivalent to 22.6 years on a single 500 MHz processor. In May 2004, the TSP of visiting all 24,978 towns in Sweden was solved: A tour of length approximately 72,500 kilometers was found and it was proven that no shorter tour exists. In March 2005, the TSP of visiting all 33,810 points in a circuit board was solved using the Concorde TSP Solver: A tour of length 66,048,945 units was found and it was proven to be optimal. The computation took approximately 15.7 CPU years, and in April 2006 an instance with 85,900 points was solved using Concorde TSP Solver, taking more than 136 CPU years [4].

In general, the TSP can be solved to optimality by formulating it as a mathematical programming problem and solving it using standard methods. Computational complexity, however, will, with problem scale, eventually defeat all such attempts. Let us look at the first-cut formulation as follows.

Parameters:

n = number of nodes ("cities" to be visited)

c_{ij} = cost of traveling from node i to node j (We assume $c_{ij} = c_{ji}$.)

Decision variables:

$$x_{ij} = \begin{cases} 1 & \text{if the salesman travels from node } i \text{ to } j, \\ 0 & \text{otherwise.} \end{cases}$$

First-cut mathematical programming model:

$$\min z = \sum_{i=1}^{n} \sum_{j=1}^{n} c_{ij} x_{ij} \tag{6.1}$$

Subject to:

$$\sum_{i \neq j} x_{ij} = 1 \quad \text{for all } j \tag{6.2}$$

$$\sum_{j \neq i} x_{ij} = 1 \quad \text{for all } i \tag{6.3}$$

$$x_{ij} \in \{0, 1\} \quad \text{for all } i, j \tag{6.4}$$

The above expressions (6.2–6.4) are assignment constraints for ensuring that each city (customer, etc.) is visited exactly once.

Will this formulation always produce a feasible solution? No! Consider the graph in Figure 6.2. Instantiating the above formulation leads to the following pattern:

$$\min z = 2x_{1,2} + 3x_{1,3} + 2x_{1,4} + 3x_{1,5} + 3x_{2,3} + 4x_{2,4} + \ldots + 5x_{4,5} \qquad (6.5)$$

s.t.

$$x_{2,1} + x_{3,1} + x_{4,1} + x_{5,1} = 1 \quad \text{for node 1}$$
$$x_{1,2} + x_{3,2} + x_{4,2} + x_{5,2} = 1 \quad \text{for node 2}$$
$$\ldots$$
$$x_{1,2} + x_{1,3} + x_{1,4} + x_{1,5} = 1 \quad \text{for node 1}$$
$$x_{2,1} + x_{2,3} + x_{2,4} + x_{2,5} = 1 \quad \text{for node 2}$$
$$\ldots$$

Solving this model will yield the solution shown in Figure 6.2.

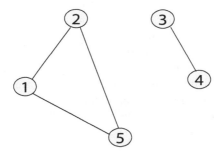

FIGURE 6.2: A disconnected (hence invalid) solution.

To prevent such infeasible solutions from being generated, we need additional constraints: the so-called *subtour elimination constraints*:

$$\sum_{i \in S} \sum_{j \in S} x_{ij} \leq |S| - 1 \quad \text{for every subset } S \text{ of } n - 1 \text{ or fewer nodes}$$

These constraints serve to prohibit the formation of cycles of lengths shorter than n. Unfortunately, while the constraints are easy to grasp and express, we actually need to introduce a number of them into the model that is exponential in the number of cities. This is not desirable from a computational standpoint, for it effectively limits the size of the problems we can handle. It is interesting to note that for large scale instances to be solved optimally, this model will need to be reformulated. Clever techniques in branch-and-bound, branch-and-cut have been introduced to reduce the effective search space. Such techniques are out of the scope of this book. The interested reader may refer to the TSP page http://www.math.uwaterloo.ca/tsp/methods/papers/index.html.

6.3.2 Heuristic Algorithms

Various heuristics that quickly yield good solutions have been introduced. Modern methods can find solutions/decisions for extremely large problems (millions of nodes) within a reasonable time that are with a high probability just 2–3% away from an optimal decision. Two categories of heuristics are in widespread use:

- Construction heuristics, and

- Iterative improvement heuristics, known as local search heuristics.

6.3.2.1 Construction Heuristics

Construction heuristics build a feasible solution by adding nodes to a partial tour one at a time and stopping when a feasible solution is found. Many construction heuristics are also called greedy heuristics because they seek to maximize the improvement at each step. One such heuristic is the nearest neighbor heuristic described as follows:

Step 1. Start with any node as the beginning of the tour.

Step 2. Find the node nearest to the last node added to the tour, add it to the tour, and make it the last node added.

Step 3. Repeat Step 2 until all nodes are contained in the tour.

Figure 6.3 provides an illustration.

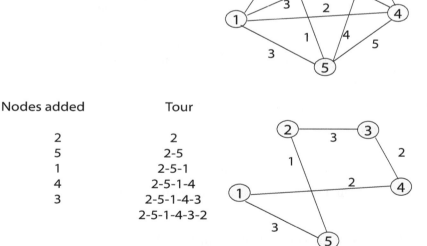

FIGURE 6.3: Example of a nearest neighbor heuristic for finding a TSP solution.

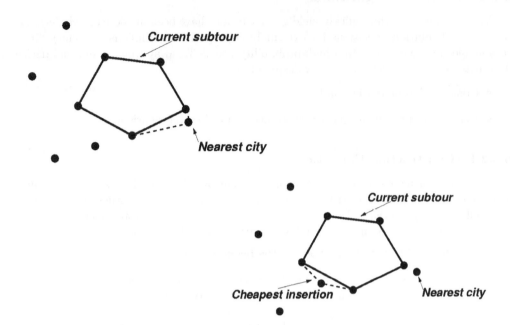

FIGURE 6.4: TSP: Cheapest versus nearest insertion heuristics.

The nearest neighbor heuristic is quite naïve. A slightly more savvy greedy heuristic that is easy to implement is known as the insertion heuristic. The basic idea is to form a subtour of increasing size starting with a subtour of two nodes chosen arbitrarily; at each step, select a node that is not yet in the subtour, and insert it into the subtour. There are different variants of insertion heuristics. Two of them are as follows.

Nearest Insertion:

- Insert the city (node) that has shortest distance to a tour city (node).

Cheapest Insertion:

- Choose the city whose insertion causes the least increase in length of the resulting subtour, and insert it at that location.

Figure 6.4 gives an illustration of these two heuristics. Can you think of other insertion heuristics?

6.3.2.2 Iterative Improvement or Local Search

The solutions constructed by the greedy heuristics given above can be quite poor in solution quality. A common method to improve the quality is to perform what is known as *iterative improvement* or local search.

Local search is a commonly used technique for solving large-scale combinatorial optimization problems. The main idea is as follows. Given an initial feasible solution (note that any permutation of nodes gives a feasible solution for TSP), we construct a *neighborhood* of feasible solutions and then search this neighborhood for a new solution. When selecting a neighbor, we may require the solution to be strictly better than the current best (if we wish to consider only improving moves), or we may use a probabilistic acceptance strategy such as simulated annealing that may accept the new solution with some probability value. More complicated heuristics such as tabu search can be seen as a kind of local search where the acceptance of a new solution depends on short- and long-term memories that are developed during the search. These and related ideas are discussed in Part III of the book, especially in Chapter 11.

There are quite a number of local search methods for the TSP. One common method is known as the *2-opt procedure*. Start with any tour (such as the tour constructed by any greedy heuristic) as the current tour. Consider the effect of removing any two edges (arcs) in the tour and replacing them with two other unique edges that form a different tour. If we find a tour with a lower cost than the current tour, then we make it the current tour. When no possible exchange of this type can produce a tour that is better than the current tour, we stop.

The following diagram gives an illustration of the 2-opt method applied to the instance given in Figure 6.5. Starting with an initial tour of cost 15, the algorithm considers all possible pairs of edges that can be removed and added to the tour. Two such possible solutions yield a resulting cost of 18, see Figure 6.6, and 11, see Figure 6.7, respectively. Hence, the second solution is chosen and replaces the current tour, and the process repeats.

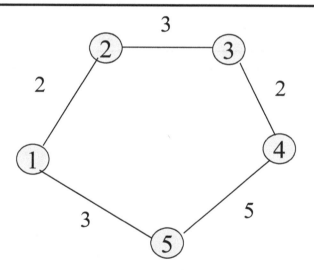

FIGURE 6.5: Initial tour is 1-2-3-4-5 with cost 15.

6.3.3 Putting Everything Together

To put everything in perspective, let us consider a larger problem instance as follows. In this instance, we have a worker who needs to service 20 jobs as plotted on a Euclidean plane, given in Figure 6.8. We assume that the distance between two locations is given by the Euclidean distance between these locations. In this plot, -101 is the depot (which provides the starting point of this tour), and 102 to 120 are the labels for the rest of the cities.

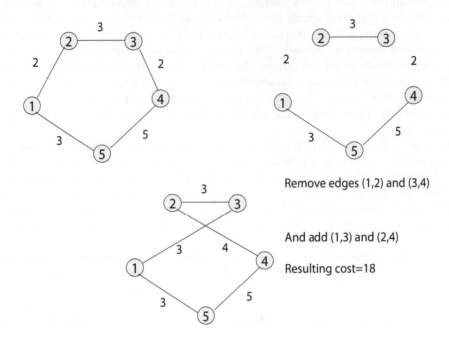

FIGURE 6.6: A first local improvement.

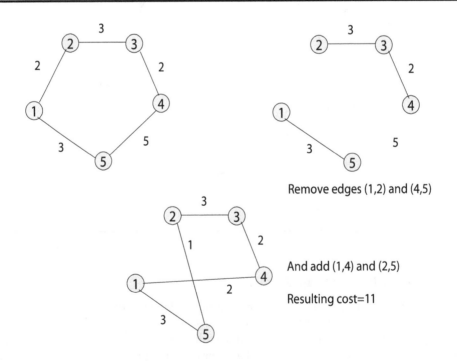

FIGURE 6.7: A second local improvement.

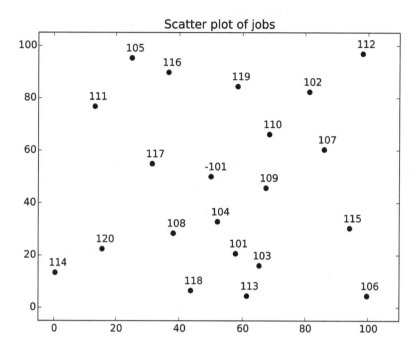

FIGURE 6.8: Scatter plot of 20 jobs. Depot at center as -101.

Figure 6.9 shows the result of a random ordering of the 20 jobs, to form a random tour. We expect it to be a very poor decision. Note that the frequent crossing of the tour path is a diagnostic suggesting that this solution is a not a good one.

Figure 6.10 shows the result of applying the simple (greedy) insertion heuristic discussed above to the problem. We see a substantial improvement over the random ordering. Notice that now the tour does not cross itself.

Figure 6.11 shows the result of applying a 2-opt procedure, described above, to the tour of Figure 6.10. This produces a modest improvement over the greedy insertion heuristic decision of Figure 6.10.

Figure 6.12 shows the best tour found by repeating the process that produced the tour of Figure 6.11 100 times with randomization. That is, in a single trial we first pick a random city as our starting point and then construct a greedy tour. Then we conduct a 2-opt improvement schedule, randomly picking the order in which the pairs of arcs to be compared are chosen. At the conclusion of the trial we note the best tour it has found and the length of the tour. We repeat this process 100 times, that is for 100 trials. Figure 6.12 represents the best results of one such experiment. It is appropriate to call this decision heuristically optimal. Notice the rather substantial, and perhaps surprising, improvement obtained over that in Figure 6.11.

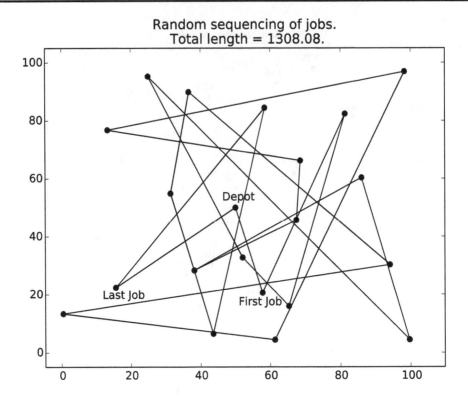

FIGURE 6.9: The 20 jobs of Figure 6.8 sequenced in a random order.

6.4   Discussion

The origins of the TSP are unclear. Mathematical problems related to the TSP were treated in the 1800s by the Irish mathematician William Hamilton, who introduced a recreational puzzle based on finding a Hamiltonian cycle of a given graph. The general form of the TSP appears to have been first studied by mathematicians during the 1930s in Vienna and at Harvard, notably by Karl Menger, who defines the problem, considers the obvious brute-force algorithm, and observes the non-optimality of the nearest neighbor heuristic.

In the 1950s and 1960s, the problem became increasingly popular in scientific circles in Europe and the USA. Notable contributions were made by George Dantzig, Delbert Ray Fulkerson, and Selmer M. Johnson at the RAND Corporation in Santa Monica, who expressed the problem as an integer linear program and developed the cutting plane method for its solution. With these new methods they solved an instance with 49 cities to optimality by constructing a tour and proving that no other tour could be shorter. In the following decades, the problem was studied by many researchers from mathematics, computer science, chemistry, physics, and other sciences.

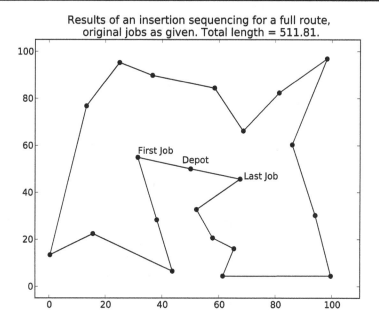

FIGURE 6.10: The 20 jobs of Figure 6.8 sequenced with a simple insertion procedure.

The TSP in its various forms appears often in real world applications. Our Example 2 is a problem that arises in practice at many banks. One of the earliest banks to use a TSP formulation for the problem, in the early days of ATMs, was the Shawmutt Bank in Boston.

6.5 For Exploration

1. Suggest other construction and/or local search heuristics for the TSP.

2. Consider the theme park routing problem mentioned in the Introduction. Suppose that you as a visitor have preferences on rides, and limited time so that you cannot afford to visit all the attractions. Design a heuristic to find a good solution for this problem.

3. Explain intuitively why the subtour elimination constraint works.

4. Consider the following insertion heuristic. Trace this heuristic against the graph given in Figure 6.1a.

   (a) Set S to be the set of all nodes in the problem.

   (b) Set T to be the initial tour of 2 nodes with the smallest distance between them.

   (c) Remove these 2 nodes (in T) from S.

   (d) While there are still cities in S:

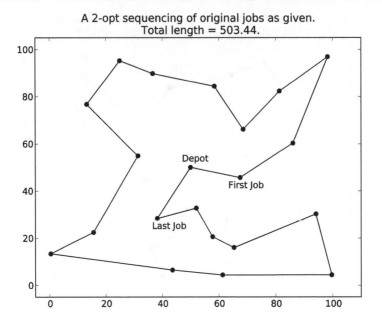

FIGURE 6.11: The 20 jobs of Figure 6.8 sequenced with a 2-opt procedure.

     i. Find the node x in S with the smallest distance to any node y in T, the current tour.

    ii. Choose the one with lowest overall distance (breaking ties randomly):
Insert x between y and its next destination z in T.
Insert x between y and its previous destination u in T.

   iii. Remove x from S.

5. On page 105 we find the following passage:

> Note that the frequent crossing of the tour path is a diagnostic suggesting that this solution is a not a good one.

Assess this statement. Is the frequent crossing of a tour path upon itself indeed a strong indication that the tour is not a very good one? Why?

6. On page 105 we find the following passage:

> We repeat this process 100 times, that is for 100 trials. Figure 6.12 represents the best results of one such experiment. It is appropriate to call this decision heuristically optimal. Notice the rather substantial, and perhaps surprising, improvement obtained over that in Figure 6.11.

The process is an example of using the method of *multiple random starts* of an algorithm for the sake of finding good decisions. We shall have much to say about this method in Part III of the book. Why use the method? That is, what reason is there to think it is appropriate as a general approach, that it will work often? It has worked for us in this example; why think it will work more generally?

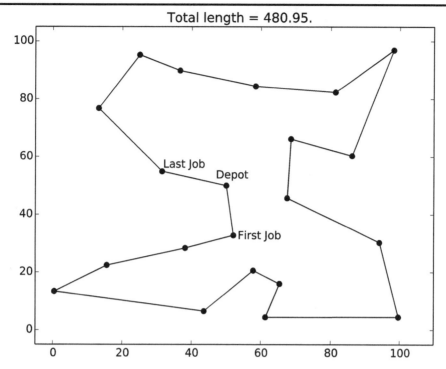

FIGURE 6.12: The best sequencing of the 20 jobs of Figure 6.8 found by 100 2-opt trials.

6.6   For More Information

The literature on the TSP is huge. Two excellent books are [114] and [4]. See [117] for the original description of the 2-opt procedure we use in the chapter and [41] for a particularly clear presentation of the procedure.

The TSP Web site is `http://www.math.uwaterloo.ca/tsp/`. The Concord TSP Solver site is `http://www.math.uwaterloo.ca/tsp/concorde/index.html`.

Chapter 7

Vehicle Routing Problems

7.1   Introduction

In the previous chapter, we discussed a sales person performing jobs scattered across a geographical region and how most efficiently to schedule them. In transportation logistics, very often there is one (or possibly more) depot where vehicles are parked. Given a set of jobs to be performed, the goal is to route the vehicles to service the jobs in a cost-optimal fashion.[1] Obviously, in settings where the cost is primarily on the number of vehicles used, if a vehicle has infinite capacity, the cheapest plan may be to simply deploy a single vehicle to serve all jobs. Realistically however, vehicles are constrained by capacity. For example, each vehicle in a courier service can only deliver packages whose total volume is constrained by the size of the vehicle. We call such problem the Capacitated Vehicle Routing Problem (CVRP) or simply the Vehicle Routing Problem (VRP).

The VRP occurs in many different ways. Here are two examples.

- **Example 1: Grocery Stores Delivery Problem:** A chain of convenience stores has twenty locations in one city. Each week goods must be delivered from a central warehouse to the stores. Items are packaged in standard sized containers. Each delivery vehicle has a capacity for 80 containers. The company would like to make all deliveries within an 8 hour shift on a single day of the week. How many vehicles are needed to fulfill this requirement?

- **Example 2: Field Service Problem:** A telco operator receives requests from households and companies for equipment repair and maintenance. There is a pool of engineers and technicians, each capable of servicing different kinds of requests. In view of manpower limitations, not all requests can be served on the same day. How can the assignment be carried out to minimize wait time of customers?

---

[1]Cost, of course, may be measured in many ways besides monetary cost, e.g., time, distance, manpower resources, etc.

7.2   Problem Definition

We shall now give a more formal definition of the VRP. Given a graph with n nodes representing customers and travel costs on the edges, the VRP involves finding a set of trips (tours), one for each vehicle, to deliver known quantities of goods to the set of customers. The objective is to minimize the number of vehicles used (this is our primary objective) and the travel costs of all trips combined (this is a secondary objective).

An example of a VRP instance and an optimal solution are given in Figure 7.1. Here, we have 14 customers on a Euclidean plane, and the depot is represented by the shaded square from which each vehicle begins its trip. Customer loads (service demands) are given next to the node, and the vehicle capacity is assumed to be 100. The optimal solution uses 5 vehicles.

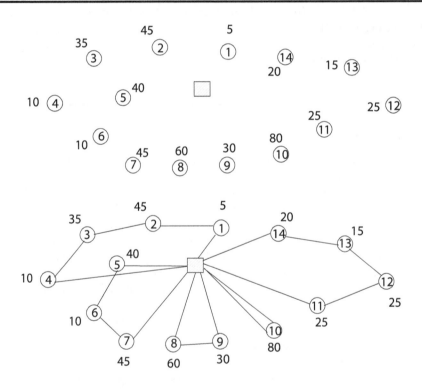

FIGURE 7.1: A VRP instance and an optimal solution

The VRP is a generalization of the Traveling Salesman Problem and (hence) is known to be a computationally hard problem. Nonetheless, great progress was made in the late 1990s for solving large scale problems with heuristics, and a survey of the literature is found in a recent edited volume by Golden, Raghavan, and Wasil [64].

To date, there are two freely available, non-commercial software packages for generating solutions to VRP instances. Both of these are focused on generating provably optimal solutions to smaller instances of the VRP. The first is the VRP code of Ralphs et al. that is included with the distribution of the open source mixed integer programming package SYMPHONY [132]. The second open source package is the CVRPSEP software of Lysgaard [119].

7.3 Solution Approaches

There are many methods to solve VRPs. In this section, we look at two broad classes of approaches: exact algorithms and heuristics. Exact algorithms will not always be able to produce an optimal solution (i.e., one with the least cost) because of problem difficulty, but if they find an optimal solution they can recognize it and report it. Heuristics, as always, are only able to report the best solution they have found, without being able to determine whether or not it is optimal.

7.3.1 Exact Algorithms

In general, the VRP can be solved to optimality by formulating it as a mathematical programming problem and solving it using standard methods. Computational complexity, however, will, with scale, defeat all such attempts. Let us look at the classical formulation as follows. Node 0 denotes the depot.

Parameters:

n = Number of customers

C = Vehicle capacity

d_i = Demand of customer i

c_{ij} = Cost (distance) of traveling from customer i to customer j

Decision variables:

$$x_{ij} = \begin{cases} 1 & \text{if edge } (i,j) \text{ is in the solution,} \\ 0 & \text{otherwise.} \end{cases}$$

k = Number of vehicles used

The following gives a classical bi-objective mathematical programming formulation, where the two goals are to first minimze the total number of vehicles used (k) and second the total distance travelled. A feasible solution is one in which the n nodes are partitioned into k disjoint routes that leave from the depot and return there, each satisfying the vehicle capacity constraint (i.e., total customer demand on each route has to be no greater than capacity C).

$$\min k \quad \text{and} \quad \sum_{i=1}^{n} \sum_{j=1}^{n} c_{ij} x_{ij} \tag{7.1}$$

Subject to:

$$\sum_{i=1}^{n} x_{oj} = 2k \tag{7.2}$$

$$\sum_{j=1}^{n} x_{ij} = 2 \quad \text{for all } i = 1, \ldots, n \tag{7.3}$$

$$\sum_{j \notin S, i \in S} x_{ij} \geq 2b(S) \quad \text{for every subset } S \text{ of 2 or more nodes} \tag{7.4}$$

$$x_{ij} = x_{ji} \quad \text{for all } i, \ j = 1, \dots, n \tag{7.5}$$

where $b(S) = \lceil (\sum_{i \in S} d_i)/C \rceil$ (intuitively this is the lower bound on the number of vehicles required to service set of customers S). The first two constrains are the degree constraints ensuring that the tours are properly set up. Constraint (7.4) can be seen as a generalization of the sub-tour elimination constraint for TSP as well as ensuring that no route has total demand exceeding capacity C.

1. Find the Traveling Salesman tour through all customers and the depot.

2. Starting at the depot and following the tour in any arbitrary orientation, partition the path into disjoint segments such that the total demand in each segment does not exceed the vehicle capacity.

FIGURE 7.2: Iterated Tour Partitioning heuristic.

1. Form sub-tours $i - 0 - i$ for $i = 1, 2, \dots, n$ (i.e., each customer is visited by a separate vehicle)

2. Compute savings $s_{ij} = c_{0i} + c_{0j} - c_{ij}$ for all $i, \ j$

3. Identify the node pair (i, j) that gives the highest saving s_{ij}.

4. Form a new subtour by connecting (i, j) and deleting arcs $(i, 0)$ and $(0, j)$ if the following conditions are satisfied:

   (a) both node i and node j have to be directly accessible from node 0;

   (b) node i and node j are not in the same tour;

   (c) forming the new subtour does not violate any of the constraints associated with the vehicles.

5. Set $s_{ij} = -\infty$, which means that this node pair is processed.

6. Go to Step 3, unless all node pairs with $s_{ij} \geq 0$ are processed.

FIGURE 7.3: Clarke-Wright Savings heuristic.

7.3.2 Heuristic Algorithms

Various heuristics that quickly yield good solutions have been introduced. Two categories of heuristics are listed as follows:

- Construction heuristics,

- Iterative improvement or local search heuristics.

7.3.2.1 Construction Heuristics

In the following, we present two such heuristics: the Iterated Tour Partitioning heuristic, in Figure 7.2, and the Clark-Wright Savings heuristic, in Figure 7.3.

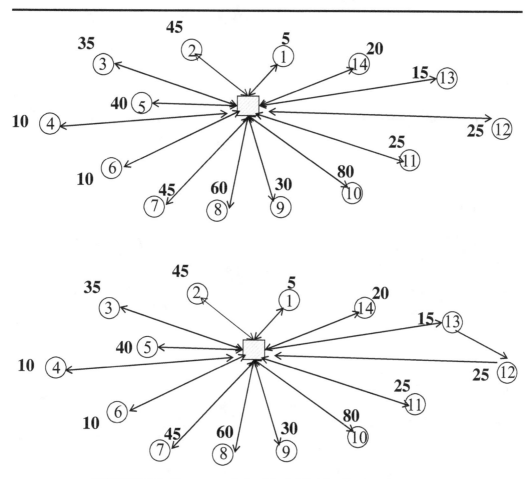

FIGURE 7.4: One step of the Clark-Wright Savings heuristic.

Referring to the problem instance given in Figure 7.1, suppose step 1 of the Iterated Tour Partitioning heuristic produces a traveling salesman tour of 1-2-...-14 in that order. Then step 2 of the heuristic simply partitions paths into segments of total demand no more than the vehicle capacity 100, thereby producing the solution which is shown in Figure 7.1.

A single iteration of the Clark-Wright Savings heuristic for Figure 7.1 is given in Figure 7.4.

7.3.2.2 Iterative Improvement or Local Search

There are a good number of local search neighborhoods for VRP. Some simple ones are (a) to perform exchange within a single route (like the 2-opt neighborhood for TSP discussed in the previous chapter) and (b) transfer of nodes between two routes, as shown in Figure 7.5. The relocate neighborhood for instance moves a node from one route to another; while the distribute neighborhood merges a short route into a long route.

More detailed discussion on local search will be found in Part III of the book.

Relocate:

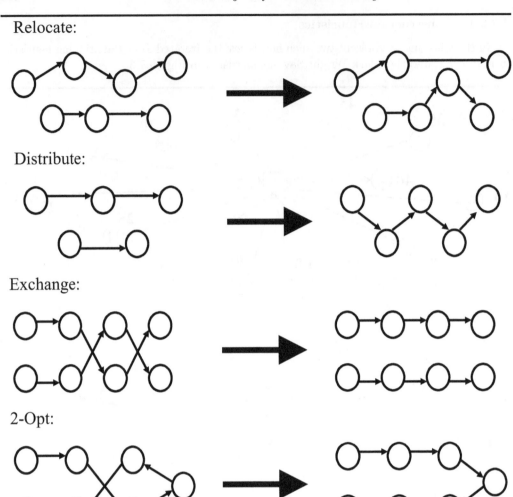

Distribute:

Exchange:

2-Opt:

FIGURE 7.5: Common local search neighborhoods for VRP.

7.4 Extensions of VRP

There are a large number of extensions for VRP. Two of the most common ones are the VRPTW (VRP with Time Windows) where customers have service durations and time windows (see for example [113]) and the PDP / PDPTW (Pickup and Delivery Problem, see for example [112]) where the job may either be a pickup job or delivery job. A pickup job will require the truck to have sufficient capacity to add the load onto the truck, while a delivery job is defined in the same way as the standard VRP.

Besides these, other common variants catering to real world requirements include the following:

1. Multi-Depot VRP, where vehicles are served out of more than a single depot.

2. Multi-Period VRP, where more than a one day plan needs to be generated and customer requests may be served in any of the days in the planning period.

3. VRP with Backhaul, where vehicles need to serve customer requests on the way back to the depot.

4. VRP with Stochastic Demands, where information about demands (such as demand loads, locations and service durations) may be uncertain.

5. Dynamic VRP, where demand information may change over time.

6. Green VRP, where the objective is to minimize environmental objectives such as carbon footprint.

7. Many real-life constraints such as driver breaks, special delivery requirement, etc.

7.5 For Exploration

1. Suggest other construction/local search heuristics for the VRP, VRPTW, and PDPTW.

2. Propose heuristics to handle the real world VRP extensions described above.

3. Read the paper written by Kant, Jacks, and Aantjes "Coca-Cola Enterprises Optimizes Vehicle Routes for Efficient Product Delivery" (*Interfaces,* Vol. 38, No. 1, 2008, pp. 40–50) [83] and answer the following questions:

   - What are the business challenges faced by the organization?

   - How is the problem different from the standard VRP described above?

   - What solution techniques could you use to solve this problem?

   - Discuss the SHORTREC decision support system (in particular the user interface shown in Figure 4 of that paper). What decision support capabilities are provided by the system and by having these decision support capabilities, how can these business challenges be resolved?

7.6 For More Information

The volume edited by Golden, Raghavan, and Wasil [64] is an up to date and comprehensive review of results for and approaches to vehicle routing problems.

The CVRPSEP package is a collection of routines, written in the C programming language, for separation of various classes of cuts in branch-and-cut algorithms for VRP based on the paper [119]. This software is OSI Certified Open Source Software.

The source code for the SYMPHONY package for solving VRPs is freely available and may be found at the Web site indicated in [132].

Chapter 8

Resource-Constrained Scheduling

8.1 Introduction

Scheduling underlies many decision making problems in diverse areas, including construction, manufacturing, and services. In general, scheduling is concerned with deciding how to commit resources over time to perform a variety of activities in order to optimize certain criteria (such as completion time, minimum cost, etc.). For example, in manufacturing a production schedule informs a production facility when to operate and on which machine equipment. The quality of a schedule makes a major impact on cost and operational efficiency.

More precisely, the solution to a scheduling problem is to determine time slots in which activities (or jobs) should be processed under given constraints. The main constraints are typically resource capacity constraints and temporal (mostly precedence) constraints between activities. In this chapter, we deal with a generic form of scheduling called the Resource-Constrained Project Scheduling Problem (RCPSP). It underlies very many scheduling problems.

The RCPSP arises in many different ways. We will discuss two examples.

- **Example 1: A Problem in Construction:** A building construction project is made up of a large number of activities which are temporally related (e.g., demolition must be completed before piling work begins, and painting can only start after plastering is completed). Each activity takes a certain duration and requires a number of resources (e.g., manpower, equipment, and machinery). The challenge is to complete the project as soon as possible given limited resource capacity.

- **Example 2: A Problem in Software Development:** Developing a software application requires a number of temporally related software engineering activities ranging from requirements analysis, to design, to coding, and to testing. Some activities can occur in parallel (e.g., two modules can be developed concurrently) while some activities have precedence relationships (e.g., testing can take place only after a module has been coded). The challenge is to complete the project as soon as possible given limited resource capacity.

8.2 Formal Definition

The RCPSP problem consists of N activities where each activity a_j, $(j = 1, \ldots, N)$ needs to be processed for a certain duration of d_j units without preemption. In addition, dummy activities a_0 and a_{N+1} with $d_0 = d_{N+1} = 0$ are introduced to represent the beginning and the completion of the project respectively.

A solution or schedule (or decision) is an assignment of start times to all activities, i.e., an array $(s(a_1), \ldots, s(a_N))$ where $s(a_i)$ represents the start time of activity a_i and $s(a_0)$ is assumed to be 0. Let $f(a_i)$ be the finish time of activity a_i. Since preemption is not allowed, we know that $s(a_i) + d_i = f(a_i)$ and the makespan is defined as the start time of the final dummy activity, which is equal to $\max_{i=1,\ldots N} f(a_i)$.

Schedules are subject to two kinds of constraints in general: temporal constraints and resource constraints. Temporal constraints restrict the time lags between activities. A minimum time lag between two different activities a_i and a_j constrains the minimum duration that must elapse between the start times of these two activities. Specifically, a zero time lag simply means that activity a_j cannot be started before activity a_i begins. For simplicity in this chapter, we assume one type of time-lag constraint, the precedence constraints, i.e., certain activities cannot start before certain other activities end. Such precedence constraints can be represented by setting the minimum time lag between these two activities to be the processing time of the first activity. Note that in general, one might also consider a maximum time lag between the start times of two different activities but in this chapter, we assume this value to be unbounded.

A resource unit is renewable and immediately available for another activity once it is no longer used by the activity that is completed. Each type of resource k, $(k = 1, \ldots, K)$ has a limited capacity C_k units. Each activity a_i requires r_{ik} units of resource of type k. Let $A(t)$ be the set of activities that are being processed at time instant t. A schedule is resource feasible if at each time instant t, the total demand for a resource k over all activities in $A(t)$ does not exceed its capacity C_k. A schedule ss is called feasible if it is both time and resource feasible. The goal in solving an RCPSP is to find a feasible schedule such that the makespan is minimized.

An example of an RCPSP instance and an optimal solution is given in Figure 8.1. For simplicity, we assume a single resource type with capacity $C = 10$. Each node represents an activity and each edge a precedence relationship. The two numbers a/b above each node stand for the processing duration and the number of units of the resource required by the activity.

Finding an optimal schedule of a given RCPSP problem is known to be computationally hard. It is interesting to note that even deciding if a schedule with a given makespan exists in a given graph is an NP-complete problem, which provides a scientific explanation for the apparent computational complexity of finding optimal tours.

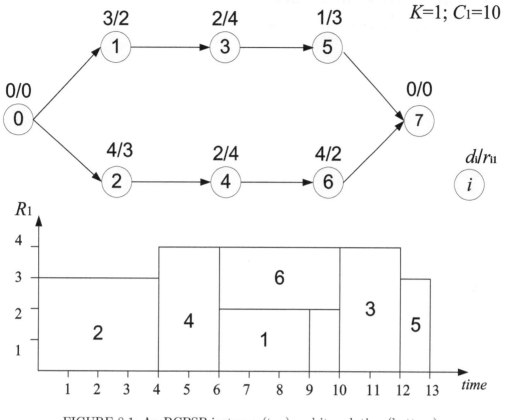

FIGURE 8.1: An RCPSP instance (top) and its solution (bottom).

8.3 Solution Approaches

There are many methods to solve the RCPSP. In this section, let us look at two broad classes of approaches: exact algorithms and basic heuristics. Exact algorithms will not always be able to produce an optimal solution (i.e., one with the least makespan cost) because of problem difficulty, but if they find an optimal solution they can recognize it and report it. Heuristics, as always, are only able to report the best solution they have found, without being able to determine whether or not it is optimal.

8.3.1 Exact Algorithms

In general, RCPSP can be solved to optimality using mathematical programming. Let us look at the first-cut formulation as follows.

Notation:

- Decision variables: F_j (finish time) for job j

- $P(j)$: set of predecessors of job j

- $A(t)$: set of active jobs j in period t

- r_{jk}: quantity of resource k required by job j

First-cut mathematical programming model:

$$\min z = F_{n+1} \tag{8.1}$$

subject to

$$F_h \leq F_j - d_j, \quad \forall j \in J, \ \forall h \in P(j) \tag{8.2}$$

$$F_j \geq 0, \quad \forall j \in J \tag{8.3}$$

$$\sum_{j \in A(t)} r_{jk} \leq C_k, \quad \forall k \in K, \forall t \tag{8.4}$$

The above model begs the question: *How to represent $A(t)$, which contains the set of jobs that are active at time t?* To do so, we need to do two things:

- Discretize time periods to $1, \ldots, H$.

- Use extra binary variables $X_{jt} = 1$ if job j ends in period t.

Hence, the final model is as follows:

$$\min z = F_{n+1} \tag{8.5}$$

subject to

$$F_h \leq F_j - d_j, \quad \forall j \in J, \ \forall h \in P(j) \ \text{Precedence constraints} \tag{8.6}$$

$$F_j = \sum_t t X_{jt}, \quad \forall j \in J \tag{8.7}$$

$$\sum_t X_{jt} = 1, \quad \forall j \in J \ \text{Determine the last period} \tag{8.8}$$

$$A_{jt} = \sum_{s=t}^{t+d_j-1} X_{js}, \quad \forall j \in J, \forall t = 1, \ldots, H \ \text{Determine active periods of tasks} \tag{8.9}$$

$$\sum_j A_{jt} r_{jk} \leq R_k, \quad \forall t = 1, \ldots, H, \ \forall k \in K \ \text{Resource capacity constraints} \tag{8.10}$$

8.3.2   Heuristic Algorithms

Schedule generation schemes (SGSs) are the core of most heuristic solution procedures for RCPSPs. The idea is to start from scratch and build a feasible schedule by stepwise extension of a partial solution (adding one job to the schedule at a time). In general, two schedule generation schemes are available, plus iterative improvement heuristics.

- Serial method (activity or job oriented): in each iteration, schedule one job at its earliest feasible starting time.

- Parallel method (time oriented): at the next finish time of a job, start a new job.

- Iterative improvement, local search.

8.3.2.1 Serial Method

The serial method is given in detail below. Notation:

- **S**: Set of scheduled jobs

- **D**: Decision set: jobs with all predecessors in **S**

- **U**: Unassigned set of jobs: these will first enter **D** and then **S**

Procedure:

1. Start with scheduled set $\mathbf{S} = \{\}$; $j = 0$.

2. Add j to **S**; Update $\mathbf{D} = \{j \in \mathbf{U} | P(j) \subseteq \mathbf{S}\}$.

3. Select job j from **D** and set j's start time as early as possible.

4. If $|\mathbf{S}| < n$, then go to Step 2.

For Step 3, one heuristic is to pick a job that has the highest priority, such as one that uses the least resource requirement, or one that has the shortest processing time, etc.

As an illustration, we consider the instance given in given in Figure 8.1, and trace the steps for constructing the schedule in Figures 8.2 and 8.3.

8.3.2.2 Parallel Method

The parallel method is given in detail below.

Procedure:

1. Start active job 0, and proceed in at most n iterations as follows.

2. In iteration g, let t_g be first finish time of any active job (i.e., the earliest time that unscheduled jobs may start) and D_g be the set of jobs that may start at time t_g.

3. Select the job from D_g with the highest priority; let it start at time t_g.

4. If jobs are still unscheduled then go to Step 2; else stop.

As an illustration, we again consider the instance given in given in Figure 8.1, and trace the steps for constructing the schedule in Figures 8.4 and 8.5. Notice that for this problem instance, the Serial method takes 6 iterations while the Parallel method takes 5 iterations to construct the schedule. The resulting schedules are different with different makespans. Even though it is generally true that the Parallel method is more efficient and effective, this is not the case all the time.

8.3.2.3 Iterative Improvement or Local Search

Unfortunately, basic local search methods do not work well in general for this complex optimization problem. We will introduce metaheuristics (tabu search, simulated annealing and genetic algorithms) in Part III of the book. Interested reader may refer to [108] for a good survey of heuristic techniques.

124

Business Analytics for Decision Making

Iteration 1

Decision set $D = \{1,2\}$
Selected job $j = 2$

Iteration 2

Decision set $D = \{1,4\}$
Selected job $j = 4$

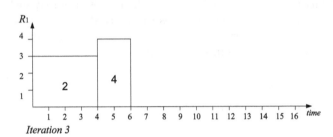

Iteration 3

Decision set $D = \{1,6\}$
Selected job $j = 1$

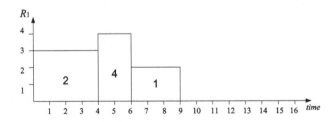

Iteration 4

Decision set $D = \{3,6\}$
Selected job $j = 3$

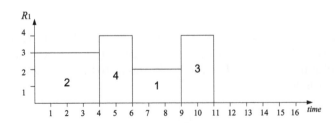

FIGURE 8.2: Tracing the Serial method.

Iteration 5

Decision set D = {5,6}
Selected job j = 5

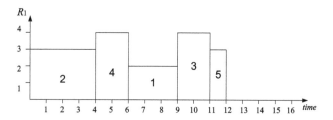

Iteration 6

Decision set D = {6}
Selected job j = 6

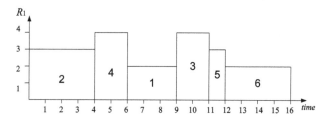

FIGURE 8.3: Tracing the Serial method (Cont'd).

8.4 Extensions of RCPSP

Hartmann and Briskorn [76] did a recent survey on variants and extensions of RCPSP. Some common extensions are given as follows:

1. RCPSP with Discounted Cash Flows, that maximizes the net present value of the project

2. Multi-Mode RCPSP, that assumes that activities have multiple possible durations, each with a different resource requirement

3. RCPSP with generalized precedence relations; an example is RCPSP/max where there is a maximum time-lag between the end time of an activity and the start time of another activity.

Iteration 1

$g = 1$
Schedule time $t = 0$
Decision set $D = \{1,2\}$
Selected job $j = 2$

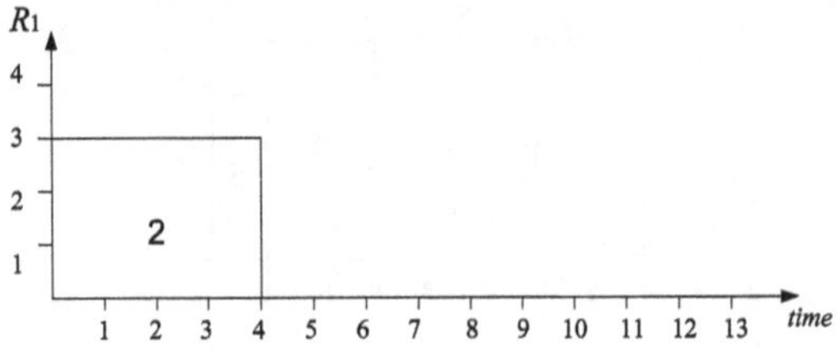

Iteration 2

$g = 2$
Schedule time $t = 4$
Decision set $D = \{1,4\}$
Selected job $j = 4$

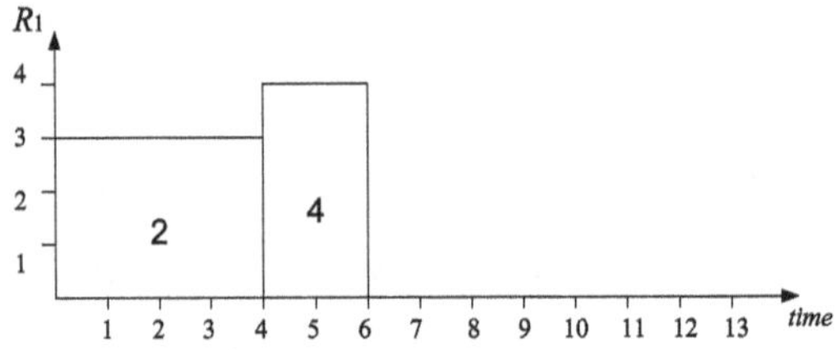

Iteration 3

$g = 3$
Schedule time $t = 6$
Decision set $D = \{1,6\}$
Selected jobs $j = 1$ and $j = 6$

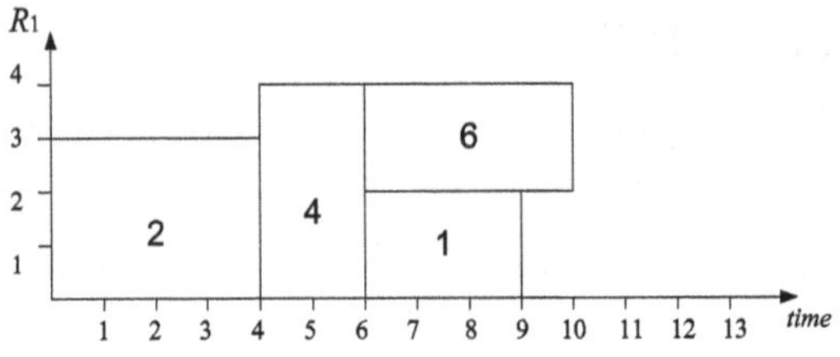

Iteration 4

$g = 4$
Schedule time $t = 10$
Decision set $D = \{3\}$
Selected job $j = 3$

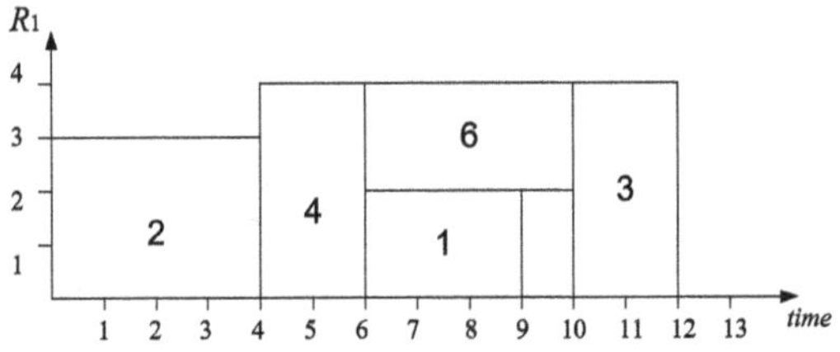

FIGURE 8.4: Tracing the Parallel method.

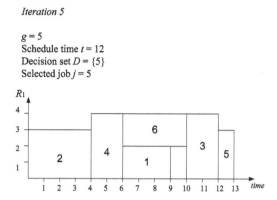

Iteration 5

$g = 5$
Schedule time $t = 12$
Decision set $D = \{5\}$
Selected job $j = 5$

FIGURE 8.5: Tracing the Parallel method (Cont'd).

8.5 For Exploration

1. Consider the problem instance given in Figure 8.1. Construct the schedule using the Serial and Parallel methods respectively.

2. Discuss the scenarios under which the Serial heuristic will outperform the Parallel heuristic, and vice versa.

3. Suggest other construction/local search heuristics for solving RCPSP.

8.6 For More Information

Much progress has been made to solve RCPSP exactly and heuristically. The reader may refer to the work of Kolisch and Hartmann [108] and a recent survey by Hartmann and Briskorn [76]. See also the Web site devoted to the problem: http://www.projectmanagement.ugent.be/?q=research/project_scheduling/rcpsp.

Chapter 9

Location Analysis

9.1   Introduction

With customers arrayed across a geographic extent, placement of service centers is contemplated with the aim of optimally serving the customers. How best to do this? Location analysis is concerned with just such questions. The need for location analysis arises in very many forms. Customers are typically people, but not necessarily; they may be other forms of life and they may be any form of non-living thing in need of servicing. Service centers may be police stations, fire stations, ambulance depots, counseling centers, retail stores, warehouses, product designs, political positions, and much else besides. The geographic extent may be conventionally spatial and two-dimensional, or it may have fewer or more dimensions. Travel times may conform to Euclidean metrics, or not. Relevant factors for optimization may include travel times or travel costs in many forms (total, maximal, minimal, average, etc.). Any number of other factors may be deemed relevant and important, including fairness, balance, and performance on ensemble attributes.

Given the enormous scope of location analysis problems, it is hardly surprising that the relevant literature is also enormous and that it has produced a very large and broad range of models. Nor can we expect the subject to be anything but open-ended. Under the circumstances, then, we essay in this chapter to introduce location analysis by focusing on a small number of representative, and widely useful, models. As we have throughout, we shall touch on model formulation and implementation, and then discuss in more depth the process of post-solution analysis in the present context, emphasizing decision sweeping and how we can use it.

Our problem focus in this chapter may be described as follows. We are given a number of areal districts—"tracts"—and we wish to locate one or more service points. We wish to minimize the total amount of travel for the collection of districts as a whole. Each district has a known geographic center. Travel distances are measured as the Euclidean distances between the known geographic centers of the various tracts.

9.2 Locating One Service Center

9.2.1 Minimizing Total Distance

To begin, we desire to find a "best" tract (areal unit) in the sense that the total distance from the geographic centers of all the tracts to the geographic center of the best tract is smaller than that for any other district. There is a simple and quite straightforward algorithm for doing this:

1. For each tract, find the total of the distances from every tract to it.

2. Find the tract with the smallest total distance score.

Let us call this the *enumeration* algorithm for finding the single best tract. With n tracts we need to sum each of the n (or $n - 1$) distances from the collection of tracts to it. This amounts to n^2 summations, which is very much a manageable number for most problems of interest. The enumeration algorithm is quite workable when we are only looking for a single service point. It will become useless from a practical standpoint when multiple service centers are being located.

Let us look at a specific example, which we shall use throughout this chapter: the 383 census tracts for Philadelphia, Pennsylvania, from the 2010 census. If we wish to locate a single service point our computational work is approximately $1 \times 383^2 = 146689$ distance calculations. If we wish to locate (as we shall) 10 service centers our work is approximately $_{383}C_{10} \times 383^2$ or about $10^{19} \times 383^2$ distance calculations, an impossible number from a practical standpoint. Even in the case of locating two service points the work is unacceptably high: $_{383}C_2 \times 383^2$ or $73153 \times 383^2 \approx 10^{10}$. The enumeration algorithm, then, can only be recommended for locating a single facility, but let us see what it tells us.

Table 9.1 on page 131 reports the top (lowest cost) 43 census tracts as single service centers. That is to say, Table 9.1 presents a sweep of the 43 best decisions (census tracts to use as a single service point). Points arising:

1. We might set 5% from optimality as our threshold for including a decision in the set of decisions of interest for the sake of deliberation. We are, after all, working with distance measures (tract centroid to tract centroid Euclidean distances) that only approximate what real people on the ground will face.

2. The ratio of #42 to #0 (the worst of this group to the best) is 29.992 / 28.542 $\approx$ 1.05. Using 5% as the solutions of interest cutoff, 43/383 or 11% of the districts would make defensible locations for locating a single service center.

No.	Sum of Distances	Tract ID
0	28.542	42101016702
1	28.542	42101017400
2	28.576	42101016600
3	28.606	42101016701
4	28.613	42101017300
5	28.622	42101020000
6	28.647	42101020101
7	28.671	42101016500
8	28.776	42101015300
9	28.803	42101017500
10	28.838	42101016800
11	28.892	42101019900
12	28.906	42101037700
13	28.915	42101016400
14	28.982	42101017201
15	29.004	42101020102
16	29.013	42101020200
17	29.065	42101014700
18	29.075	42101015200
19	29.100	42101020300
20	29.196	42101019800
21	29.208	42101014600
22	29.227	42101014800
23	29.377	42101017202
24	29.430	42101017601
25	29.431	42101016901
26	29.434	42101014500
27	29.444	42101014000
28	29.492	42101016200
29	29.576	42101015600
30	29.591	42101980500
31	29.614	42101019501
32	29.645	42101013900
33	29.648	42101016300
34	29.655	42101020500
35	29.668	42101020400
36	29.711	42101017100
37	29.732	42101017602
38	29.749	42101019700
39	29.756	42101014100
40	29.791	42101019502
41	29.822	42101015102
42	29.992	42101013800

TABLE 9.1: The 43 best single locations for minimizing total distance.

9.2.2   Weighting by Population

The Philadelphia census tracts are hardly uniform in population, as the descriptive table below shows. (There is even a single tract with no population!)

count	383
mean	3972.6
std	1711.9
min	0
25%	2795.5
50%	3913
75%	5116.5
max	8322

In this regard, the census tracts are quite typical of other location analysis problems: different areal units are distinguishable on size of customer base (and much else). It makes sense, then, to seek a minimum cost location, where cost is the sum of the population-weighted distances. In the distance-only calculations in the previous section we used an array of pairwise distances, with element $d(i, j)$ representing the distance from tract i to tract j. Now we use an array of weights, with element $w(i, j)$ representing the distance from tract i to tract j times the population of track j. Thus, $w(i, j)$ is the cost of assigning tract j to service center (or *hub*) i.

Table 9.2, page 133, reports the sweep of the top 43 decisions (lowest cost census tracts) as single service centers, weighted by population. Points arising:

1. The ratio of #42 to #0 (the worst of this group to the best) is 122163.138/116121.793 ≈ 1.05. Using 5% as the decisions of interest cutoff, 43/383 or 11% of the districts would again make defensible locations for locating a single service center.

2. There is considerable overlap between the weighted and unweighted results. Table 9.3 135 identifies the 33 tracts in common between Tables 9.1 (unweighted results) and 9.2 (weighted results).

3. Figure 9.1, on page 134, shows the locations of the top 12 tracts from Table 9.2.

9.3   A Naïve Greedy Heuristic for Locating n Centers

For the remainder of this chapter we will be concerned with the more general problem of locating n (> 1) service centers (or hubs) and we will work with weighted distances. Continuing to use the Philadelphia 2010 census data—which is both generally available and representative, as well as interesting in its own right—our distances are weighted by total population in a tract. Many other factors could be of interest, such as population between 20 and 35 years old at the time of the census, occurrences of demand, and so on.

The enumeration algorithm discussed in the previous section is deterministic and optimal. It is deterministic because given a fixed data set, it will always return the same answers (answers presented in part in Tables 9.1 and 9.2). It is optimal because given the data it does find the least cost solution. That there are many nearly optimal decisions is a feature of the problem. Indeed it is a feature typical of these kinds of problems.

We explore in this section and the next two heuristics for locating n centers. The subject of this section is a deterministic greedy procedure that leverages the enumeration algorithm.

No.	Sum of Distances	Tract ID
0	116121.793	42101019900
1	116141.768	42101017500
2	116374.623	42101020000
3	116386.492	42101017400
4	116720.317	42101019800
5	117044.649	42101020101
6	117201.416	42101016500
7	117218.681	42101016400
8	117382.421	42101019501
9	117430.195	42101020300
10	117462.859	42101016600
11	117553.554	42101017601
12	117629.602	42101019502
13	117806.663	42101016702
14	117889.380	42101020102
15	117894.080	42101019700
16	117951.933	42101017602
17	118001.983	42101017300
18	118213.736	42101980500
19	118536.243	42101016701
20	118774.744	42101037700
21	118776.394	42101016300
22	119175.157	42101020200
23	119211.530	42101016200
24	119220.023	42101038300
25	119501.240	42101015300
26	119695.687	42101020400
27	119997.283	42101028700
28	120107.969	42101016800
29	120356.232	42101017702
30	120381.104	42101017201
31	120428.962	42101015600
32	120498.018	42101028400
33	120876.837	42101014600
34	120888.009	42101028300
35	120980.300	42101014500
36	121079.066	42101014700
37	121556.955	42101028800
38	121592.962	42101020500
39	121644.000	42101015200
40	121683.902	42101016100
41	121948.734	42101015700
42	122163.138	42101028500

TABLE 9.2: The 43 best single locations for minimizing total weighted distance.

FIGURE 9.1: Philadelphia census tract map with high quality tracts highlighted, and #0, the putative best, separately distinguished.

No.	Tract ID
0	42101014500
1	42101014600
2	42101014700
3	42101015200
4	42101015300
5	42101015600
6	42101016200
7	42101016300
8	42101016400
9	42101016500
10	42101016600
11	42101016701
12	42101016702
13	42101016800
14	42101017201
15	42101017300
16	42101017400
17	42101017500
18	42101017601
19	42101017602
20	42101019501
21	42101019502
22	42101019700
23	42101019800
24	42101019900
25	42101020000
26	42101020101
27	42101020102
28	42101020200
29	42101020300
30	42101020400
31	42101020500
32	42101037700
33	42101980500

TABLE 9.3: Thirty-three tracts are within 5% of the best tract on unweighted distances and on weighted distances.

To locate n centers, our greedy heuristic (see [37, chapter 6]) first uses the enumeration method to find the single best service point. It essentially repeats the procedure for the second service point by enumerating the costs of adding each of the remaining tracts as the second service point *given that the first service point is in place*. The cost of adding a new service point, tract i, for any tract j is just the minimum of $w(i,j)$ (the weighted cost of traveling from j to i) and the cost of traveling from j to any service points already established, particularly the first.

It should be clear why calling this a greedy heuristic is apt. The procedure is rather clearly both greedy and a heuristic. It should also be apparent that the computational cost is low, basically n times the cost of placing the first service center.

Table 9.4 on page 137 reports the sweep of the top 43 decisions (lowest cost census tracts) having in common their first 9 tracts, weighted by population. The table gives the tenth tract in each case in the right most column. Points arising:

1. In the best found solution, number 0, the 10 districts are (added in order): 42101019900, 42101033400, 42101036900, 42101025200, 42101008400, 42101002300, 42101035302, 42101029200, 42101014800, 42101035701.

   Every other decision in the table has in common the first 9 of these tracts.

2. The ratio of the worst of the 43 to the best is $33455.587/33032.923 \approx 1.01$, indicating ample density of decisions of interest, if we use the 5% cutoff as we did above.

3. We would like to know how well the greedy heuristic is performing. To that end we could (and should if we can) compare it with the result of an exact optimizer. Such an optimizer, however, will only deliver one (optimal) solution at best. This prevents us, absent the heuristic, from having access during deliberation to high quality but non-optimal solutions. Remember: the optimal solution is only optimal with respect to the data and the model.

4. An alternative (a complementary one) to seeking an optimal solution for comparison is to explore an alternative heuristic. Perhaps there are better ways to sweep the decisions and explore the solutions of interest. The next section presents just such an alternative.

9.4   Using a Greedy Hill Climbing Heuristic

Greedy hill climbing is a quite generally applicable metaheuristic (family of heuristics). It will often give satisfactory results and it is conceptually important as a basis for metaheuristics based on local search. Unlike the naïve greedy heuristic of the previous section, which is deterministic, greedy hill climbing is inherently stochastic. That said, greedy hill climbing is plenty naïve on its own account, but that is a topic we take up in the sequel. See especially Chapter 11.

We encountered greedy hill climbing briefly in Chapter 2 and gave a high-level description of it in Figure 2.8 on page 38. Figure 9.2 presents a more detailed description of the heuristic.

To restate our problem, let us assume we wish to locate 10 (more generally, n) service centers. A decision in this case, as above, is simply a list of 10 (n) distinct tracts in which the service centers are to be located. We call these the *hubs*. The value of a decision is the total weighted travel cost to the hubs, assuming each tract is assigned to its lowest cost hub.

In our version of greedy hill climbing for this problem, the neighborhood exploration step (step (3) in Figure 9.2) works as follows. We introduce a parameter, *number_of_tries*. We used 140 and 200 to get the results reported below. For each neighborhood exploration step undertaken (step (3) in Figure 9.2) we randomly remove one of the tracts from the incumbent decision. Then we randomly introduce a new tract (resulting in a complete decision again with 10 (n) hubs) and evaluate the decision. We do this *number_of_tries* times and then (step (4)) we examine the values of the new decisions. If none improve on the value of our incumbent solution, we stop the run; otherwise, we accept the best of the new decisions and we continue the run (step (2)).

This single run greedy hill climbing procedure is wrapped within another procedure that calls it a number of times. That number is set with a second parameter, *number_of_runs*, which equals 43 in our present example. It sets the number of times the greedy hill climbing

No.	Sum of Distances	Last added tract
0	33032.923	42101035701
1	33082.596	42101021300
2	33096.745	42101021000
3	33132.855	42101035601
4	33136.939	42101021400
5	33150.859	42101021200
6	33158.323	42101035702
7	33177.402	42101017800
8	33184.679	42101017900
9	33204.558	42101035900
10	33210.938	42101035500
11	33212.727	42101021500
12	33214.981	42101021700
13	33227.871	42101021100
14	33234.697	42101035800
15	33235.276	42101020900
16	33286.497	42101027600
17	33300.711	42101017701
18	33302.304	42101017702
19	33317.732	42101016100
20	33318.921	42101027500
21	33333.235	42101016000
22	33338.137	42101036000
23	33342.410	42101018001
24	33347.458	42101028200
25	33348.150	42101027800
26	33358.153	42101006200
27	33374.746	42101026700
28	33374.770	42101035602
29	33375.070	42101980100
30	33386.129	42101018800
31	33387.846	42101027700
32	33395.896	42101027401
33	33397.372	42101018002
34	33401.740	42101031700
35	33417.766	42101006300
36	33419.953	42101019200
37	33431.178	42101006600
38	33442.319	42101031900
39	33448.564	42101031800
40	33453.866	42101006000
41	33454.471	42101027000
42	33455.587	42101006700

TABLE 9.4: Top 43 single locations having the first 9 of 10 tracts in common, using the naïve greedy algorithm with weighted distances.

1. Begin with a random decision; make it the incumbent decision.

2. Evaluate the incumbent decision.

3. Explore the neighborhood of the current decision, evaluating the neighborhood decisions.

4. If no decision is found that produces an improvement over the current decision, stop; otherwise continue.

5. If an improvement is found in the neighborhood, make the best found decision the (new) incumbent decision and evaluate it.

6. Go to step (2).

FIGURE 9.2: Pseudocode description of one run of the greedy hill climbing metaheuristic.

heuristic is to be run in a single *trial* of the procedure. In a trial, each of the *number_of_runs* runs begins with a random start (step (1) in Figure 9.2). We collect and report results from the totality of runs undertaken in the trial.

Figure 9.3 presents pseudocode that is specific to the greedy hill climbing procedure we use in this chapter. The actual Python code is in the file *location_greedy_climb.py* in the folder for this chapter at the book's Web site.

Table 9.5, on page 141, reports the results of a trial of 43 runs (43 executions of the procedure implementing Figure 9.3). It is, thus, a sweep of the top 43 decisions (lowest cost census tracts) having in common that they were found by independent runs, starting randomly, of our greedy hill climbing heuristic, with parameter *number_of_tries* set to 140. The third column of the table lists the ten tracts in the associated decision. The tracts are identified by their array indices for the sake of saving space.[1] Points arising:

1. The greedy hill climbing results, even with only 43 runs, noticeably outperform the deterministic greedy procedure of the previous section, whose best decision has a value of 33032.923 (Table 9.4). Of the 43 results in Table 9.5, 34 or 79% have superior values.

2. The best value obtained by this trial of 43 runs of the greedy hill climbing heuristic is 31770.088,. With $33032.923/31770.088 \approx 1.04$, it constitutes a 4% improvement. This is small, but definitely an improvement.

3. Upping *number_of_tries* to 200 and *number_of_runs* to 500, gives us a trial that takes longer to run, but yields improved results. See Table 9.6 on page 142. The best result from that trial is 31203.451, yielding $33032.923/31203.451 \approx 1.06$ for a 6% improvement over the best result from the greedy procedure reported in Table 9.4.

4. The greedy hill climbing procedure is, as announced, a *heuristic* solver, not an *exact* solver. In consequence, we cannot know without trying whether increasing the search effort—by increasing values of the *number_of_tries* or *number_of_runs* parameters—will yield substantial improvements. In practice, we rely on experience and good judgment

---

[1]The actual tract IDs may be recovered from **arraytractindices**, a dictionary available by unpickling the Python pickle file *weights_and_dicts.p*, which in turn is available on the book's Web site, in the folder for this chapter. The tract IDs are numbers created by the U.S. Bureau of the Census. Our array indices are to these numbers sorted in increasing value.

1. *incumbent_decision* ← a randomly created decision

   (Alternatively, begin with a decision chosen with a purpose in mind, such as to explore in the neighborhood of a proposed or known good solution.)

2. *number_of_tries* ← a value chosen with experience

3. *incumbent_decision_value* ← evaluate *incumbent_decision*

4. *improvement* ← *true*

5. *While improvement == true*

   (a) *improvement* ← *false*

   (b) Explore the neighborhood of the *incumbent_decision*, by obtaining *number_of_tries* decisions in the neighborhood of the *incumbent_decision*; evaluate each of the new decisions.

   (c) If an improvement is found in the neighborhood, make the best found decision the (new) incumbent decision and evaluate it.

      i. *incumbent_decision* ← best found decision

      ii. *incumbent_decision_value* ← evaluate *incumbent_decision*

      iii. *improvement* ← *true*

FIGURE 9.3: More specific pseudocode description of the greedy hill climbing metaheuristic.

in conducting the meta-search to set parameters value (here *number_of_tries* and *number_of_runs*) for our heuristic procedure. As with any other randomized heuristic, the outcome of any single trial or run is random and will vary with other trials or runs. Replication is necessary for confidence in any general performance estimate. That, however, is not our immediate concern.

5. Note in particular that *all* of the decisions reported in Table 9.6 (page 142) have solution values superior to the best solution value from Table 9.5 (page 141).

6. Table 9.7 on page 143 presents 43 top results from a trial of 500 runs of a *threshold accepting algorithm*. Threshold accepting is a metaheuristic we discuss in some detail in Chapter 11 and use elsewhere in the book.

7. Table 9.8 on page 144 presents 43 top results from a trial of 120 runs of a *simulated annealing* procedure. Simulated annealing is a metaheuristic we discuss in some detail in Chapter 11 and use elsewhere in the book.

8. This is the best overall decision we have found at the time of the writing:

No.	Sum of Weighted Distances	Tract Array Indices
0	31057.108	[44, 81, 102, 125, 173, 194, 242, 282, 291, 341]

It was found by a run of simulated annealing. See Figure 9.4 for a map.

9. All of our results produced by metaheuristics (including greedy hill climbing as a courtesy)—see in particular Tables 9.5, 9.6, 9.7 and 9.8—show a remarkable density

of good decisions that are distinct from and better than the value of the best decision found by the naïve greedy method of §9.3. Randomized metaheuristics have, in this case at least, consistently improved upon results returned by a deterministic heuristic.

10. As visual inspection indicates, the high quality solutions obtained by decision sweeping are quite diverse in detail. See for example column 3, "Tract Array Indices," of Table 9.5. In all, 175 different tract indices appear in the table. Comparing the best and second best decisions in the table, different in value by $31770.736/31770.088 = 1.00002$ or 0.002%, the two decisions have only 3 tracts in common. The observation holds generally for our results.

The upshot of these examples (using real data) is that our (heuristic) optimization procedures have apparently succeeded in finding good decisions. Of equal or greater importance is that these procedures have afforded successful decision sweeping, yielding a significant number of high-quality, very distinct decisions.

Taking the contents of the several as our consideration set, we now face the problem of choosing a decision from among these solutions of interest. They are more or less indistinguishable from one another in terms of their associated objective values. Focusing now on Tables 9.6, 9.7 and 9.8, the worst:best ratio is $31701.183/31064.461 \approx 1.02$. How we might go about making a good choice among these decisions is our subject in the next section.

9.5   Discussion

The technical name of the variety of location problem we have addressed in this chapter is the *p-median problem*. Although it is a constrained optimization problem, we have not formulated it as a mathematical program. We leave that task to Chapter 18.

Our principal aim in this chapter has been to illustrate—with real data on a real location problem—decision sweeping with metaheuristics and to establish in the present case that local search metaheuristics (in particular greedy hill climbing, threshold accepting algorithms, and simulated annealing) are not only able to find high quality decisions, but—what we think is even more important—are able to find quantitatively substantial corpora of high quality decisions.

With this embarrassment of riches, which we shall see again and again, how are we to decide what decision, if any, to implement? Picking the best found solution or picking at random among the best found are strategies with obvious merit. There may even be situations in which this is appropriate. For the most part, however, productive decision sweeping of a model should be seen as an opportunity to deliberate with information not already in the model.

Chapter 16 discusses a principled and quite general response to the availability of a plurality of solutions: create an index or scoring mechanism—in the form of a mathematical model—to evaluate and compare the plurality of items, e.g., decisions of interest, on multiple criteria.

No.	Sum of Weighted Distances	Tract Array Indices
0	31770.088	[39, 81, 102, 129, 170, 189, 204, 275, 354, 366]
1	31770.736	[17, 40, 102, 126, 141, 219, 253, 284, 354, 374]
2	31889.880	[87, 102, 106, 126, 137, 255, 308, 317, 351, 366]
3	31931.503	[39, 44, 86, 102, 187, 281, 300, 354, 361, 374]
4	31991.851	[40, 129, 142, 186, 204, 219, 310, 348, 355, 366]
5	32004.844	[99, 140, 194, 204, 221, 242, 255, 281, 300, 366]
6	32038.497	[27, 87, 125, 159, 166, 187, 327, 354, 365, 370]
7	32048.275	[76, 78, 90, 125, 152, 187, 282, 348, 367, 374]
8	32085.226	[40, 75, 90, 152, 212, 254, 284, 300, 348, 374]
9	32086.113	[3, 30, 42, 81, 101, 125, 159, 254, 281, 312]
10	32137.030	[3, 42, 53, 80, 102, 171, 185, 196, 282, 300]
11	32314.216	[8, 86, 91, 122, 137, 187, 282, 290, 349, 367]
12	32315.105	[44, 51, 160, 172, 184, 194, 291, 300, 309, 319]
13	32323.892	[8, 85, 101, 125, 171, 216, 268, 275, 365, 366]
14	32336.504	[3, 29, 42, 76, 90, 164, 185, 253, 283, 380]
15	32352.537	[32, 39, 42, 80, 99, 164, 254, 275, 350, 374]
16	32375.947	[17, 102, 106, 187, 224, 227, 301, 348, 355, 372]
17	32450.858	[23, 87, 125, 170, 184, 216, 246, 311, 355, 366]
18	32474.279	[39, 141, 163, 199, 214, 224, 301, 308, 312, 337]
19	32492.859	[11, 69, 80, 242, 255, 282, 286, 310, 332, 375]
20	32506.754	[90, 151, 204, 222, 253, 312, 328, 355, 373, 379]
21	32513.371	[3, 17, 86, 117, 161, 173, 188, 242, 284, 332]
22	32530.165	[8, 89, 153, 200, 221, 257, 268, 301, 309, 366]
23	32552.179	[7, 101, 127, 170, 184, 265, 282, 317, 368, 379]
24	32579.286	[3, 31, 81, 90, 125, 160, 256, 270, 307, 367]
25	32669.079	[42, 81, 127, 157, 254, 275, 288, 308, 310, 348]
26	32679.986	[85, 126, 136, 214, 246, 253, 268, 281, 310, 332]
27	32705.577	[99, 107, 129, 151, 185, 205, 224, 253, 275, 373]
28	32713.976	[71, 78, 153, 199, 221, 231, 255, 307, 351, 367]
29	32752.471	[27, 81, 101, 117, 156, 173, 184, 348, 363, 366]
30	32754.277	[4, 39, 43, 101, 125, 139, 171, 185, 222, 355]
31	32846.034	[104, 126, 142, 161, 188, 193, 217, 265, 282, 370]
32	32859.892	[3, 44, 80, 100, 160, 253, 275, 300, 307, 353]
33	32869.155	[12, 40, 68, 152, 162, 174, 284, 287, 312, 315]
34	32918.423	[98, 106, 153, 176, 219, 301, 309, 312, 367, 373]
35	33094.915	[40, 86, 126, 135, 166, 174, 189, 214, 313, 355]
36	33118.760	[17, 100, 127, 153, 157, 182, 194, 284, 289, 327]
37	33243.389	[23, 81, 102, 105, 164, 175, 211, 341, 357, 373]
38	33462.119	[68, 87, 98, 113, 190, 242, 287, 309, 333, 347]
39	33607.242	[31, 101, 108, 129, 139, 223, 257, 270, 308, 374]
40	33853.244	[16, 72, 141, 163, 255, 281, 291, 293, 323, 347]
41	34100.216	[17, 121, 153, 160, 168, 192, 224, 261, 287, 355]
42	34221.807	[102, 127, 153, 174, 196, 203, 223, 283, 312, 334]

TABLE 9.5: Locating 10 service centers: Greedy hill climbing results from 43 runs, sorted by total population-weighted distance.

No.	Sum of Weighted Distances	Tract Array Indices
0	31203.451	[3, 42, 81, 101, 129, 174, 184, 194, 242, 282]
1	31244.406	[3, 40, 44, 102, 129, 140, 224, 255, 282, 348]
2	31334.119	[40, 42, 81, 101, 129, 173, 188, 193, 282, 291]
3	31365.693	[42, 76, 99, 129, 172, 195, 242, 282, 341, 373]
4	31387.927	[39, 44, 76, 99, 129, 173, 184, 194, 282, 290]
5	31430.672	[26, 34, 42, 81, 101, 125, 187, 193, 242, 291]
6	31438.537	[34, 39, 42, 76, 101, 126, 255, 279, 291, 354]
7	31489.512	[8, 42, 76, 99, 126, 174, 184, 197, 282, 291]
8	31507.429	[39, 75, 90, 129, 140, 170, 184, 194, 282, 366]
9	31526.861	[39, 101, 141, 212, 255, 281, 299, 327, 354, 373]
10	31528.269	[40, 102, 126, 140, 212, 224, 254, 354, 355, 374]
11	31537.680	[5, 26, 32, 40, 42, 81, 125, 253, 311, 351]
12	31551.196	[3, 39, 44, 81, 99, 173, 185, 283, 301, 347]
13	31565.727	[42, 76, 99, 125, 159, 173, 194, 282, 287, 341]
14	31578.033	[44, 81, 101, 129, 194, 243, 254, 268, 282, 373]
15	31580.264	[3, 39, 44, 81, 99, 187, 275, 299, 309, 354]
16	31585.004	[26, 32, 39, 99, 125, 151, 187, 327, 354, 370]
17	31590.251	[39, 101, 129, 152, 174, 221, 282, 341, 347, 366]
18	31600.086	[42, 101, 126, 140, 193, 221, 242, 255, 284, 291]
19	31600.215	[76, 125, 163, 173, 188, 242, 284, 291, 348, 367]
20	31603.697	[44, 86, 122, 137, 242, 254, 282, 291, 310, 349]
21	31616.700	[17, 81, 102, 173, 183, 194, 242, 284, 300, 374]
22	31625.909	[26, 33, 98, 106, 129, 140, 172, 224, 351, 370]
23	31628.233	[31, 44, 76, 78, 90, 125, 172, 282, 291, 312]
24	31628.482	[3, 8, 42, 90, 129, 151, 193, 255, 282, 315]
25	31632.111	[40, 44, 80, 129, 163, 170, 194, 291, 308, 341]
26	31632.767	[5, 33, 40, 44, 76, 101, 125, 255, 282, 354]
27	31633.584	[42, 81, 91, 125, 173, 185, 192, 242, 291, 308]
28	31637.684	[86, 98, 129, 212, 219, 242, 255, 281, 354, 373]
29	31638.818	[8, 81, 99, 173, 183, 199, 291, 299, 308, 367]
30	31644.874	[101, 125, 141, 194, 242, 247, 284, 327, 341, 366]
31	31648.596	[16, 39, 99, 129, 174, 184, 284, 291, 319, 347]
32	31651.574	[23, 26, 33, 40, 76, 129, 173, 184, 310, 370]
33	31663.792	[39, 101, 129, 142, 173, 216, 309, 341, 347, 366]
34	31666.216	[26, 44, 81, 102, 129, 173, 185, 291, 357, 379]
35	31669.163	[8, 75, 99, 129, 152, 255, 274, 309, 354, 366]
36	31679.825	[76, 125, 173, 185, 197, 242, 291, 308, 311, 369]
37	31686.032	[42, 76, 106, 125, 173, 183, 198, 291, 309, 311]
38	31689.658	[3, 8, 44, 81, 90, 172, 183, 282, 301, 347]
39	31689.930	[8, 44, 85, 99, 126, 174, 184, 197, 282, 291]
40	31692.710	[80, 98, 126, 170, 184, 194, 212, 242, 284, 373]
41	31694.852	[44, 102, 106, 141, 187, 219, 282, 300, 349, 373]
42	31701.183	[23, 39, 99, 126, 170, 282, 317, 319, 341, 366]

TABLE 9.6: Locating 10 service centers: Top 43 greedy hill climbing results from 500 runs and a sampled neighborhood of 200, sorted by total population-weighted distance.

No.	Sum of Weighted Distances	Tract Array Indices
0	31057.11	[44, 81, 102, 125, 173, 194, 242, 282, 291, 341]
1	31057.11	[44, 81, 102, 125, 173, 194, 242, 282, 291, 341]
2	31057.11	[44, 81, 102, 125, 173, 194, 242, 282, 291, 341]
3	31057.11	[44, 81, 102, 125, 173, 194, 242, 282, 291, 341]
4	31057.11	[44, 81, 102, 125, 173, 194, 242, 282, 291, 341]
5	31057.11	[44, 81, 102, 125, 173, 194, 242, 282, 291, 341]
6	31057.11	[44, 81, 102, 125, 173, 194, 242, 282, 291, 341]
7	31057.11	[44, 81, 102, 125, 173, 194, 242, 282, 291, 341]
8	31070.69	[3, 42, 81, 102, 125, 173, 194, 242, 282, 341]
9	31070.69	[3, 42, 81, 102, 125, 173, 194, 242, 282, 341]
10	31070.69	[3, 42, 81, 102, 125, 173, 194, 242, 282, 341]
11	31070.69	[3, 42, 81, 102, 125, 173, 194, 242, 282, 341]
12	31078.05	[3, 42, 81, 101, 129, 173, 194, 242, 282, 341]
13	31148.02	[40, 44, 102, 129, 140, 224, 255, 282, 291, 354]
14	31148.02	[40, 44, 102, 129, 140, 224, 255, 282, 291, 354]
15	31148.02	[40, 44, 102, 129, 140, 224, 255, 282, 291, 354]
16	31176.12	[40, 44, 102, 129, 151, 221, 255, 282, 291, 354]
17	31176.12	[40, 44, 102, 129, 151, 221, 255, 282, 291, 354]
18	31176.12	[40, 44, 102, 129, 151, 221, 255, 282, 291, 354]
19	31189.71	[3, 40, 42, 102, 129, 151, 221, 255, 282, 354]
20	31189.71	[3, 40, 42, 102, 129, 151, 221, 255, 282, 354]
21	31207.19	[102, 129, 140, 173, 194, 224, 242, 282, 341, 366]
22	31207.19	[102, 129, 140, 173, 194, 224, 242, 282, 341, 366]
23	31225.62	[27, 39, 42, 81, 99, 129, 255, 291, 354, 365]
24	31240.12	[27, 40, 42, 81, 102, 125, 255, 291, 354, 365]
25	31240.12	[27, 40, 42, 81, 102, 125, 255, 291, 354, 365]
26	31245.18	[44, 102, 106, 129, 140, 172, 224, 282, 291, 312]
27	31245.18	[44, 102, 106, 129, 140, 172, 224, 282, 291, 312]
28	31263.30	[3, 40, 42, 99, 129, 141, 219, 255, 282, 354]
29	31267.50	[39, 42, 81, 99, 129, 172, 291, 307, 312, 380]
30	31267.50	[39, 42, 81, 99, 129, 172, 291, 307, 312, 380]
31	31273.60	[27, 39, 42, 81, 99, 129, 254, 291, 351, 365]
32	31273.60	[27, 39, 42, 81, 99, 129, 254, 291, 351, 365]
33	31273.60	[27, 39, 42, 81, 99, 129, 254, 291, 351, 365]
34	31287.68	[27, 39, 42, 81, 99, 129, 172, 291, 312, 365]
35	31305.21	[8, 42, 81, 99, 129, 172, 199, 291, 309, 312]
36	31316.11	[32, 39, 44, 81, 99, 129, 172, 281, 312, 374]
37	31328.87	[33, 39, 99, 129, 151, 221, 255, 281, 354, 370]
38	31356.20	[27, 34, 81, 99, 129, 173, 193, 242, 341, 366]
39	31394.82	[27, 34, 40, 102, 129, 140, 224, 255, 354, 366]
40	31398.81	[39, 99, 129, 151, 221, 255, 305, 354, 355, 373]
41	31402.79	[27, 34, 39, 99, 129, 151, 221, 255, 354, 366]
42	31438.35	[33, 40, 102, 129, 140, 172, 224, 281, 312, 370]
43	31442.95	[39, 101, 129, 140, 172, 224, 307, 312, 366, 380]
44	31468.43	[39, 99, 129, 151, 221, 254, 281, 286, 351, 374]
45	31490.78	[27, 34, 39, 101, 129, 140, 172, 224, 312, 366]
46	31499.02	[39, 101, 129, 140, 224, 254, 307, 351, 366, 380]
47	31499.62	[39, 101, 129, 140, 172, 224, 305, 312, 355, 373]
48	31702.20	[44, 81, 129, 231, 239, 242, 255, 282, 291, 348]
49	31911.30	[11, 40, 71, 81, 254, 281, 286, 310, 351, 374]

TABLE 9.7: Locating 10 service centers: Results from 50 runs, sorted by total population-weighted distance, using a threshold accepting algorithm.

0	31064.461	[44, 81, 101, 129, 173, 194, 242, 282, 291, 341]
1	31069.513	[44, 81, 99, 129, 173, 194, 242, 282, 291, 341]
2	31078.046	[3, 42, 81, 101, 129, 173, 194, 242, 282, 341]
3	31079.212	[3, 42, 81, 101, 125, 173, 194, 242, 282, 341]
4	31111.674	[3, 42, 81, 102, 125, 173, 193, 242, 282, 341]
5	31118.026	[40, 44, 81, 102, 129, 173, 193, 282, 291, 341]
6	31120.326	[44, 76, 101, 129, 173, 194, 242, 282, 291, 341]
7	31148.367	[3, 40, 44, 81, 99, 129, 173, 193, 282, 341]
8	31152.800	[3, 40, 42, 76, 102, 125, 173, 194, 282, 341]
9	31175.188	[3, 42, 81, 101, 129, 173, 184, 193, 242, 282]
10	31178.622	[40, 44, 102, 126, 140, 224, 255, 282, 291, 354]
11	31179.538	[3, 40, 42, 81, 102, 125, 173, 184, 282, 332]
12	31185.272	[3, 40, 81, 102, 125, 173, 194, 282, 341, 367]
13	31186.088	[39, 44, 99, 129, 151, 221, 255, 282, 291, 354]
14	31197.774	[40, 44, 99, 129, 140, 224, 255, 282, 291, 354]
15	31201.454	[44, 76, 98, 125, 173, 194, 242, 282, 291, 341]
16	31217.076	[3, 39, 42, 101, 129, 151, 221, 255, 282, 354]
17	31218.613	[102, 129, 140, 172, 194, 224, 242, 282, 341, 366]
18	31226.950	[3, 40, 42, 102, 125, 151, 255, 282, 327, 354]
19	31227.592	[44, 76, 99, 173, 184, 194, 242, 282, 299, 373]
20	31228.138	[3, 39, 42, 101, 126, 151, 221, 255, 282, 354]
21	31229.492	[3, 40, 42, 81, 102, 125, 172, 194, 308, 341]
22	31233.478	[3, 40, 42, 81, 102, 129, 170, 188, 282, 354]
23	31238.866	[44, 102, 106, 129, 140, 224, 254, 282, 291, 351]
24	31239.367	[81, 99, 129, 173, 194, 242, 284, 291, 341, 367]
25	31242.826	[3, 42, 81, 99, 125, 173, 194, 242, 308, 341]
26	31248.543	[39, 44, 101, 129, 140, 187, 224, 282, 291, 354]
27	31254.064	[40, 44, 99, 129, 151, 187, 221, 282, 291, 354]
28	31255.750	[8, 44, 81, 99, 129, 173, 184, 282, 291, 332]
29	31255.841	[40, 44, 101, 129, 151, 187, 221, 282, 291, 354]
30	31257.575	[3, 40, 42, 99, 151, 255, 282, 299, 327, 354]
31	31261.005	[27, 39, 42, 81, 99, 129, 255, 354, 365, 373]
32	31279.946	[23, 81, 102, 125, 170, 242, 282, 341, 367, 373]
33	31285.351	[3, 39, 44, 99, 126, 140, 221, 255, 282, 354]
34	31288.591	[17, 81, 102, 125, 173, 184, 194, 242, 284, 373]
35	31289.389	[3, 42, 75, 102, 106, 129, 140, 172, 282, 312]
36	31293.505	[23, 40, 42, 102, 129, 140, 224, 255, 282, 291]
37	31293.568	[17, 23, 39, 81, 99, 173, 282, 291, 299, 341]
38	31295.377	[87, 102, 129, 173, 194, 222, 242, 282, 341, 366]
39	31296.243	[17, 23, 81, 102, 125, 173, 242, 282, 291, 341]
40	31300.909	[44, 101, 129, 151, 193, 221, 242, 255, 282, 291]
41	31301.310	[27, 39, 42, 81, 101, 125, 253, 291, 351, 365]
42	31302.005	[27, 34, 39, 42, 81, 99, 129, 253, 291, 351]

TABLE 9.8: Locating 10 service centers: Top 43 results from a trial of 120 runs of a simulated annealing heuristic, sorted by total population-weighted distance.

FIGURE 9.4: Best found decision.

9.6 For Exploration

1. Discuss the following statement, found on page 139 above:

   > In practice, we rely on experience and good judgment in conducting the meta-search to set parameters value (here *number_of_tries* and *number_of_runs*) for our heuristic procedure.

   Imagine that you must conduct an analysis for the sort of problem discussed in this chapter. How would you conduct the meta-search for good parameter values, given limited time and computational resources? How would you explain your reasoning and why your recommendations should be accepted?

2. Discuss and critically assess the following passage, found on page 140:

   > The upshot of these examples (using real data) is that our (heuristic) optimization procedures have apparently succeeded in finding good decisions. Of equal or greater importance is that these procedures have afforded successful decision sweeping, yielding a significant number of high-quality, very distinct decisions.

3. The following passage occurs on page 140:

   > As visual inspection indicates, the high quality solutions obtained by decision sweeping are quite diverse in detail. See for example column 3, "Tract Array Indices," of Table 9.5. In all, 175 different tract indices appear in the table. Comparing the best and second best decisions in the table, different in value by $31770.736/31770.088 = 1.00002$ or 0.002%, the two decisions have only 3 tracts in common. The observation holds generally for our results.

   Examine Tables 9.5, 9.6, 9.7, and 9.8 and elaborate upon the passage. Characterize the variations among the different tables.

4. Is there a systematic difference between the naïve greedy algorithm and greedy hill climbing in *which* decisions they find? Or do they tend to find the same decisions? What about greedy hill climbing compared to simulated annealing? How would you design an experiment to answer these questions definitively (more or less)? See §9.7 for information on the data with census tract IDs behind Tables 9.5 and 9.8.

5. Follow up the suggestion in Chapter 16 and build a realistic multiattribute decision model as in equation (16.4). Census files for all of Pennsylvania, with a rather complete roster of demographic information by 2010 census tract can be found in Data/PA_Tracts/ on the book's Web site. See §9.7 for information on the data with census tract IDs behind Tables 9.5 and 9.8.

6. Follow up the suggestion in Chapter 16 and design a negotiation game based on realistic model of preferences for at least two players, as in equation (16.4). Census files for all of Pennsylvania, with a rather complete roster of demographic information by 2010 census tract can be found in Data/PA_Tracts/ on the book's Web site. See §9.7 for information on the data with census tract IDs behind Tables 9.5 and 9.8.

9.7 For More Information

The p-median problem, considered in this chapter and revisited as a mathematical program in Chapter 18, is only one of several useful location analysis problems. A substantial literature is devoted to location problems in their various forms. See [37, 46, 111, 135] for overviews and textbook treatments. Location problems in general and the p-median problem in particular have attracted much work with metaheuristics. See, for example, Chiyoshi, F. and Galvão, R. D [31] on simulated annealing for the p-median problem and [1] for a greedy hill climber.

We note with pleasure the treatment by Alp, Erkut, and Drezner of the p-median problem with genetic algorithms (a metaheuristic we discuss in the sequel). In particular we note this passage in their paper, which makes a point in close coherence with the thrust of this chapter.

> While it is not surprising that near-optimal solutions are quite similar to one another, note that the similarity is not extreme. On average 8 of the 100 locations are different between a pair of solutions at termination. Hence, we have a large number of very good solutions to the problem, and this may be more useful than one optimal solution. [1, page 40]

Edwards and Barron [43] are invaluable for the sort of multiattribute modeling discussed in §9.5. For depth of background Keeney and Raiffa [84] and especially Dawes [38] are not to be missed.

The book's Web site is `http://pulsar.wharton.upenn.edu/~sok/biz_analytics_rep`. The Philadelphia census files are in `Data/Tracts_2010_Phila/` on the book's Web site. Census files for all of Pennsylvania, with a rather complete roster of demographic information by 2010 census tract can be found in `Data/PA_Tracts/`, also on the book's Web site.

The data for the trial of 43 runs behind Table 9.5, but with the decisions given with census tract IDs instead of our array indices, is present in the file `tractIDs_greedy_hill_140.csv`. The data for the trial of 120 runs behind Table 9.8, but with the decisions given with census tract IDs instead of our array indices, is present in the file `tractIDs_simulated_annealing.csv`. The files may be found on the book's Web site in the `Chapter_Files/Location_Analysis/` folder.

Chapter 10

Two-Sided Matching

10.1   Quick Introduction: Two-Sided Matching Problems

This section gives a very quick introduction to two-sided matching problems. Subsequent sections tell the story in more detail, at times repeating some information presented here.

A *match* μ between two sets or sides $\mathcal{A}$ and $\mathcal{B}$ can be represented nicely in tabular form. We pick one set to be the row (player), R, and the other to be the column (player), C, and we impose an arbitrary ordering on them individually. Then in the table element $(r_i, c_j)=1$ if there is a match—a pairing up—between $r_i \in R$ and $c_j \in C$; otherwise the entry is 0.

In the simple case of what we call a *conventional marriage matching problem* there are n agents on each side and each agent is matched to exactly one agent on the other side. The table then looks like this:

	1	2	3	4	5
1	0	0	0	1	0
2	0	1	0	0	0
3	0	0	0	0	1
4	1	0	0	0	0
5	0	0	1	0	0

That is, this match table or matrix represents one of 5!=120 possible matches between the 5 agents on each side. Taking the column's perspective, a more compact form of this table would be

$$(\quad 4 \quad 2 \quad 5 \quad 1 \quad 3 \quad) \tag{10.1}$$

which we call a *match vector*. Its interpretation is direct: column 1 is matched with row 4, column 2 with row 2, column 3 with row 5, column 4 with row 1, and column 5 with row 3. Notice that in this special case, the match vector is a *permutation* of the numbers 1 through 5. A matrix that is all 0s except for a single 1 in each row and column, as in the match table above, is thus called a *permutation matrix*. For those familiar with matrix multiplication notice that:

$$
(1 \quad 2 \quad 3 \quad 4 \quad 5) \times \begin{pmatrix} 0 & 0 & 0 & 1 & 0 \\ 0 & 1 & 0 & 0 & 0 \\ 0 & 0 & 0 & 0 & 1 \\ 1 & 0 & 0 & 0 & 0 \\ 0 & 0 & 1 & 0 & 0 \end{pmatrix} = (4 \quad 2 \quad 5 \quad 1 \quad 3) \qquad (10.2)
$$

Points arising:

1. The square matrix arrangement is characteristic of the two-sided matching problem known as the (conventional) *marriage problem.*

2. In the marriage problem, then, a match, μ, is essentially a permutation, of which there are $n!$. This is too many matches to search exhaustively for any but the smallest problems. Even 20! is larger than 10^{18}.

3. The entries in the permutation matrix are 0–1 decision variables. Equivalently, the entries in the associated match vector (e.g., expression (10.1)) are decision variables whose values cover the index values of the rows. So far, we have not specified an objective function for these decision variables (but we will, soon). For the moment, consider that we seek a (solution) procedure for finding a decision (setting of the decision variables in either the match array, or the match vector) that has nice properties, which are yet to be specified.

4. The representation is richer and more flexible than is perhaps evident at first glance. For example, if we have n "men" and n "women" we could add two dummies to each list and specify that if a man (or woman) ranked the dummy above any woman (man), then the man (woman) would prefer not to be "married" at all to being married to a woman (man) ranked below the dummy. And so on.

5. In the marriage problem we say that each agent has a quota of 1. We can easily let one side have agents with quotas larger than 1; we can even let different agents have different quotas. When we do this, we get what is called an *admissions problem.* Think of students (quotas of 1) applying for admission to schools (quotas of hundreds or even thousands).

6. Matching problems, even the special case of marriage problems, occur in practice quite often, as we shall see.

10.2 Narrative Description of Two-Sided Matching Problems

Consider, prototypically, the following problem. We have n men and n women, all of whom would like to get married. We consider no other individuals than those before us, the n men and the n women. Our job is to find a match, μ, in which each man has be paired with one woman and each woman has been paired with one man. The men have rated the women and the women have rated the men. How should we pair them up, assign them to each other, make the match μ? This is called the *marriage problem* (or something similar) in an extensive literature. Cheerfully yielding to modern sentiments, we prefer to call it the *conventional marriage problem,* and shall later consider variations on it. For the present, however, the mathematical properties of the conventional problem hold plenty of interest for us.

The (conventional) marriage problem is prototypical in two ways. First, there are many purposes, besides marrying off men and women, for which the problem is apt, such as matching roommates in a dormitory, partners at a dance, workers for a task, and so on. Second, the marriage problem is an instance of the much larger class of *2-sided matching problems* in which individuals from two collections are to be paired up. For this more general class we do not require that the two collections be equally sized or even non-overlapping (although most authors do require the latter). Further, we do not require that all of the individuals in either set be matched. And so on.

Indefinitely many complications may legitimately arise in two-sided matching and do in real-world applications. Student applicants and schools are matched, but the schools have a limit on how many students they are willing to admit. Similarly for interns and hospitals. In a variant of the marriage problem, some individuals may prefer to remain single than to be paired with individuals below a certain level of attractiveness. And so on.

And why limit matching to two sides? The interesting possibilities are manifold. But first things first.

The first question we need to ask in dealing with the (conventional) marriage problem is: On what basis should we create matches? Here the key thing to note is that this is (assumed to be) a strategic problem, in distinction to a non-strategic (aka: parametric) problem. That is, we assume that the men and women (and more generally the agents in any n-sided matching problem) have a degree of autonomy in their preferences and their actions. In particular, it will not do just to pair them up willy-nilly. Unhappy couples will find other partners, undoing our match.

Reflections of just this sort led Gale and Shapley, the authors of the seminal paper on two-sided matching [53], to propose *stability* as a requirement for any adequate match. Part of what makes the paper seminal is that an extensive literature since agrees with, or rather presupposes, the requirement of stability.

A match μ is *unstable* if there are two matched couples, call them (r_1, c_1) and (r_2, c_2) such that r_1 prefers to be matched with c_2 rather than c_1 and c_2 prefers to be matched with r_1 rather than r_2. The presumption is they—r_1 and c_2—will tend to get their way, making the match unstable. If a match is not unstable—if there are no such pairs of frustrated couples, more generally called *blocking pairs*—then the match is said to be *stable*.

Note that here we are foreshadowing the notational conventions we shall use later. Instead of men and women to be matched, we shall often prefer to have two sets to be matched, the row set, of which the r_is are members, and the column set, containing the c_is. In any particular case, we may think of the men as being assigned to the row set and the women to the column set, or vice versa. (It will matter which is which.)

We have thus bumped up against issues of problem representation, and shall now address it explicitly in the next section. To foreshadow what is to come, our aim for this chapter is to explore in some depth the concept of stability in the context of the marriage problem. What are its operating characteristics? Does it capture everything we want or need to assess a match or matching procedure? These and other key questions for any application of the marriage problem, and more generally matching problems, will exercise us for the remainder of the chapter.

10.3 Representing the Problem

In the original paper [53], Gale and Shapley assumed that each of the men individually ranked in order of preference each of the women and vice versa. The ranks were 1, 2, ...n, with 1 being the best, most preferred, value and n the least. We shall abide by these assumptions until further notice. (The literature varies on this crucial but inconsequential convention.)

Tables are natural and convenient representation forms for preference ranking information. In an *individual preference table* the rows correspond to individuals of one side or another and the columns to their rank scores for the individuals on the other side. For example, this table

Men	Women			
	1	2	3	4
1	1	2	3	4
2	1	4	3	2
3	2	1	3	4
4	4	2	3	1

shows the "men's" preferences for women in an $n = 4$ marriage problem. Man 2 ranks woman 1 the highest, woman 4 the next highest, then woman 3, and woman 2 the lowest. "Women's" preferences are indicated in a second table

Women	Men			
	1	2	3	4
1	3	4	2	1
2	3	1	4	2
3	2	3	4	1
4	3	2	1	4

where we see that woman 2 likes man 2 the most, followed by man 4, then man 1, and finally man 3.

A *ranking table* combines the individual preference tables. Using the examples above, the resulting ranking table when the men are assigned to be row players and the women to be column players is [53, Example 2]:

	1	2	3	4
1	1,3	2,3	3,2	4,3
2	1,4	4,1	3,3	2,2
3	2,2	1,4	3,4	4,1
4	4,1	2,2	3,1	1,4

If instead we assign the women the role of row players and the men that of column players we get this as our ranking table:

	1	2	3	4
1	3,1	4,1	2,2	1,4
2	3,2	1,4	4,1	2,2
3	2,3	3,3	4,3	1,3
4	3,4	2,2	1,4	4,1

(The one is easily transformed into the other by reversing the pairs in each cell and then taking the transpose.)

We can also represent a matches as a table: 1 if the pair is matched, 0 if not. For example

	Women			
Men	1	2	3	4
1	0	0	1	0
2	0	0	0	1
3	1	0	0	0
4	0	1	0	0

indicates matches between row player (in this case, man) 1 and column player (here, woman) 3. Compressing this we say (1,3). The other matches are (2,4), (3,1) and (4,2).

Notice that a more compact representation of a match is possible (as discussed in §10.1). Choosing to take the column perspective, we can represent the match above as a vector or 1-dimensional table: [3,4,1,2] or

3	4	1	2

This has the additional advantage of letting us see immediately how many possible matches there are for a given ranking matrix: $n!$ where n is the number of agents to a side. In consequence, the search space of matches for a given ranking matrix of size n grows very rapidly with n

n	$n!$
1	1
2	2
3	6
4	24
5	120
6	720
7	5040
8	40320
9	362880
10	3628800
11	39916800
12	479001600

so that anything larger than 12 or so cannot as a practical matter be generated and visited (Knuth's felicitous terms) on a contemporary personal computer.

These are just the counts for the number of possible matches for a given ranking matrix. Given n, how many ranking matrices are there? On the row side, each player can independently choose among $n!$ preference rankings for itself. Since there are n row players, there are $(n!)^n$ possible individual preference matrices for the rows. Similarly for the columns, yielding $(n!)^{2n}$ possible ranking matrices of size n. This number grows even faster than $n!$ (one wants to add another "!").

n	$(n!)^{2n}$
2	16
3	46,656
4	11,007,531,4176
5	6.1917364224e+20

Of course there will be "duplications" of sorts among the $(n!)^{2n}$ possibilities. With $n=2$, for example, we will have both

	1	2
1	1,1	2,2
2	1,1	2,2

and

	1	2
1	2,2	1,1
2	2,2	1,1

Why not just switch the women's names (column IDs) and call these the same? An argument could be made for this, but we are unpersuaded and shall not pursue this prospect for—rather modest—reduction of complexity.[1]

With so many possible matches for a given ranking table, $n!$, and given the problem of choosing a match, how are we to find a good one? Well, before we can say how to find a good one, we perhaps should say something about what would count as a good one. To that now, plus an algorithm for finding good matches on this criterion.

10.4   Stable Matches and the Deferred Acceptance Algorithm

As we noted above, Gale and Shapley in their paper [53] propose *stability* as a, even the, requirement for any adequate match. Again,

> A match μ is *unstable* if there are two couples, call them (r_1, c_1) and (r_2, c_2) such that r_1 prefers being matched with c_2 rather than c_1 and c_2 prefers to be matched with r_1 rather than r_2. The presumption is they—r_1 and c_2—will tend to get their way, making the match unstable. If a match is not unstable—if there are no such frustrated couples—then the match is said to be *stable*.

In the related literature that has extended this work, stability has continued to figure centrally in evaluation of procedures for two-sided matching. It is easy to see why. Besides its obvious stability property, Gale and Shapley showed that for the marriage problem (and others) there always exists at least one stable matching. In addition, they presented a fast algorithm for finding a stable match. The algorithm, called the *deferred acceptance algorithm*, requires at most $n^2 - 2n + 2$ passes, each of which is relatively cheap. So, while $n!$ is huge (for $n > 12$ or so), $n^2 - 2n + 2$ is not. Even for $n = 1000$, $n^2 - 2n + 2$ is still under a million, at 998002.

The procedure, more precisely, the deferred acceptance algorithm as applied to the marriage problem, is quite simple. Here is Gale and Shapley's original description of it.

> To start, let each boy propose to his favorite girl. Each girl who receives more than one proposal rejects all but her favorite from among those who have proposed to her. However, she does not accept him yet, but keeps him on a string to allow for the possibility that someone better may come along later.

> We are now ready for the second stage. Those boys who were rejected now propose to their second choices. Each girl receiving proposals chooses her favorite from the group consisting of the new proposers and the boy on her string, if any. She rejects all the rest and again keeps the favorite in suspense.

---

[1]Nevertheless the mathematical problem of defining what counts as a duplication and why, and then counting the number of possible ranking matrices, excluding duplications, is itself a delightful one.

We proceed in the same manner. Those who are rejected at the second stage propose to their next choices, and the girls again reject all but the best proposal they have had so far.

Eventually (in fact, in at most $n^2 - 2n + 2$ stages) every girl will have received a proposal, for as long as any girl has not been proposed to there will be rejections and new proposals, but since no boy can propose to the same girl more than once, every girl is sure to get a proposal in due time. As soon as the last girl gets her proposal the "courtship" is declared over, and each girl is now required to accept the boy on her string. [53, pages 12–3]

10.5 Once More, in More Depth

Again, now with more technical detail, in a two-sided matching problem we are given two sets (sides) of agents, X and Y, and are asked to find a match, μ, consisting of a decision (in or out?) for each pair $(x, y), x \in X, y \in Y$. It is helpful to view a match as represented by a table, $\mathbf{M}$, of size $|X| \times |Y|$, based upon arbitrary orderings of X and Y. The element $m_{i,j}$ of $\mathbf{M}$ equals 1, if $x_i \in X$ is matched with $y_j \in Y$; otherwise the element is 0. Thus the element $m_{i,j}$ of $\mathbf{M}$ represents the *pair* (x_i, y_j). Matchings pair up agents from X and Y, which need not in general be the size.

Particular matching problems come with particular requirements on μ (or $\mathbf{M}$) as well as X and Y. For example in the conventional marriage matching problem (as discussed above), we require that $|X| = |Y| = n$; the number of "men" equals the number of "women." We further require of any (valid) match that each man (or member of X (Y)) be paired (or matched) with exactly one woman (or member of Y (X)), and vice versa. In terms of $\mathbf{M}$, this means that there is one 1 in each row and one 1 in each column. $\mathbf{M}$ is thus a permutation matrix and the number of possible valid matches is $n!$. In *admissions* matching problems, which are used to model, for example, interns applying to hospitals and students applying to schools, one side of the problem, say X, is much larger than the other. There are more doctors than hospitals, more students than schools. Unlike conventional marriage problems, however, one side will have *quotas* larger than 1. Each doctor and each student will have a quota of 1, but each hospital and each school may have a much larger quota and admit many doctors or students. Thus, in a valid match for an admissions problem, each agent on one side (X or Y) is paired to one or more agents on the counter party side, not to exceed the counter party's quota. With students as X, and schools as Y, $\mathbf{M}$ will have one 1 in each row, and each column will have a number of 1s not to exceed the quota of the corresponding school.

Many other variations are possible and are met in practice for two-sided matching problems, a topic we take up in the sequel. For now we focus on a more technical question. How do and how should centralized market agencies produce matches? In practice, some form of, variation on, the deferred acceptance algorithm of Gale and Shapley [53] is used to find a *stable* match, which is then used. A match is *unstable* if there is a pair of matched pairs—(x_i, y_j) and (x_k, y_l)—such that x_i prefers to be matched with y_l over being matched with y_j and y_l prefers to be matched with x_i to being matched with x_k. Stable matches are the ones that are not unstable.

Although the procedure was discovered and used independently before, the deferred acceptance algorithm (DAA) was first published in a paper by Gale and Shapley [53], quoted above. Figure 10.1 (after [53]) presents the DAA in pseudocode.

1. Each $x \in X$ ranks each $y \in Y$, and each $y \in Y$ ranks each $x \in X$.

2. *Matched* $\longleftarrow \emptyset$, *Unmatched* $\longleftarrow \emptyset$.

3. For each y, *string.y* $\longleftarrow$ [].

4. Each $x \in X$ proposes to its most preferred y, appending x to *string.y*.

5. Each y with *length*(*string.y*) > 1 (i.e., with more than one proposal), retains in the string its most preferred member of the string, and removes the rest, adding them to *Unmatched*.

6. Each x remaining on some string is added to *Matched*.

7. Do while *Unmatched* $\neq \emptyset$:

   (a) *Matched* $\longleftarrow \emptyset$, *Unmatched* $\longleftarrow \emptyset$.

   (b) Each $x \in$ *Unmatched* proposes to its most preferred y, among the ys that have not already rejected x, appending x to *string.y*.

   (c) Each y with *length*(*string.y*) > 1 (i.e., with more than one proposal), retains in the string its own most preferred member of the string, and removes the rest, adding them to *Unmatched*.

   (d) Each x remaining on some string is added to *Matched*.

8. Stop. Each x is matched to a distinct y, which has x as the sole member of its string.

FIGURE 10.1: Pseudocode for the deferred acceptance algorithm (DAA) for the simple marriage matching problem, Xs proposing to Ys.

10.6   Generalization: Matching in Centralized Markets

In the usual case, markets are *distributed*, with buyers and sellers mostly on their own in finding each other and in negotiating terms of trade. Distributed markets may fail in one way or another, however. A common response is to create a *centralized* market organized by a third party whose responsibility it is to set the conditions of trade, for example the price, based on the bids and asks from the buyers and sellers. Many electricity markets are organized in this way. Deregulated electricity producers, for example, offer supply schedules to a third party, often called the *independent systems operator* or ISO, who aggregates the supply schedules, observes the market demand, and sets the price of electricity (for a given period of time).

Quite a number of labor markets are similarly centralized, most famously, markets in which physicians are matched to hospitals for internships [136]. Roughly speaking, the individual doctors submit their rankings of hospitals, the hospitals submit their rankings of doctors, and a third party organization undertakes to match doctors with hospitals. This is an example of a *two-sided matching* problem, which problems are the subject of this chapter.

As Roth notes [136], two-sided matching models are often natural for representing markets, in which agents need to be paired up and in which price alone is insufficient to govern the exchanges. Thus, matching markets are inherently different from the usual commodity markets of textbook economics.

> The price does all the work [in a commodity market], bringing the two of you [buyer and seller] together at the price at which supply equals demand. On the NYSE, the price decides who gets what.

> But in *matching markets,* prices don't work that way. [For example]...In the working world, courtship often goes both ways, with employers offering good salaries, perks, and prospects for advancement, and applicants signaling their passion, credentials, and drive. College admissions and labor markets are more than a little like courtship and marriage: each is a two-sided matching market that involves searching and wooing on both sides. A market involves matching whenever price isn't the only determinant of who gets what. [137, pages 5–6]

Men and women want to find partners, workers want to find employment and employers want to find workers, and so on. Moreover, many of these markets, decentralized or free markets, experience failure and unsatisfactory performance in practice. They experience unraveling, e.g., offers to students are made earlier and earlier; congestion, e.g., offerers have insufficient time to make new offers after candidates have rejected previous offers; and participants engage in disruptive strategic behavior, so that behaving straightforwardly with regard to one's preferences becomes risky, e.g., in scheduling offers and responding to them [17, 136, 137, 138]. In consequence, there is a large and growing number of applications of two-sided matching in which decentralized markets have been replaced by centralized ones, in which a coordinating agency undertakes periodic matching between two sides of a specific market. [136] lists over 40 labor markets, mostly in medical fields; schools in New York and Boston are using centralized markets to match students to schools; [137] describes a number of applications, including kidney exchanges; see also [17].

In brief, matching markets occur naturally and just as naturally often fail, leading (again naturally) to institution of centralized exchanges for them. How can, how do, the exchanges do their jobs? Using the deferred acceptance algorithm or some variant.

10.7 Discussion: Complications

A presumption in matching problems (as distinguished from assignment problems, Chapter 5, which are treated in operations research and employ non-strategic decision making) is that both sides consist of agents who have interests of their own and capacities to act on them. Consequently, matches are ordinarily evaluated in terms of *stability.* Matching problems are inherently strategic, or game-theoretic, and stability is the accepted equilibrium concept. This is in contrast to most or all of the other problems we consider in this book. A match is said to be stable, again, if there is no pair of matched couples in it containing individuals who would prefer to be matched to each other but are not. The thought is that, with stability, unraveling will not happen and without stability it is difficult to prevent unraveling and concomitant market failure. Requiring matches to be stable in the first place will, it is thought, prevent breakup and reformation among pairs and its attendant costs.

The point of departure for this section is the observation that two-sided matches can be evaluated—and for many applications should be evaluated—according to several objectives, particularly stability, equity, and social welfare. For present purposes, by stability we mean the count of unstable pairs of matched couples in a match. This should be minimized and at 0 the match is stable.[2] By equity we mean the sum of the absolute differences in the preference scores of each matched pair. We will be scoring preference on a ranking 1 to n scale (1 = most preferred, n = least preferred), so this too should be minimized. Finally, by social welfare we mean the sum of the agent scores in the match. Again, since scoring is from low to high, this quantity should also be minimized. (To illustrate, if agents i and j are paired in a match, and i's preference score for j is 5 and j's preference score for i is 3, then the matched pair's contribution to social welfare is the sum, 8, and their contribution to equity is the absolute value of the difference of the scores, $|5 - 3| = 2$.)

Given that we would consider designing or even centralizing a matching market (as is widely done in practice), the question arises of how best to provide the market operators and users with match options that honor these three objectives, and inform the decision makers about them.

Several points by way of additional background on the deferred acceptance algorithm. First, as proved by Gale and Shapley [53], under the special assumptions they made (e.g., preference ranking by agents), the stable marriage problem and the admissions problem (see above) have stable matches and the DAA will find one and will find one quickly ($O(n^2)$). Second, the DAA is asymmetric. One side proposes, the other disposes. Focusing now on the marriage problem, if the men propose, they obtain a stable match that is male optimal in the sense that no man in this match does better in any other stable match. Conversely, the match is female pessimal in the sense that no woman is worse off in any other stable match. And vice versa if the women propose [53, 73, 107].

This asymmetry is a general characteristic of the DAA in its various forms. It occasions the important question of whether better matches exist and can be found. To this end, we will want to look at stable matches that may be preferable to the matches found by the DAA. And as announced above, we want to examine both social welfare and equity.

These issues could be neatly resolved by finding all of the stable solutions for a given problem and comparing them with respect to equity, social welfare, and whatever other measures of performance are relevant. Predictably, however, this is an intractable problem. Irving and Leather [79] have shown that the maximum number of stable matches for the simple marriage matching problem grows exponentially in n (see also [73, 75]). Further, they provide a lower bound on the maximum by problem size. Remarkably, for a problem as small as $n = 32$, the lower bound is $104, 310, 534, 400$ [79]. Further, they establish that the problem of determining the number of stable matches is #P-complete, and hence in a theoretically very difficult complexity class. These are, of course, extreme-case results, but very little is known about average cases. So we are left to rely upon heuristics, and indeed on metaheuristics.

The first complication for the deferred acceptance algorithm thus is the fact that there can be very many stable matches and those found by the algorithm may not be very attractive on grounds other than mere stability. Simulation results with artificial data, e.g., [74, 91, 99, 110], indicate that indeed both evolutionary algorithms and agent-based models can find stable and nearly stable decisions that are superior on other grounds to those found by the deferred acceptance algorithm. Unfortunately, real world matching data is hard to come by and it remains to be seen whether real world matching problems can yield attractive alternative matches with these methods.

There is a second complication pertaining to the deferred acceptance algorithm: beyond the simple cases discussed by Gale and Shapley [53], when such real world considerations as "two body problems" arise (e.g., two partners who wish to be placed in internships near one another, siblings (perhaps more than two) who wish to attend the same school or schools

---

[2]If the count of unstable pairs is 1, there is no guarantee that if the two pairs rematch by exchanging partners the resulting match would be stable. In fact, it could have a higher count of unstable pairs [107].

near one another, etc.), then the deferred acceptance algorithm may fail in various ways and there may not even be any stable matches to be found. Such problems are handled somewhat in an ad hoc fashion in existing applications [137]. Whether and to what extent metaheuristics or other algorithms can contribute usefully to more complex existing or potential new matching markets is an intriguing possibility that lies on the frontier of current knowledge.

10.8   For More Information

The original paper by Gale and Shapley [53] was published in a journal aimed at high school teachers of mathematics. It remains a delightful exemplar of accessible mathematical insight and clarity. Alvin Roth has made something of a career designing matching markets, for which he shared the Nobel Prize with Lloyd Shapley. Roth's popular account of matching markets [137] is readable and quite accessible, although lacking in technical details and fine points of matching problems (e.g., there is no mention of asymmetry and pessimal matches).

At an intermediate level of mathematical discussion, Knuth's book [107] is a gem. Roth and Sotomayor's book [138] is an essential read from an economic perspective, while the book by Gusfield and Irving [73] is essential from a computer science perspective.

See [5] for a related treatment using agent-based modeling.

Part III

Metaheuristic Solution Methods

Chapter 11

Local Search Metaheuristics

11.1   Introduction

Metaheuristics may usefully be categorized as either *local search* metaheuristics (the subject of this chapter) or *population based* metaheuristics (the subject of the next chapter). In either case *meta*heuristics are heuristics that may be realized in many different ways as particular heuristics. A goal of this chapter, and the next, is to clarify this statement. That is something easily and best done with specific examples before us. Let us proceed, then, to discuss the first of the four local search metaheuristics that are the main subject of this chapter.

11.2   Greedy Hill Climbing

To illustrate greedy hill climbing, bring to mind a Simple Knapsack model. Given a particular decision (setting of the decision variables, indicating which items are in and which are out of the knapsack), we might seek to improve the decision by looking for a better one in the immediate neighborhood of the incumbent decision. We might do this, for example, by randomly generating an incumbent decision and setting the provisional decision by copying the incumbent, and then by the following *neighborhood exploration* procedure:

1. Randomly select a locus in the incumbent decision

   For example, at position 9.

2. Change the value at the selected locus, creating thereby a *candidate* decision to replace the incumbent decision.

   For the Simple Knapsack, from 0 to 1 or 1 to 0 as the case may be.

3. Evaluate the candidate decision.

4. Compare the evaluated value of the candidate decision with that of the incumbent decision. If the candidate's value represents an improvement over the incumbent, make the candidate decision the provisional incumbent decision; otherwise, retain the existing incumbent.

We might repeat this procedure a fixed number of times, say 100, after which time we accept the current and continue in this vein until we no longer see any improvement over the provisional incumbent. That incumbent, present at the end of the procedure, will be our heuristic choice for an optimal solution to the constrained optimization problem. Figure 11.1 presents this procedure more generally in high-level pseudocode. (We are assuming our optimization problem is one of maximization.)

1. Initialization

    (a) Select a model to be optimized, *Model*.

    (b) Set *number_of_tries* to a positive value.

    (c) Set *incumbent_decision* to an arbitrary decision for *Model*.

    (d) Set *value_inc* by evaluating *Model* on *incumbent_decision*.

    (e) Set *improvement?* to True.

2. While *improvement?*

    (a) Set *improvement?* to False

    (b) For *number_of_tries* times:

        i. Set *candidate_decision* to an arbitrary decision in the neighborhood of *incumbent_decision*.

        ii. Set *value_cand* by evaluating *Model* on *candidate_decision*.

        iii. If *incumbent_decision* is feasible, *candidate_decision* is feasible, and *value_cand* > *value_inc*, then set *incumbent_decision* to *candidate_decision* and set *improvement?* to True.

        iv. If *incumbent_decision* is not feasible and *candidate_decision* is feasible, then set *incumbent_decision* to *candidate_decision* and set *improvement?* to True.

        v. If *incumbent_decision* is not feasible and *candidate_decision* is not feasible, and *candidate_decision* is closer to being feasible than *incumbent_decision*, then set *incumbent_decision* to *candidate_decision* and set *improvement?* to True.

    (c) If *improvement?* is True:

        i. Set *value_inc* by evaluating *Model* on *incumbent_decision*.

3. Report *incumbent_decision* as heuristically optimal.

FIGURE 11.1: Pseudocode for a version of a greedy hill climbing metaheuristic.

Points arising:

1. The core idea of greedy hill climbing (for optimization) is to seek improvements for an incumbent decision by looking at candidate decisions that are very similar to it. Such solutions are said to be "in the neighborhood" of the incumbent solution. Search continues, possibly replacing incumbent solutions many times, until there is no apparent candidate in the neighborhood that is better than the present incumbent. Metaphorically, this is a hill climbing procedure: accepting an improving candidate decision in the neighborhood is like taking a step upwards on a hill. It is also a greedy procedure because it always accepts a candidate if it is better than the incumbent.

2. Any instance of greedy hill climbing is clearly a heuristic solver for its constrained optimization model. If the procedure happens upon an optimal decision, that decision will become and remain the incumbent decision, but this is only because the outlook of the procedure is limited to its neighborhood, however defined. Any decision encountered that is best in its neighborhood (said to be *locally optimal*) will also become and remain the incumbent decision. Greedy hill climbing cannot distinguish locally and globally optimal decisions. This is what makes it a heuristic solver.

3. In our version we repeat the neighborhood search procedure a fixed number of times or tries, with *number_of_tries* = 100 in the example. This is significant on two counts.

   (a) Fixing *number_of_tries* at 100 is considerably less greedy, and much more patient, that fixing it at, say, 10 or even 1. It is an empirical matter what yields the best results.

   (b) The fact that we have a parameter, *number_of_tries*, that can be set to different values illustrates how the procedure described in Figure 11.1 is a *meta*heuristic: The procedure is in fact not a single procedure, but a schema or template or pattern that may be followed by many particular procedures, differing in detail.

4. Greedy hill climbing will often given satisfactory results. At the very least it serves as a benchmark heuristic (metaheuristic) against which other methods can be compared.

11.2.1 Implementation in Python

Figure 11.1 (page 164) presents a greedy hill climbing metaheuristic in pseudocode, Figure 11.2 (page 168) presents an implementation in the programming language Python (version 3.4). Pseudocode, as in Figure 11.1, is intended to be comprehensible by people who do not program. Programming languages, including Python, must be "comprehensible"— executable—on a computer. Our understanding of any procedure is deepened when we can comprehend together both its pseudocode and its implementation. To that end, we now compare Figures 11.1 and 11.2.

The initialization portion of the pseudocode

1. Initialization

   (a) Select a model to be optimized, *Model*.

   (b) Set *number_of_tries* to a positive value.

   (c) Set *incumbent_decision* to an arbitrary decision for *Model*.

   (d) Set *value_inc* by evaluating *Model* on *incumbent_decision*.

   (e) Set *improvement?* to True.

corresponds to lines 1–6 of the Python implementation:

```
1 def doGreedyHillClimbRun(themodel,nTries):
2     # Obtain a random decision for starting the run:
3     incumbentDecision = themodel.getRandomDecision()
4     (feasible_inc,obj_inc,slacks_inc) = \
5         themodel.evaluateDecision(incumbentDecision)
6     improvement = True
```

Line 1 announces the beginning of the definition (**def**) of a function, called **doGreedyHillClimbRun**. This function corresponds closely to the pseudocode in Figure 11.1. The model to be optimized, *Model*, and the *number_of_tries* parameter are determined outside of the function and are given to the function as **themodel** and **nTries**. Line 3 corresponds to "Set *incumbent_decision* to an arbitrary decision for *Model*." in the pseudocode. **themodel** is an object whose class definition is stated in the Python module *comodel.py*, which is available and described in the book's Web site. It instantiates the problem specific model. **getRandomDecision()** is a *method* (procedure) attached to the model object. Calling it, as we do in line 3, causes the model object to execute the method and thereby produce a random, or arbitrary, decision, which is then assigned to the variable **incumbentDecision** (*incumbent_decision* in the pseudocode). Notice that because **themodel** is supplied exogenously to the function, it may be any kind of model at all, so long as it supports the requisite methods needed by the **doGreedyHillClimbRun** function. Again, it is the repository of problem specific information, e.g., that there are 383 districts, that they are certain distances from each other, and so on.

Lines 4–5 of the code correspond to "Set *value_inc* by evaluating *Model* on *incumbent_decision*" in the pseudocode. The evaluation is effected by another method attached to the model, **evaluateDecision()**. The implementation at this point becomes more specific than the pseudocode, as inevitably it must. **evaluateDecision()** returns three values, which should be understood as constituting elements of *value_inc* in the pseudocode: **feasible_inc** is a Boolean (true or false) value indicating whether the evaluated decision is feasible or not; **obj_inc** is a scalar number holding the value of the objective function for the evaluated decision; and **slacks_inc** is a vector representing the (right-hand-side − left-hand-side) difference in values for the constraints in the model. The implementation assumes that all constraints are ≤ constraints, so a decision is feasible when and only when every value in **slacks_inc** is non-negative.

Finally for this section of the pseudocode, line 6 of the Python code implements the "Set *improvement?* to True" step of the pseudocode.

Now we consider the main loop of the pseudocode for our greedy hill climbing metaheuristic:

2. While *improvement?*

   (a) Set *improvement?* to False
   (b) For *number_of_tries* times:
       i. Set *candidate_decision* to an arbitrary decision in the neighborhood of *incumbent_decision*.
       ii. Set *value_cand* by evaluating *Model* on *candidate_decision*.
       iii. If *incumbent_decision* is feasible, *candidate_decision* is feasible, and *value_cand* > *value_inc*, then set *incumbent_decision* to *candidate_decision* and set *improvement?* to True.
       iv. If *incumbent_decision* is not feasible and *candidate_decision* is feasible, then set *incumbent_decision* to *candidate_decision* and set *improvement?* to True.

v. If *incumbent_decision* is not feasible and *candidate_decision* is not feasible, and *candidate_decision* is closer to being feasible than *incumbent_decision*, then set *incumbent_decision* to *candidate_decision* and set *improvement?* to True.

(c) If *improvement?* is True:

i. Set *value_inc* by evaluating *Model* on *incumbent_decision*.

This portion of the pseudocode corresponds to lines 8–36 of the Python implementation, shown in Figure 11.2 on page 168. We will focus on the highlights. Line 12 implements step 2(b)i of the pseudocode, "Set *candidate_decision* to an arbitrary decision in the neighborhood of *incumbent_decision*", with **newdecision** implementing *candidate_decision* and the model method **getRandomNeighbor(incumbentDecision)** serving as the procedure to draw a candidate decision from the neighborhood of the incumbent.

Lines 13–14 implement step 2(b)ii of the pseudocode, "Set *value_cand* by evaluating *Model* on *candidate_decision*." Notice that a triple is returned, as before.

Lines 17–20 implement step 2(b)iii of the pseudocode, "If *incumbent_decision* is feasible, *candidate_decision* is feasible, and *value_cand* > *value_inc*, then set *incumbent_decision* to *candidate_decision* and set *improvement?* to True."

Lines 21–23 implement step 2(b)iv of the pseudocode, "If *incumbent_decision* is not feasible and *candidate_decision* is feasible, then set *incumbent_decision* to *candidate_decision* and set *improvement?* to True."

Lines 24–29 implement step 2(b)v of the pseudocode, "If *incumbent_decision* is not feasible and *candidate_decision* is not feasible, and *candidate_decision* is closer to being feasible than *incumbent_decision*, then set *incumbent_decision* to *candidate_decision* and set *improvement?* to True." In the Python implementation, "closer to being feasible" is implemented by obtaining the sums of the negative slack values for the two solutions, and counting the candidate as closer to feasible if its sum is greater than that of the incumbent's.

Completing our discussion of the current chunk, lines 30–32 implement steps 2c–2(c)i of the pseudocode. Finally, lines 34–36 implement step 3 of the pseudocode, "Report *incumbent_decision* as heuristically optimal."

11.2.2 Experimenting with the Greedy Hill Climbing Implementation

The pseudocode describes the procedure for doing one run of the metaheuristic. Normally, we want to undertake multiple runs because the outcome of any one run is randomly conditioned and will vary. By undertaking a number of runs we can expect to get a better picture of the performance of the implementation than we would with any single run. We say that a *trial* is an undertaking of multiple runs. The function definition at lines 37–43 in Figure 11.2 implements a trial procedure and returns a sorted list of results for the runs it undertakes. This affords experimentation with the metaheuristic. We discuss a small experiment in this section.

The experiments were run with variations on the following script, present in the file *exercise_ghc.py*. The requisite files are available on the book's Web site.

```
1 import greedy_hill_climber as ghc
2 import comodel
3 import pickle
4
5 def exercise_gap(numTries,numRuns):
6     ###### Get the model. ###############################
7     [info,objcs,lhscs,rhsvs] = pickle.load(open('gap1_c5_15_1.p','rb'))
```

```
1 def doGreedyHillClimbRun(themodel,nTries):
2     # Obtain a random decision for starting the run:
3     incumbentDecision = themodel.getRandomDecision()
4     (feasible_inc,obj_inc,slacks_inc) = \
5         themodel.evaluateDecision(incumbentDecision)
6     improvement = True
7     # Go until no further improvements are found:
8     while improvement:
9         improvement = False
10        for atry in range(nTries):
11            # Find a decision in the neighborhood of the incumbent.
12            newdecision = themodel.getRandomNeighbor(incumbentDecision)
13            (feasible_new,obj_new,slacks_new) = \
14                themodel.evaluateDecision(newdecision)
15            # If it is an improvement make it the new incumbent,
16            # and record that an improvement was found.
17            if feasible_inc and feasible_new:
18                if obj_new > obj_inc:
19                    incumbentDecision = newdecision.copy()
20                    improvement = True
21            if feasible_new and not feasible_inc:
22                incumbentDecision = newdecision.copy()
23                improvement = True
24            if not feasible_new and not feasible_inc:
25                negs_new = slacks_new * (slacks_new < 0)
26                negs_inc = slacks_inc * (slacks_inc < 0)
27                if negs_new.sum() > negs_inc.sum():
28                    incumbentDecision = newdecision.copy()
29                    improvement = True
30            if improvement:
31                (feasible_inc,obj_inc,slacks_inc) = \
32                    themodel.evaluateDecision(incumbentDecision)
33    # Return the information from the best found decision in the run.
34    if slacks_inc is not None:
35        slacks_inc = slacks_inc.tolist()
36    return (feasible_inc,obj_inc,slacks_inc,incumbentDecision.tolist())
37 def doGreedyHillClimbTrial(themodel,nTries,nRuns):
38    found = []
39    for i in range(nRuns):
40        if i % 1000 == 0:
41            print("starting run ", i)
42        found.append(doGreedyHillClimbRun(themodel,nTries))
43    return sorted(found,reverse=True)
```

FIGURE 11.2: From *greedy_hill_climber.py*, an implementation of a greedy hill climbing metaheuristic in Python. (Line numbers added.)

```
8     model = comodel.GAP(objcs,lhscs,rhsvs)
9     ###### Do the trial of numRuns runs. #####################
10    found = ghc.doGreedyHillClimbTrial(model,numTries,numRuns)
11    evalCount = model.getEvalCounter()
12    return [found,evalCount]
13 #%%
14 [results, withCount] = exercise_gap(1,200000)
```

With the values numTries = 1 and numRuns = 200000, and the model GAP 1-c5-15-1, as shown above, doGreedyHillClimbTrial is called and it in turn runs doGreedyHillClimbRun numRuns = 200000 times. In each of these runs the number of tries, numTries is set to 1. In a typical trial with these variable settings, the procedure halts after having done 630,744 decision evaluations (calls to evaluateDecision()). This quantity, the number of times a decision is evaluated, is a good proxy for the amount of computational work done to complete the trial and is generally accepted as such because typically the decision evaluation procedure is the most expensive step undertaken. Table 11.1 shows information from the best 12 decisions found in this particular trial. Notice that only 1 feasible solution has been found and its objective value, at 304, is rather far from the optimal value of 336.

0	True	304.0
1	False	343.0
2	False	341.0
3	False	340.0
4	False	339.0
5	False	339.0
6	False	339.0
7	False	338.0
8	False	338.0
9	False	338.0
10	False	338.0
11	False	338.0

TABLE 11.1: Objective function values of the best 12 decisions found in a trial of greedy hill climbing on the GAP 1-c5-15-1 model with numTries = 1 and numRuns = 200000, and 630,744 decision evaluations.

Setting the values numTries = 10 and numRuns = 10000 and doing another trial yields a typical result (with these settings) shown in Table 11.2. We see that for roughly the same computational effort, allocating more tries and fewer runs yields both an improved best decision found and 9 additional feasible decisions each of which has a better objective value than the best decision found in the previous trial. (Although some of the found decisions are tied on objective value, in fact they were distinct decisions.)

Continuing in the direction of our success, we undertake a third and for now final trial, setting the values numTries = 50 and numRuns = 2000. Table 11.3 reports the results of a typical case. Remarkably, with less computational effort (only 517,550 decision evaluations) we have found a nearly optimal solution as well as several feasible solutions that are better than the best on the previous trial.

The lesson here is general and important. It is *not* that greedy hill climbing will find an optimal solution with enough fiddling; it might, but you can hardly rely on it. Instead, the lesson to hand is that greedy hill climbing in particular, and metaheuristics in general, have adjustable parameters (numTries and numRuns for instance) whose settings can signifi-

0	True	315.0
1	True	312.0
2	True	310.0
3	True	309.0
4	True	308.0
5	True	307.0
6	True	307.0
7	True	306.0
8	True	305.0
9	True	305.0
10	True	304.0
11	True	304.0

TABLE 11.2: Objective function values of the best 12 decisions found in a trial of greedy hill climbing on the GAP 1-c5-15-1 model with `numTries` = 10 and `numRuns` = 10000, and 644,082 decision evaluations.

0	True	334.0
1	True	317.0
2	True	316.0
3	True	316.0
4	True	315.0
5	True	315.0
6	True	315.0
7	True	313.0
8	True	313.0
9	True	313.0
10	True	312.0
11	True	312.0

TABLE 11.3: Objective function values of the best 12 decisions found in a trial of greedy hill climbing on the GAP 1-c5-15-1 model with `numTries` = 50 and `numRuns` = 2000, and 517,550 decision evaluations.

cantly affect performance. There is no way short of experience of estimating good parameter settings.

In practice we engage in *tuning* the parameters by running, more or less systematically (more is preferred), a series of experiments designed to find good quality values, parameter values that afford good performance by the metaheuristic. How best to do this is a large topic and it shall be with us in the sequel. Its comprehensive exploration is something that goes beyond the scope of this book. As a practical matter, however, one might tune the parameters of a metaheuristic by testing some sensible settings and then proceeding under a greedy hill climbing regime.

11.3 Simulated Annealing

Simulated annealing is greedy hill climbing with a twist. Greedy hill climbing always accepts a candidate decision if it is better than the incumbent and always rejects a candidate

1. Initialization

    (a) Select a model to be optimized, *Model*.

    (b) Set *number_of_tries* to a positive value.

    (c) Set *incumbent_decision* to an arbitrary decision for *Model*.

    (d) Set *value_inc* by evaluating *Model* on *incumbent_decision*.

    (e) Set *temperature_start* to an initial "temperature."

    (f) Set *repetition_schedule* to fix the number of iterations at each temperature step.

    (g) Set *cooling_schedule* to the temperature value for each temperature step.

2. For each temperature step k:

    (a) For each iteration i:

        i. Probe the neighborhood for *number_of_tries* tries:
            A. Set *candidate_decision* to an arbitrary decision in the neighborhood of *incumbent_decision*.
            B. Set *value_cand* by evaluating *Model* on *candidate_decision*.

        ii. Set *best_of_tries* to *value_cand* for the best candidate found in the *number_of_tries* neighborhood probes.

        iii. Set Δ to *best_of_tries* - *value_inc*.

        iv. If *incumbent_decision* is feasible and *candidate_decision* is feasible
            A. If *value_cand* > *value_inc* or *random* < $exp(-\Delta/cooling\_schedule(k))$ then set *incumbent_decision* to *candidate_decision*. (*random* is a uniform random draw from $[0, 1]$.)

        v. If *incumbent_decision* is not feasible and *candidate_decision* is feasible
            A. Set *incumbent_decision* to *candidate_decision*.

        vi. If *incumbent_decision* is not feasible and *candidate_decision* is not feasible, and *candidate_decision* is closer to being feasible than *incumbent_decision*,
            A. Set *incumbent_decision* to *candidate_decision*.

        vii. If a new *incumbent_decision* has been found, set *value_inc* by evaluating *Model* on the new *incumbent_decision*.

3. Report *incumbent_decision* as heuristically optimal.

FIGURE 11.3: High-level pseudocode for simulated annealing.

if it is equal to or worse than the incumbent. Simulated annealing, in contrast, always accepts a candidate decision if it is better than the incumbent and *sometimes* (usually, not always) rejects a candidate if it is equal to or worse than the incumbent. Why do this? Local search in its greedy hill climbing incarnation risks failing to enter a neighborhood containing an optimum decision, or even a decision close to being optimal, e.g., because no run happens to alight on an optimal neighborhood or because the procedure finds its way to a locally optimal decision in a neighborhood adjacent to an optimal neighborhood. This is apparent, although for any given problem it is generally not possible to assess this risk, except empirically.

As a heuristic, recall, simulated annealing has no way of determining whether a given decision is globally optimal or not. Simulated annealing, and indeed every other metaheuristic besides greedy hill climbing, is characterized by an approach to avoid being stuck on a local optimum. The approach simulated annealing uses is to accept as its new incumbent any decision it finds that is a definite improvement on the incumbent, and to accept with a small probability a candidate that is not much worse than the incumbent. That is the basic idea. In detail it is somewhat more subtle, as Figure 11.3, "High-level pseudocode for simulated annealing," indicates.

11.4   Running the Simulated Annealer Code

The Python code in *simulated_annealer.py,* available on the book's Web site, implements a detailed version of the pseudocode of Figure 11.3. Table 11.4 presents results on a trial using simulated annealing on the Philadelphia data and problem discussed in Chapter 9. Compare these results with Table 9.8 on page 144.

Notice that in 50 runs the algorithm found the apparently optimal decision 7 times and also found a very nearly optimal decision 4 times. The apparently optimal decision is that displayed in Figure 9.4 on page 145. The worst decision reported from these 50 trials has a ratio to the best of $31556.097 / 31057.108 = 1.016$, that is, it is within 2% of the heuristically optimal decision.

As a second example, Table 11.5 on page 174 shows the results of a trial of 50 runs on the GAP 1-c5-15-1 model introduced in Chapter 5. Notice that an optimal decision has been found as well as a number of high quality feasible but non-optimal decisions.

11.5   Threshold Accepting Algorithms

Threshold accepting algorithms (TAAs) are extremely simple local search heuristics. In fact, we would be pressed to imagine one that is simpler. Put verbally, a TAA begins with an arbitrary solution to the problem at hand. It then iteratively explores solutions in the neighborhood of the current solution. If a newly found solution is better than the current solution, *or is at least not much worse,* it replaces the current solution. *Not much worse* is operationalized by a threshold value, T. If the new solution is better than the current solution or if it is worse by less than T, the new solution is accepted and becomes the current solution. This iterative exploration continues until a certain number of trials has been completed. After each iteration the algorithm typically reduces the size of T until at

	ObjVal	Decision
0	31057.108	[44, 81, 102, 125, 173, 194, 242, 282, 291, 341]
1	31057.108	[44, 81, 102, 125, 173, 194, 242, 282, 291, 341]
2	31057.108	[44, 81, 102, 125, 173, 194, 242, 282, 291, 341]
3	31057.108	[44, 81, 102, 125, 173, 194, 242, 282, 291, 341]
4	31057.108	[44, 81, 102, 125, 173, 194, 242, 282, 291, 341]
5	31057.108	[44, 81, 102, 125, 173, 194, 242, 282, 291, 341]
6	31057.108	[44, 81, 102, 125, 173, 194, 242, 282, 291, 341]
7	31070.693	[3, 42, 81, 102, 125, 173, 194, 242, 282, 341]
8	31070.693	[3, 42, 81, 102, 125, 173, 194, 242, 282, 341]
9	31070.693	[3, 42, 81, 102, 125, 173, 194, 242, 282, 341]
10	31070.693	[3, 42, 81, 102, 125, 173, 194, 242, 282, 341]
11	31148.016	[40, 44, 102, 129, 140, 224, 255, 282, 291, 354]
12	31148.016	[40, 44, 102, 129, 140, 224, 255, 282, 291, 354]
13	31148.016	[40, 44, 102, 129, 140, 224, 255, 282, 291, 354]
14	31148.016	[40, 44, 102, 129, 140, 224, 255, 282, 291, 354]
15	31148.016	[40, 44, 102, 129, 140, 224, 255, 282, 291, 354]
16	31148.016	[40, 44, 102, 129, 140, 224, 255, 282, 291, 354]
17	31148.016	[40, 44, 102, 129, 140, 224, 255, 282, 291, 354]
18	31161.601	[3, 40, 42, 102, 129, 140, 224, 255, 282, 354]
19	31176.121	[40, 44, 102, 129, 151, 221, 255, 282, 291, 354]
20	31176.121	[40, 44, 102, 129, 151, 221, 255, 282, 291, 354]
21	31176.121	[40, 44, 102, 129, 151, 221, 255, 282, 291, 354]
22	31176.121	[40, 44, 102, 129, 151, 221, 255, 282, 291, 354]
23	31189.706	[3, 40, 42, 102, 129, 151, 221, 255, 282, 354]
24	31207.187	[102, 129, 140, 173, 194, 224, 242, 282, 341, 366]
25	31207.187	[102, 129, 140, 173, 194, 224, 242, 282, 341, 366]
26	31207.187	[102, 129, 140, 173, 194, 224, 242, 282, 341, 366]
27	31214.623	[3, 39, 42, 101, 129, 140, 224, 255, 282, 354]
28	31223.038	[3, 40, 42, 81, 102, 125, 170, 188, 282, 354]
29	31225.617	[27, 39, 42, 81, 99, 129, 255, 291, 354, 365]
30	31238.866	[44, 102, 106, 129, 140, 224, 254, 282, 291, 351]
31	31238.866	[44, 102, 106, 129, 140, 224, 254, 282, 291, 351]
32	31245.179	[44, 102, 106, 129, 140, 172, 224, 282, 291, 312]
33	31245.179	[44, 102, 106, 129, 140, 172, 224, 282, 291, 312]
34	31269.576	[33, 81, 99, 129, 173, 193, 242, 281, 341, 370]
35	31269.576	[33, 81, 99, 129, 173, 193, 242, 281, 341, 370]
36	31269.576	[33, 81, 99, 129, 173, 193, 242, 281, 341, 370]
37	31287.677	[27, 39, 42, 81, 99, 129, 172, 291, 312, 365]
38	31320.723	[27, 40, 42, 81, 102, 125, 172, 291, 312, 365]
39	31323.358	[3, 39, 42, 101, 129, 140, 172, 224, 282, 312]
40	31328.868	[33, 39, 99, 129, 151, 221, 255, 281, 354, 370]
41	31328.868	[33, 39, 99, 129, 151, 221, 255, 281, 354, 370]
42	31350.099	[33, 40, 102, 129, 151, 221, 255, 281, 354, 370]
43	31402.794	[27, 34, 39, 99, 129, 151, 221, 255, 354, 366]
44	31440.086	[39, 81, 99, 129, 170, 188, 281, 286, 354, 374]
45	31442.946	[39, 101, 129, 140, 172, 224, 307, 312, 366, 380]
46	31466.450	[39, 101, 129, 140, 224, 254, 305, 351, 355, 373]
47	31468.426	[39, 99, 129, 151, 221, 254, 281, 286, 351, 374]
48	31499.620	[39, 101, 129, 140, 172, 224, 305, 312, 355, 373]
49	31556.097	[8, 99, 129, 140, 224, 254, 274, 309, 351, 366]

TABLE 11.4: Results from a trial of 50 runs simulated annealing on the Philadelphia districting problem.

	Feasible?	ObjVal	Slacks	Decision
0	True	334.0	[1.0, 1.0, 0.0, 1.0, 5.0]	[4, 1, 3, 2, 0, 4, 3, 1, 0, 1, 3, 3, 0, 0, 2]
1	True	330.0	[2.0, 1.0, 0.0, 2.0, 9.0]	[4, 1, 3, 2, 0, 3, 0, 1, 4, 1, 3, 3, 0, 0, 2]
2	True	326.0	[1.0, 9.0, 0.0, 1.0, 0.0]	[1, 4, 3, 2, 4, 3, 0, 1, 0, 3, 4, 3, 0, 0, 2]
3	True	323.0	[3.0, 2.0, 13.0, 0.0, 0.0]	[1, 1, 3, 4, 0, 4, 0, 2, 0, 3, 3, 3, 1, 0, 2]
4	True	323.0	[2.0, 2.0, 12.0, 0.0, 4.0]	[1, 1, 3, 4, 0, 2, 0, 1, 4, 3, 3, 3, 0, 0, 2]
5	True	322.0	[1.0, 2.0, 2.0, 6.0, 0.0]	[1, 1, 2, 4, 0, 4, 0, 1, 0, 2, 3, 3, 0, 3, 2]
6	True	321.0	[3.0, 2.0, 11.0, 0.0, 0.0]	[1, 1, 3, 4, 0, 4, 0, 1, 0, 3, 3, 3, 2, 0, 2]
7	True	319.0	[3.0, 3.0, 2.0, 0.0, 12.0]	[4, 2, 3, 4, 0, 1, 0, 1, 0, 3, 3, 3, 1, 0, 2]
8	True	318.0	[12.0, 2.0, 3.0, 0.0, 4.0]	[1, 1, 3, 4, 0, 2, 0, 2, 4, 3, 3, 3, 1, 0, 2]
9	True	318.0	[3.0, 3.0, 2.0, 1.0, 12.0]	[4, 4, 2, 3, 0, 1, 0, 1, 0, 2, 3, 3, 1, 0, 2]
10	True	316.0	[1.0, 9.0, 2.0, 2.0, 0.0]	[1, 4, 2, 3, 4, 3, 0, 1, 0, 2, 4, 3, 0, 0, 2]
11	True	316.0	[1.0, 2.0, 8.0, 0.0, 0.0]	[4, 2, 3, 4, 0, 2, 0, 1, 0, 3, 3, 3, 0, 4, 1]
12	True	315.0	[1.0, 2.0, 12.0, 0.0, 2.0]	[4, 4, 3, 1, 4, 2, 0, 1, 0, 3, 3, 3, 0, 0, 2]
13	True	315.0	[1.0, 1.0, 0.0, 3.0, 5.0]	[1, 1, 2, 2, 4, 2, 0, 1, 0, 1, 4, 3, 0, 0, 3]
14	True	314.0	[4.0, 2.0, 6.0, 0.0, 5.0]	[4, 0, 2, 3, 4, 2, 1, 1, 0, 3, 4, 3, 1, 0, 2]
15	True	313.0	[3.0, 2.0, 3.0, 0.0, 12.0]	[4, 4, 3, 1, 0, 2, 0, 2, 0, 3, 3, 3, 1, 0, 2]
16	True	313.0	[2.0, 7.0, 2.0, 2.0, 2.0]	[4, 4, 2, 3, 0, 3, 0, 1, 1, 2, 4, 3, 0, 0, 2]
17	True	313.0	[1.0, 3.0, 7.0, 1.0, 7.0]	[1, 4, 2, 4, 0, 3, 0, 2, 0, 1, 3, 3, 0, 3, 2]
18	True	312.0	[3.0, 8.0, 7.0, 2.0, 2.0]	[4, 4, 2, 3, 0, 3, 0, 2, 0, 1, 4, 3, 1, 0, 2]
19	True	308.0	[3.0, 2.0, 11.0, 0.0, 2.0]	[4, 4, 2, 3, 0, 2, 0, 1, 0, 3, 4, 3, 2, 0, 1]
20	True	308.0	[1.0, 2.0, 8.0, 0.0, 2.0]	[4, 4, 2, 3, 4, 2, 0, 1, 0, 3, 3, 2, 0, 0, 1]
21	True	306.0	[5.0, 5.0, 1.0, 0.0, 4.0]	[4, 1, 2, 2, 3, 3, 0, 2, 4, 3, 3, 0, 4, 0, 1]
22	True	306.0	[3.0, 7.0, 8.0, 3.0, 2.0]	[4, 1, 2, 4, 0, 2, 0, 1, 0, 2, 4, 3, 1, 0, 3]
23	True	303.0	[3.0, 1.0, 13.0, 3.0, 2.0]	[4, 1, 2, 4, 0, 2, 0, 2, 0, 1, 4, 3, 1, 0, 3]
24	True	303.0	[3.0, 1.0, 13.0, 3.0, 2.0]	[4, 1, 2, 4, 0, 2, 0, 2, 0, 1, 4, 3, 1, 0, 3]
25	True	302.0	[4.0, 2.0, 1.0, 3.0, 5.0]	[0, 3, 2, 4, 0, 2, 0, 1, 2, 3, 4, 3, 4, 0, 1]
26	True	302.0	[1.0, 0.0, 1.0, 5.0, 2.0]	[4, 3, 2, 2, 1, 3, 0, 2, 0, 1, 1, 3, 0, 0, 4]
27	True	300.0	[2.0, 1.0, 7.0, 1.0, 3.0]	[1, 4, 2, 3, 0, 1, 3, 2, 0, 0, 1, 3, 4, 4, 2]
28	True	299.0	[4.0, 2.0, 10.0, 0.0, 2.0]	[4, 0, 2, 4, 4, 3, 0, 1, 0, 3, 2, 3, 2, 3, 1]
29	True	298.0	[4.0, 5.0, 1.0, 0.0, 2.0]	[1, 1, 2, 2, 4, 3, 0, 2, 4, 3, 1, 3, 4, 3, 0]
30	True	297.0	[1.0, 6.0, 0.0, 1.0, 2.0]	[4, 1, 3, 4, 1, 2, 0, 1, 0, 2, 4, 2, 0, 0, 3]
31	True	296.0	[3.0, 2.0, 4.0, 4.0, 2.0]	[1, 4, 2, 4, 0, 3, 0, 1, 0, 2, 1, 2, 4, 0, 3]
32	True	295.0	[5.0, 1.0, 7.0, 3.0, 2.0]	[1, 1, 2, 0, 1, 3, 3, 2, 4, 3, 3, 4, 4, 0, 2]
33	True	295.0	[3.0, 5.0, 3.0, 3.0, 2.0]	[0, 1, 4, 4, 0, 2, 1, 1, 0, 2, 2, 3, 4, 0, 3]
34	True	294.0	[3.0, 9.0, 2.0, 0.0, 5.0]	[2, 1, 2, 3, 0, 1, 0, 2, 0, 3, 1, 3, 4, 0, 4]
35	True	294.0	[3.0, 9.0, 2.0, 0.0, 0.0]	[0, 1, 4, 3, 1, 4, 0, 2, 0, 3, 1, 3, 2, 0, 2]
36	True	294.0	[3.0, 4.0, 4.0, 2.0, 5.0]	[0, 1, 2, 3, 0, 3, 0, 2, 0, 1, 1, 3, 4, 2, 4]
37	True	292.0	[5.0, 0.0, 0.0, 2.0, 2.0]	[4, 1, 2, 4, 4, 2, 0, 1, 1, 3, 2, 0, 2, 0, 3]
38	True	292.0	[2.0, 5.0, 4.0, 3.0, 2.0]	[4, 0, 2, 4, 4, 3, 3, 2, 0, 3, 1, 3, 0, 2, 1]
39	True	290.0	[4.0, 0.0, 1.0, 0.0, 5.0]	[4, 1, 2, 3, 4, 2, 3, 1, 2, 3, 1, 4, 1, 0, 0]
40	True	290.0	[1.0, 5.0, 2.0, 2.0, 5.0]	[4, 1, 2, 1, 4, 2, 0, 2, 0, 3, 2, 4, 0, 0, 3]
41	True	289.0	[4.0, 2.0, 0.0, 0.0, 0.0]	[1, 1, 2, 4, 0, 3, 3, 1, 2, 3, 4, 4, 2, 3, 0]
42	True	289.0	[2.0, 9.0, 2.0, 2.0, 2.0]	[4, 1, 2, 4, 0, 1, 0, 2, 2, 3, 1, 4, 0, 0, 3]
43	True	288.0	[3.0, 0.0, 9.0, 3.0, 5.0]	[0, 1, 2, 4, 0, 1, 3, 2, 0, 1, 4, 2, 4, 0, 3]
44	True	288.0	[3.0, 0.0, 3.0, 0.0, 0.0]	[4, 2, 2, 1, 4, 3, 1, 2, 0, 3, 4, 3, 4, 3, 0]
45	True	286.0	[4.0, 5.0, 2.0, 0.0, 2.0]	[4, 1, 4, 1, 3, 3, 3, 2, 0, 3, 4, 0, 2, 0, 2]
46	True	285.0	[4.0, 3.0, 7.0, 1.0, 1.0]	[0, 1, 2, 3, 4, 0, 3, 2, 1, 4, 1, 3, 4, 0, 2]
47	True	285.0	[4.0, 3.0, 7.0, 1.0, 1.0]	[0, 1, 2, 3, 4, 0, 3, 2, 1, 4, 1, 3, 4, 0, 2]
48	True	285.0	[3.0, 9.0, 1.0, 3.0, 0.0]	[4, 1, 2, 2, 4, 3, 3, 2, 0, 3, 4, 3, 4, 1, 0]
49	True	276.0	[4.0, 3.0, 9.0, 2.0, 0.0]	[4, 1, 2, 3, 4, 3, 3, 2, 1, 2, 1, 4, 4, 0, 0]

TABLE 11.5: Results from a trial of 50 runs of simulated annealing on the GAP 1-c5-15-1 model.

the very end of the run T is very close to zero. The entire process may be repeated multiple times, starting from different initial solutions. The best solution found during the entire process is offered as the (heuristically) optimal solution to the problem.

Expressed in pseudocode, TAAs are trivially different from simulated annealing. Where simulated annealing probabilistically accepts a worse decision based on the state of cooling, TAAs deterministically accept a worse decision based on the state of cooling.

The Python code in *threshold_accepter.py*, implements a TAA. It is available on the book's Web site. Table 11.6 on page 176 shows the results of a trial of 50 runs on the GAP 1-c5-15-1 model introduced in Chapter 5. Notice that an optimal decision has not been found and that the decisions in the table are generally inferior to those in Table 11.5 for simulated annealing.

11.6 Tabu Search

Tabu search is a fourth important local search metaheuristic. Introduced by Fred Glover [59] and conceived as a metaheuristic from the outset, tabu search has been articulated and developed with a large number of variations, some of them quite complex. It has, in various forms, also seen very widespread and generally successful application on COModels, especially of the combinatorial (integer variables) type.

As a starting point, for fixing ideas, we might think of tabu search as a threshold accepting algorithm augmented by a *tabu list* that prevents acceptance of a new incumbent decision if doing so would result in, or risk, returning to—"cycling back to"—a decision that was itself previously an incumbent. The core idea is to maintain, as a heuristic, a time decaying list of previous moves in the search and to prevent reversal of the moves for a number of steps in the search process. The motivation, again, is to avoid previously explored positions. As it happens, tabu search has proved very productive and successful. This applies to the considerable body of variations on the method; no one variation has been dominant. Because these variations offer a degree of complexity beyond the scope of this book and because there are so many of them, we content ourselves with noting with approval the method and commending it for further study by the reader.

11.7 For Exploration

1. Discuss the following passage, which appears on page 165.

> Fixing *number_of_tries* at 100 is considerably less greedy, and much more patient, that fixing it at, say, 10 or even 1. It is an empirical matter what yields the best results.

Why is it that "Fixing *number_of_tries* at 100 is considerably less greedy, and much more patient, that fixing it at, say, 10"? How should experiments be designed to investigate this empirical matter? On what basis should comparison be made between, for example, results from setting *number_of_tries* at 100 and setting it to 1? Would you expect to find that results vary by type of model (GAP, traveling salesman, etc.) used in the comparison? Why or why not?

	Feasible?	ObjVal	Slacks	Decision
0	True	326.0	[1.0, 8.0, 0.0, 2.0, 2.0]	[4, 4, 3, 2, 4, 3, 0, 1, 0, 1, 3, 3, 0, 0, 2]
1	True	326.0	[1.0, 8.0, 0.0, 2.0, 2.0]	[4, 4, 3, 2, 4, 3, 0, 1, 0, 1, 3, 3, 0, 0, 2]
2	True	323.0	[3.0, 2.0, 13.0, 0.0, 0.0]	[1, 1, 3, 4, 0, 4, 0, 2, 0, 3, 3, 3, 1, 0, 2]
3	True	322.0	[1.0, 2.0, 2.0, 6.0, 0.0]	[1, 1, 2, 4, 0, 4, 0, 1, 0, 2, 3, 3, 0, 3, 2]
4	True	321.0	[3.0, 2.0, 7.0, 1.0, 0.0]	[1, 1, 3, 2, 0, 3, 0, 2, 0, 3, 4, 3, 1, 0, 4]
5	True	321.0	[1.0, 2.0, 10.0, 0.0, 0.0]	[1, 1, 2, 2, 0, 3, 0, 1, 0, 3, 4, 3, 0, 3, 4]
6	True	319.0	[3.0, 2.0, 7.0, 1.0, 2.0]	[4, 4, 3, 2, 0, 3, 0, 2, 0, 3, 4, 3, 1, 0, 1]
7	True	318.0	[3.0, 3.0, 2.0, 1.0, 12.0]	[4, 4, 2, 3, 0, 1, 0, 1, 0, 2, 3, 3, 1, 0, 2]
8	True	317.0	[5.0, 7.0, 0.0, 0.0, 5.0]	[4, 1, 0, 2, 4, 3, 0, 1, 0, 3, 3, 4, 1, 3, 2]
9	True	317.0	[4.0, 2.0, 12.0, 0.0, 0.0]	[4, 0, 3, 4, 0, 2, 1, 1, 0, 3, 3, 3, 1, 4, 2]
10	True	315.0	[2.0, 7.0, 0.0, 1.0, 2.0]	[4, 1, 4, 2, 4, 3, 0, 1, 0, 0, 3, 3, 1, 3, 2]
11	True	313.0	[3.0, 2.0, 3.0, 0.0, 12.0]	[4, 4, 3, 1, 0, 2, 0, 2, 0, 3, 3, 3, 1, 0, 2]
12	True	313.0	[3.0, 2.0, 3.0, 0.0, 12.0]	[4, 4, 3, 1, 0, 2, 0, 2, 0, 3, 3, 3, 1, 0, 2]
13	True	313.0	[2.0, 2.0, 1.0, 4.0, 0.0]	[1, 1, 2, 2, 4, 3, 3, 2, 0, 0, 3, 3, 1, 0, 4]
14	True	312.0	[5.0, 0.0, 1.0, 0.0, 2.0]	[4, 1, 2, 2, 0, 3, 0, 2, 1, 3, 3, 0, 1, 3, 4]
15	True	312.0	[3.0, 7.0, 8.0, 0.0, 5.0]	[4, 1, 3, 3, 0, 3, 0, 1, 0, 2, 4, 4, 1, 0, 2]
16	True	308.0	[4.0, 2.0, 3.0, 0.0, 5.0]	[4, 0, 3, 1, 4, 2, 3, 2, 0, 3, 4, 3, 1, 0, 2]
17	True	307.0	[1.0, 1.0, 1.0, 0.0, 5.0]	[4, 3, 2, 2, 0, 1, 0, 2, 0, 3, 4, 4, 0, 3, 1]
18	True	306.0	[3.0, 7.0, 8.0, 3.0, 2.0]	[4, 1, 2, 4, 0, 2, 0, 1, 0, 2, 4, 3, 1, 0, 3]
19	True	304.0	[3.0, 0.0, 1.0, 1.0, 0.0]	[0, 1, 2, 2, 0, 3, 3, 2, 1, 0, 3, 4, 1, 3, 4]
20	True	302.0	[4.0, 0.0, 0.0, 3.0, 1.0]	[0, 1, 2, 2, 0, 2, 0, 1, 1, 4, 3, 4, 4, 0, 3]
21	True	302.0	[3.0, 1.0, 5.0, 1.0, 2.0]	[4, 1, 3, 4, 0, 2, 0, 2, 0, 1, 4, 2, 1, 0, 3]
22	True	301.0	[5.0, 5.0, 1.0, 0.0, 2.0]	[0, 0, 2, 2, 0, 3, 3, 2, 4, 3, 1, 4, 4, 3, 1]
23	True	300.0	[3.0, 0.0, 3.0, 2.0, 0.0]	[4, 3, 3, 4, 0, 2, 0, 2, 1, 0, 1, 3, 1, 4, 2]
24	True	299.0	[6.0, 6.0, 0.0, 0.0, 4.0]	[0, 1, 0, 4, 1, 3, 0, 1, 4, 3, 2, 3, 2, 3, 2]
25	True	299.0	[4.0, 2.0, 8.0, 0.0, 3.0]	[0, 0, 2, 4, 3, 2, 3, 1, 0, 2, 3, 3, 4, 4, 1]
26	True	299.0	[3.0, 0.0, 2.0, 1.0, 3.0]	[0, 4, 2, 3, 0, 2, 1, 2, 0, 4, 3, 3, 2, 0, 1]
27	True	296.0	[4.0, 2.0, 2.0, 0.0, 7.0]	[4, 0, 2, 4, 3, 3, 0, 1, 0, 3, 2, 3, 4, 2, 1]
28	True	296.0	[3.0, 1.0, 2.0, 5.0, 5.0]	[0, 2, 3, 4, 3, 3, 0, 1, 0, 1, 1, 4, 4, 0, 2]
29	True	296.0	[0.0, 5.0, 7.0, 1.0, 5.0]	[4, 1, 2, 1, 4, 3, 3, 2, 0, 0, 4, 3, 0, 3, 2]
30	True	295.0	[3.0, 0.0, 7.0, 2.0, 0.0]	[4, 1, 2, 4, 1, 0, 0, 2, 0, 1, 3, 3, 3, 4, 2]
31	True	294.0	[4.0, 0.0, 3.0, 0.0, 7.0]	[4, 1, 4, 0, 1, 2, 3, 2, 0, 1, 3, 3, 4, 3, 2]
32	True	294.0	[1.0, 5.0, 12.0, 0.0, 2.0]	[4, 1, 2, 1, 3, 3, 0, 2, 0, 3, 2, 3, 0, 0, 4]
33	True	293.0	[5.0, 0.0, 2.0, 0.0, 2.0]	[0, 4, 2, 4, 3, 3, 1, 2, 2, 3, 3, 0, 4, 0, 1]
34	True	292.0	[2.0, 4.0, 7.0, 3.0, 2.0]	[4, 1, 2, 4, 0, 3, 0, 2, 3, 1, 1, 4, 0, 0, 2]
35	True	291.0	[1.0, 6.0, 1.0, 2.0, 0.0]	[2, 1, 2, 4, 1, 2, 0, 1, 0, 3, 4, 4, 0, 0, 3]
36	True	290.0	[5.0, 5.0, 2.0, 1.0, 3.0]	[4, 1, 0, 1, 3, 3, 3, 2, 0, 4, 2, 3, 4, 0, 2]
37	True	287.0	[4.0, 8.0, 1.0, 0.0, 0.0]	[0, 1, 2, 2, 3, 3, 1, 2, 4, 3, 1, 3, 4, 4, 0]
38	True	287.0	[3.0, 5.0, 2.0, 0.0, 0.0]	[2, 1, 2, 4, 4, 3, 1, 2, 0, 3, 4, 3, 1, 3, 0]
39	True	286.0	[4.0, 1.0, 2.0, 1.0, 2.0]	[4, 1, 2, 4, 3, 3, 3, 2, 2, 1, 3, 4, 1, 0, 0]
40	True	285.0	[4.0, 5.0, 2.0, 0.0, 0.0]	[4, 1, 2, 1, 4, 3, 0, 2, 2, 3, 4, 3, 4, 3, 0]
41	True	284.0	[5.0, 2.0, 7.0, 3.0, 2.0]	[0, 1, 2, 0, 4, 3, 3, 2, 4, 3, 1, 3, 4, 1, 2]
42	True	283.0	[2.0, 5.0, 2.0, 3.0, 0.0]	[4, 1, 2, 1, 4, 2, 3, 2, 0, 0, 2, 4, 4, 0, 3]
43	True	283.0	[1.0, 9.0, 2.0, 3.0, 4.0]	[0, 1, 2, 4, 0, 3, 3, 4, 0, 3, 3, 2, 0, 1, 2]
44	True	282.0	[3.0, 5.0, 2.0, 3.0, 1.0]	[0, 1, 2, 1, 0, 2, 3, 2, 0, 4, 2, 4, 4, 0, 3]
45	True	279.0	[4.0, 4.0, 2.0, 1.0, 0.0]	[4, 1, 2, 3, 1, 3, 1, 2, 2, 3, 4, 4, 4, 0, 0]
46	True	278.0	[4.0, 0.0, 1.0, 0.0, 2.0]	[4, 1, 2, 0, 1, 3, 3, 2, 0, 3, 2, 1, 2, 3, 4]
47	True	277.0	[3.0, 5.0, 4.0, 3.0, 0.0]	[4, 1, 2, 1, 4, 3, 3, 2, 0, 3, 3, 4, 4, 2, 0]
48	True	268.0	[3.0, 9.0, 4.0, 1.0, 0.0]	[4, 1, 3, 4, 1, 3, 3, 2, 0, 3, 1, 2, 2, 4, 0]
49	False	-1.0	[4.0, 8.0, -1.0, 0.0, 0.0]	[4, 1, 2, 0, 4, 2, 1, 2, 0, 2, 1, 4, 4, 3, 3]

TABLE 11.6: Results from a trial of 50 runs of a threshold accepting algorithm on the GAP 1-c5-15-1 model.

2. Discuss the following passage, which appears on page 165.

> [T]he procedure described in Figure 11.1 is a *meta*heuristic: The procedure is in fact not a single procedure, but a schema or template or pattern that may be followed by many particular procedures, differing in detail.

Design several variations on the greedy hill climbing procedure. Are some variations—some versions of the metaheuristic—more suited than others for certain applications? How should experiments be designed to compare the different versions?

3. Discuss the following passage, which appears on page 170.

> In practice we engage in *tuning* the parameters by running, more or less systematically (more is preferred), a series of experiments designed to find good quality values, parameter values that afford good performance by the metaheuristic. How best to do this is a large topic and it shall be with us in the sequel. Its comprehensive exploration is something that goes beyond the scope of this book. As a practical matter, however, one might tune the parameters of a metaheuristic by testing some sensible settings and then proceeding under a greedy hill climbing regime.

Beyond discussion, devise and carry out a parameter tuning exercise as suggested. What do you find?

4. The following passage appears on page 175.

> Table 11.6 on page 176 shows the results of a trial of 50 runs on the GAP 1-c5-15-1 model introduced in Chapter 5. Notice that an optimal decision has not been found and that the decisions in the table are generally inferior to those in Table 11.5 for simulated annealing.

What can we conclude from this finding about the relative merits of simulated annealing and threshold accepting algorithms? Design an experimental regime that could resolve the issue.

5. Describe a threshold accepting algorithm in pseudocode. Note: The Python code in *threshold_accepter.py*, is available on the book's Web site.

11.8 For More Information

Simulated annealing has received a great deal of attention every since its originating publications [103] (see also [82]). Applications have been extensive and include essentially every kind of optimization problem traditionally associated with operations research and management science, as well as multiple specialist areas, such as political redistricting [23, 120]. See [129] for a recent review. The general reputation of simulated annealing is that it can be effective at finding good decisions for difficult optimization problems, but it requires a substantial amount of computational effort to do so.

Simulated annealing has, from its inception, benefitted from continual theoretical investigation. See reviews in [31, 40, 129] and the reviews they cite. Despite extensive theoretical progress, application guidance remains primarily in the purview of experience.

The chapters by Dowsland [40] and by Nikolaev and Jacobson [129] are very useful and accessible surveys of simulated annealing. These are excellent points of departure from this chapter.

Many other local search metaheuristics exist and are flourishing in applications. The volumes in which [40] and [129] appear also review many of these methods. We will mention just a few.

As noted above, threshold accepting algorithms, originally introduced in Dueck and Scheuer [41], are closely related to simulated annealing, but are even simpler. Essentially, the threshold accepting metaheuristic is a deterministic simplification of simulated annealing. Where simulated annealing accepts a candidate with a declining probability based on a declining temperature, threshold accepting heuristics accept a candidate with a declining unfavorable Δ, based on a declining temperature. The book by Winker [156] is thorough and very thoughtful in applying threshold accepting algorithms to econometrics and statistics. The Web page at `http://comisef.wikidot.com/concept:thresholdaccepting` (accessed 2015-06-06) is succinct and usefully clear. Korevaar et al. [109] describe a substantial and quite successful real-world application of this metaheuristic, which is ongoing. This said, there is comparatively little literature on application or theory for the method.

Tabu search, on the other hand, is in the same league as simulated annealing with regard to theoretical exploration and to applications. It may be described as a form of *guided search*: in tabu search greedy hill climbing is augmented with a judicious visiting of worsening decisions, as in simulated annealing and threshold accepting algorithms, along with a memory of recently visited positions. These latter become (at least temporarily) tabu, thus focusing the search on new areas not experienced to be weak. Glover introduced the metaheuristic [59] and has covered it comprehensively in subsequent publications, e.g., [58, 60, 61]. Gendreau and Potvin [56] provide an up to date review and their article is an excellent point of departure from this chapter.

Systematic treatment of tuning metaheuristics—searching among their parameter values for felicitous settings—is beyond the scope of this book. This has, however, become an important research area with much spillover to practice. Standard reference works are [14] and [15]. Chapter 8 of [156] addresses tuning threshold accepting algorithms in depth with many of the points having more general applicability.

Chapter 12

Evolutionary Algorithms

12.1   Introduction

Evolutionary algorithms (EAs) constitute a loosely defined family of heuristic problem solving procedures—they are metaheuristics—that are inspired by Darwinian evolution in biology. The basic notion of an evolutionary algorithm is remarkably simple. A population of decisions is created and maintained throughout a run of the algorithm. Individuals in the population are assessed with regard to their performance on the objective at hand. Better performing individuals, said to have higher *fitness*, are more likely to remain in the population over time and to have their "descendants" present in the population over time. Descendants are formed by perturbing individuals in the population and, typically, exchanging information among them, through what are called *genetic operators*. Over time the constitution of the population changes—evolves—and typically better and better decisions are found to the objective at hand. Figure 12.1 outlines this process in high-level pseudocode.

Several characteristics are core to the concept of an evolutionary algorithm.

1. Decisions for the problem at hand can be rigorously characterized.

   For example, a decision for a Simple Knapsack problem with n decision variables may be specified as a vector of length n, all of whose entries are either 0s or 1s.

2. For every possible decision for the problem we can specify how to evaluate it and give it a score with regard to an appropriate *measure of performance* for the problem.

   We call this score the *fitness* of the solution, by analogy to fitness as understood in biological evolution. In the case of the Simple Knapsack problem the fitness of a decision is simply its objective function score, z. Recall expression (1.1)

$$\max z = \sum_{i=1}^{n} c_i x_i \qquad (12.1)$$

1. Create a number of decisions for the problem to be solved, thereby creating a *population* of decisions by analogy with biological processes.

2. Evaluate each member of the population and obtain its *fitness score*.

3. Choose which individuals will be represented in the next generation.

   Do this with bias towards individuals (decisions) with higher fitnesses. Keeping the population size constant, some individuals will be represented by more than one ("daughter") solution in the next generation.

4. Alter the decisions in the next generation.

   Apply genetic operators such as computer program analogs of mutation and recombination in biology. Alterations are performed in a randomized fashion, resulting in a next generation population that resembles but is different from the current generation population.

5. If the stopping condition (such as number of generations) has not been met, return to step (2), using the next generation population as the new current generation.

FIGURE 12.1: Evolutionary algorithms: high-level pseudocode.

on page 7, in Figure 1.2. z for any single decision $[x_1, x_2, \ldots, x_n]$ is $z = \sum_{i=1}^{n} c_i x_i$.

3. The algorithm processes a *population* of decisions to the problem at hand. Under processing by the algorithm, the population of decisions changes—it evolves.

   This is a distinctive aspect of EAs. Many other metaheuristics process only a single solution at a time. We discussed several of these, along with the concept of a local search metaheuristic, in Chapter 11.

4. New decisions are normally generated by modifying existing decisions.

   The modifiers are called *genetic operators* by analogy with biological concepts. We will see examples below.

5. Decisions that are comparatively fitter than other decisions in the population at a given stage are more likely to be retained in the population or be modified by the genetic operators and hence have descendants in subsequent stages of the population.

To solve a problem, such as a Simple Knapsack problem, with an evolutionary algorithm, we obtain an initial population of decisions (usually by a random process) and we run the algorithm on the population for a number of iterations. When we halt the process we can hope to have one or more high quality decisions for the problem. It's that simple, although the details may be manifold. Key points:

1. Evolutionary algorithms do not optimize, they *meliorize*. Given an existing population of decisions, they seek better decisions, with no guarantee that they will find optimal or even good decisions. We will say that they *heuristically optimize*.

2. The evolutionary process is *blind* in its search. The genetic operators make changes to decisions without any knowledge of whether the changes are beneficial. By its nature, evolution blindly tries new combinations without foreknowledge of whether they will

be successful.[1] In biology as well as in evolutionary algorithms, it is typically the case that nearly all new combinations tried are unsuccessful.

3. Evolutionary algorithms, including genetic algorithms (GAs) of many sorts, have met with excellent success, and are widely used in practice, especially on difficult problems for which more conventional methods do not perform well. (Simple Knapsack problems, as we have said repeatedly, *are* well-solved by conventional methods. Even so, there remains scope for handling them with evolutionary algorithms and other metaheuristics.) Applications of evolutionary algorithms abound in the STEM fields (science, technology, engineering, and mathematics) including business (and finance), as well as design in several fields, including architecture.

In this chapter we look into two rather different prototypical varieties of evolutionary algorithms: evolutionary programs (EPs) and genetic algorithms (GAs). In fact, EPs are sufficiently distinct that the high-level pseudocode of Figure 12.1 does not capture them very well. So that is where we shall begin, starting in the next section.

12.2 EPs: Evolutionary Programs

Evolutionary programming (EP, which also abbreviates *evolutionary program*, an algorithm that implements evolutionary programming) is a family of loosely related EAs (evolutionary algorithms). It was originally developed and named by Lawrence Fogel [52] in the 1960s (see [49, 50, 51] for useful discussions and background materials). Since then, EP has been developed by many different people, and many different varieties or "flavors" have been created, with successful applications in diverse areas. The Fogels founded a company in 1993, Natural Selection, Inc., whose mission is to use evolutionary computation to solve real-world problems. The company continues to do business today and its Web site is a useful source of information on applications of EC (http://www.natural-selection.com/).

The evolutionary computation community generally recognizes four main varieties of evolutionary algorithms: (1) evolutionary programs and (2) genetic algorithms, which we study in this chapter, and (3) evolution strategies and (4) genetic programming, which we will not discuss in any detail. It is generally acceded that the different methods have in many ways come to overlap and that there is little of value to be gained in attempting to make precise distinctions among them. We agree. As we wish to emphasize, the practically useful and theoretically interesting space of metaheuristics is enormous and far from fully specified. Innovations are constantly appearing and we expect this will continue indefinitely. Our strategy—here, in Chapter 11, and indeed throughout the book—is to present and discuss a small number of prototypical approaches that have good use in real applications. The reader, having mastered these examples, will be in good position to explore alternatives.

12.2.1 The EP Procedure

In the prototypical evolutionary program, a population of decisions is maintained and processed over multiple generations, as is characteristic of evolutionary algorithms. What is distinctive in EP (compared to typical genetic algorithms, etc.) is that during each generation, the population is expanded by copying the incumbent decisions to form a larger

---

[1]This is not to deny the possibility that certain structures may evolve in part because they have done relatively better at engendering new solutions.

1. Initialization:

    (a) Choose a model and set parameter values from Table 12.1.

    (b) Create a *currentPopulation*.

2. Do *numGens* times:

    (a) Reproduction: Expand *currentPopulation* by copying it (1 + *offspringCount*) times into *nextPopulation*.

    (b) Mutation: In *nextPopulation*, subject to mutation each of the *offspringCount* copies of the decisions in *currentPopulation*

    (c) Culling duplicates: Remove all duplicate decisions from *nextPopulation* and put the results into *nextPopulationCulled*.

    (d) Evaluation: Evaluate the decisions in *nextPopulationCulled*.

    (e) Selection: Identify the *popSize* best decisions found in the evaluation of *nextPopulationCulled* and remove all other decisions in the population.

    (f) Replacement: Set (replace) the *currentPopulation* with the *nextPopulationCulled*.

3. Collect statistics and report out.

FIGURE 12.2: High-level pseudocode for an evolutionary program.

population. The new copies are subjected to mutation, the expanded population is evaluated (that is, each decision in the expanded population is evaluated), and the highest scoring decisions are selected for the population of the next generation. Normally, there is a parameter that sets the fixed size of the population, *popSize*. Processing during a generation creates a much larger population from the incumbent population of *popSize* decisions. After this larger population is evaluated the best *popSize* decisions are taken to create the new incumbent population for the next generation.

Our EP, like EPs generally, has four main parameters as shown in Table 12.1.

Parameter	Example Value	Description
popSize	200	Population size. Called μ in the literature.
offspringCount	2	Number of offspring per decision per generation. Called λ in the literature.
r	1	Mutation rate. The mean of the Poisson distribution we use to determine the number of mutations for a single decision.
numGens	2000	Number of generations for a run.

TABLE 12.1: Parameters for an evolutionary program.

Figure 12.2 above presents high-level pseudocode for evolutionary programs. Points arising with respect to the pseudocode in Figure 12.2:

1. Step 1a, "Choose a model and set parameter values from Table 12.1," is quite straightforward and needs no further comment.

2. Step 1b, "Create a *currentPopulation*," is the preparatory step of creating an initial population of decisions. Normally this is done by collecting *popSize* random decisions. Of course, it is possible to "seed" this initial population with decisions obtained in some other way, for example from another heuristic.

3. Step 2, "Do *numGens* times," signifies the main loop of the procedure. In the body of this procedure we undertake one generation of learning in the evolutionary programming style. In the complete EP we repeat this multiple times, governed by the *numGens* parameter.

4. Step 2a is "Reproduction: Expand *currentPopulation* by copying it $(1 + offspringCount)$ times into *nextPopulation*." This is the first of five essential steps undertaken during one generation of EP learning. Here we combine the *currentPopulation* with *offspringCount* copies of it into a single population, *nextPopulation*, that is $(1 + offspringCount)$ times the size of the *currentPopulation*.

5. Step 2b is "Mutation: In *nextPopulation*, subject to mutation each of the *offspringCount* copies of the decisions in *currentPopulation*." *nextPopulation* consists of $(1 + offspringCount)$ copies of each of the decisions in *currentPopulation*. We leave one copy of each decision intact and subject to mutation each of the remaining *offspringCount* copies. Of course, whether a given decision is actually changed, and if so how much it is changed, when it is subjected to mutation is a matter of chance and how the mutation procedure is implemented.

6. Step 2c, "Culling duplicates: Remove all duplicate decisions from *nextPopulation* and put the results into *nextPopulationCulled*." This removes any duplicate decisions existing after mutation in *nextPopulation*. It is an optional step in EP generally. Our experience supports retaining it. We encourage the reader to experiment.

7. In step 2d, "Evaluation: Evaluate the decisions in *nextPopulationCulled*," we (finally!) evaluate the decisions under consideration. The Python code we use for EP, *evolutionary_programming.py* (available on the book's Web site), calculates a net fitness score for each decision. If the decision is feasible, its fitness is its objective function value, which is assumed to be positive. If the decision is infeasible, its fitness is a negative number obtained by summing the negative slack values on the constraints for the decision in question. The code is assuming maximization and $\leq$ constraints. When these assumptions cannot be met it will be necessary to make minor modifications to the code.

8. Step 2e, "Selection: Set the new *currentPopulation* to the *popSize* best decisions found in the evaluation of *nextPopulationCulled*," completes the body of our EP learning procedure. We create a new *currentPopulation* by taking the best *popSize* (or fewer if necessary) decisions from *nextPopulationCulled*. We have now created a population for a new generation, one that we can hope is on average somewhat improved from the one that commenced the current generation.

As a further note, our flavor of EP uses what is called $(\mu + \lambda)$ selection. In $(\mu + \lambda)$ selection there is a population of μ individuals. Each individual produces λ offspring each generation, with these offspring being subject to change by mutation. The μ parents and the $\mu \times \lambda$ daughters produced by mutation jointly compete on fitness to be taken for the next generation. Thus, each generation a total of $\mu(1+\lambda)$ individuals compete to be present in the next generation. Under $(\mu + \lambda)$ selection, these $\mu(1 + \lambda)$ individuals are evaluated with respect to fitness and ranked. The best-scoring μ individuals are selected for the next generation. In consequence, $(\mu + \lambda)$ selection is a form of *ranking selection*. There are other forms; this should be kept in mind, although it is beyond the scope of this book.

12.2.2 Applying the EP Code to the Test Problems

We have been discussing three principal test problems: Simple Knapsack 100 (stored in *simpleKS100_20150527.p* on the book's Web site), GAP 1-c5-15-1 (stored in *gap1_c5_15_1.p* on the book's Web site), and Philadelphia 2010 Census Tracts (stored in *Philly2010CensusTractsWeightedDistances.p* on the book's Web site). How does our EP code perform on these problems?

Table 12.2 on page 185, shows the fifty best decision values for a representative test run of the EP code on the Simple Knapsack 100 problem. The run has $popSize = 200$, $offspringCount = 2$, r (mutation rate) $= 1$, and $numGens = 1,200$. Note that the bang-for-buck heuristic decision has an objective value of 560.664 and slack of 2.258. The EP run did not find the bang-for-buck decision, but it did find nine decisions that have a higher objective function value. Interestingly, and surely not systematically, the bang-for-buck solution has much more slack than any of the 50 best EP decisions. This illustrates an important general point for applications. Different heuristics may well find different decisions; they may even be systematically different in doing so.

Finally on this problem, notice the high density of high quality decisions. The best found decision has an objective value of 562.068. The 200th best decision (not shown in the table) has an objective value of 558.379. Their ratio is $562.068/558.379 = 1.0066$, a less than 1% difference.

Table 12.3 on page 186, shows the fifty best decision values for a representative test run of the EP code on the GAP 1-c5-15-1 problem. The run also has $popSize = 200$, $offspringCount = 2$, r (mutation rate) $= 1$, and $numGens = 1,200$. Note that the run found the optimal decision we found otherwise, as well as very many high quality decisions. The ratio of the best objective to the 200th best found objective value is $336/320 = 1.05$.

Table 12.4 on page 187, shows the fifty best decision values for a representative test run of the EP code on the Philadelphia 2010 Census Tracts problem with 10 hubs to be located. The run also has $popSize = 200$, $offspringCount = 2$, r (mutation rate) $= 1$, and $numGens = 1,200$. (Recall that under our formulation the Philadelphia 2010 Census Tracts model has no constraints, so every decision produced is feasible.)

Note that the run found the heuristically optimal decision we found otherwise (page 139), as well as very many nearly optimal decisions. The ratio of the best objective to the 200th best found objective value to the best found value (we are minimizing) is $31132.144/31057.108 = 1.0024$. Again, we see a very large number of nearly equal objective values.

12.2.3 EP Discussion

We have now seen that EP can do a fine job of obtaining good decisions for our test models. This is not an aberration: EP can be expected to perform well on a broad range of problems. Because not a lot is known about how to match problems with particular metaheuristics, an empirical, trial-and-error approach is apt. We can recommend EP as an essential element in any tool kit for heuristic optimization. Any such kit should include several different tools, including the local search metaheuristics we discussed in Chapter 11. The analyst should approach any real world decision model of import with multiple tools.

Decision ID	Feasible?	Objective Value	Slack
0	True	562.068	0.423
1	True	561.469	0.163
2	True	561.425	0.873
3	True	561.306	0.717
4	True	561.144	0.882
5	True	560.917	1.060
6	True	560.834	0.036
7	True	560.831	0.049
8	True	560.707	0.471
9	True	560.660	1.247
10	True	560.531	0.001
11	True	560.530	0.172
12	True	560.523	1.348
13	True	560.481	0.657
14	True	560.472	0.462
15	True	560.409	0.731
16	True	560.302	0.737
17	True	560.229	0.055
18	True	560.223	0.514
19	True	560.191	0.807
20	True	560.142	0.596
21	True	560.117	1.114
22	True	560.066	0.423
23	True	560.039	0.458
24	True	559.971	1.057
25	True	559.967	0.015
26	True	559.922	0.473
27	True	559.912	1.425
28	True	559.853	0.187
29	True	559.833	0.020
30	True	559.825	0.126
31	True	559.818	0.987
32	True	559.729	0.866
33	True	559.710	0.842
34	True	559.709	0.097
35	True	559.680	0.146
36	True	559.663	0.447
37	True	559.608	0.432
38	True	559.608	0.614
39	True	559.595	0.563
40	True	559.591	0.643
41	True	559.590	0.245
42	True	559.580	2.212
43	True	559.555	0.790
44	True	559.533	0.248
45	True	559.522	1.247
46	True	559.508	0.649
47	True	559.491	1.825
48	True	559.472	0.072
49	True	559.454	0.512

TABLE 12.2: Simple Knapsack 100: Fifty best decision values from a representative run of evolutionary programming.

Decision ID	Feasible?	Objective Value	Slacks
0	True	336.000	[1.0, 2.0, 0.0, 0.0, 1.0]
1	True	335.000	[3.0, 2.0, 0.0, 0.0, 0.0]
2	True	335.000	[1.0, 2.0, 0.0, 0.0, 3.0]
3	True	334.000	[1.0, 1.0, 0.0, 1.0, 5.0]
4	True	334.000	[3.0, 7.0, 0.0, 0.0, 5.0]
5	True	334.000	[1.0, 2.0, 0.0, 0.0, 3.0]
6	True	333.000	[3.0, 2.0, 0.0, 0.0, 8.0]
7	True	332.000	[1.0, 5.0, 0.0, 0.0, 5.0]
8	True	332.000	[2.0, 0.0, 0.0, 0.0, 5.0]
9	True	331.000	[1.0, 2.0, 0.0, 0.0, 3.0]
10	True	331.000	[4.0, 3.0, 0.0, 0.0, 4.0]
11	True	331.000	[3.0, 7.0, 0.0, 0.0, 1.0]
12	True	331.000	[1.0, 1.0, 0.0, 1.0, 1.0]
13	True	330.000	[1.0, 2.0, 0.0, 2.0, 0.0]
14	True	330.000	[3.0, 7.0, 0.0, 0.0, 3.0]
15	True	330.000	[2.0, 1.0, 0.0, 2.0, 9.0]
16	True	330.000	[2.0, 2.0, 0.0, 1.0, 7.0]
17	True	330.000	[3.0, 1.0, 0.0, 7.0, 0.0]
18	True	329.000	[11.0, 1.0, 0.0, 1.0, 0.0]
19	True	329.000	[3.0, 3.0, 0.0, 0.0, 0.0]
20	True	329.000	[3.0, 7.0, 0.0, 0.0, 3.0]
21	True	329.000	[2.0, 0.0, 0.0, 0.0, 1.0]
22	True	329.000	[1.0, 5.0, 0.0, 0.0, 1.0]
23	True	329.000	[2.0, 1.0, 0.0, 0.0, 4.0]
24	True	328.000	[3.0, 3.0, 0.0, 0.0, 12.0]
25	True	328.000	[2.0, 6.0, 0.0, 0.0, 9.0]
26	True	328.000	[2.0, 0.0, 0.0, 0.0, 3.0]
27	True	328.000	[3.0, 1.0, 0.0, 1.0, 8.0]
28	True	328.000	[3.0, 17.0, 0.0, 0.0, 0.0]
29	True	327.000	[2.0, 1.0, 0.0, 2.0, 5.0]
30	True	327.000	[2.0, 2.0, 0.0, 1.0, 7.0]
31	True	327.000	[3.0, 7.0, 0.0, 0.0, 0.0]
32	True	327.000	[11.0, 5.0, 0.0, 0.0, 0.0]
33	True	327.000	[12.0, 0.0, 0.0, 0.0, 0.0]
34	True	327.000	[1.0, 1.0, 0.0, 0.0, 0.0]
35	True	327.000	[1.0, 8.0, 0.0, 2.0, 0.0]
36	True	326.000	[3.0, 0.0, 0.0, 6.0, 5.0]
37	True	326.000	[3.0, 7.0, 0.0, 0.0, 3.0]
38	True	326.000	[6.0, 2.0, 0.0, 3.0, 0.0]
39	True	326.000	[1.0, 2.0, 6.0, 0.0, 0.0]
40	True	326.000	[1.0, 9.0, 0.0, 1.0, 0.0]
41	True	326.000	[4.0, 0.0, 0.0, 0.0, 8.0]
42	True	326.000	[11.0, 3.0, 0.0, 0.0, 0.0]
43	True	326.000	[1.0, 1.0, 0.0, 0.0, 12.0]
44	True	326.000	[3.0, 5.0, 0.0, 0.0, 8.0]
45	True	326.000	[1.0, 2.0, 8.0, 1.0, 0.0]
46	True	326.000	[1.0, 6.0, 0.0, 0.0, 5.0]
47	True	326.000	[2.0, 1.0, 0.0, 2.0, 7.0]
48	True	326.000	[1.0, 8.0, 0.0, 2.0, 2.0]
49	True	325.000	[11.0, 3.0, 0.0, 0.0, 2.0]

TABLE 12.3: GAP 1-c5-15-1: Fifty best decision values from a representative run of evolutionary programming.

Decision ID	Feasible?	Objective Value
0	True	31057.108
1	True	31064.461
2	True	31067.549
3	True	31068.084
4	True	31068.084
5	True	31069.513
6	True	31069.513
7	True	31073.624
8	True	31074.790
9	True	31078.245
10	True	31078.267
11	True	31078.944
12	True	31081.160
13	True	31081.160
14	True	31083.098
15	True	31083.742
16	True	31084.716
17	True	31086.987
18	True	31087.139
19	True	31088.283
20	True	31088.305
21	True	31088.685
22	True	31088.685
23	True	31088.707
24	True	31089.220
25	True	31089.242
26	True	31089.877
27	True	31089.891
28	True	31090.323
29	True	31090.323
30	True	31090.412
31	True	31090.650
32	True	31090.650
33	True	31090.672
34	True	31090.672
35	True	31090.967
36	True	31091.852
37	True	31091.883
38	True	31091.883
39	True	31092.529
40	True	31094.949
41	True	31095.156
42	True	31095.156
43	True	31095.881
44	True	31096.280
45	True	31096.302
46	True	31096.302
47	True	31097.067
48	True	31097.120
49	True	31097.327

TABLE 12.4: Philadelphia 2010 Census Tracts: Fifty best decision values from a representative run of evolutionary programming.

Here are several points—about EP, but having larger significance—before turning to genetic algorithms.

1. In the EP we have presented, and in any EP very much like it, once a good decision is found it is automatically retained in the population until it fails to rank in the top *popSize* discovered decisions. (We are neglecting the fine point of the possibility of ties.) In particular, once the best decision that the run will discover is found, whenever it occurs in a run, EP as we have presented it will retain this decision in the working populations.

2. Extending the previous point, it is obvious that EP as presented will retain in its generation-to-generation working populations the best *popSize* decisions it encounters in a run. In consequence, at the end of a run the current population contains the *popSize* best decisions discovered during the run. We have seen the fruits of this in Figures 12.2–12.4. These figures present the objective function and constraint slack information for the best *popSize* decisions, but of course the associated decisions are available as well. Taken together they go far to enable a solution pluralism—and in particular a decision sweeping—approach to model analytics. This is a point we shall take up in detail in part IV.

3. The best *popSize* decisions delivered at the end of a run of EP will typically all be feasible. When the model has constraints that are not encoded in its representation, as occurs in our Simple Knapsack 100 and GAP 1-c5-15-1 problems, we will often be interested in high quality infeasible decisions as well. It is also worth considering whether there might be additional ways to capture feasible decisions of high quality. An example might be feasible decisions with objective values above a certain value that have relatively large amounts of slack in their constraints. This is a subject we will explore in some detail in the sequel, especially in Chapter 13.

12.3 The Basic Genetic Algorithm (GA)

12.3.1 The GA Procedure

The basic GA is a prototypical evolutionary algorithm, well suited for many purposes. It belongs in any study of evolutionary algorithms (EAs) and is very often a good starting point for a business analytics investigation. Figure 12.3 presents the Basic BA in pseudocode.

The basic GA pseudocode differs in detail from that for EP in Figure 12.2, page 182. The two varieties of EA plainly bear a strong family resemblance to each other. Both pseudocode descriptions are simple with the basic GA perhaps the simpler of the two.

We will now elaborate the basic GA and discuss it in some detail as implemented in *basic_genetic_algorithm.py*, which is available on the book's Web site. This affords an opportunity to discuss genetic operators and other matters of general interest when implementing or using EAs. Figure 12.4 presents a more elaborated pseudocode version of the basic GA. It will be the main reference point for the discussion that follows immediately.

1. Step 1, "Initialization," does just that. It undertakes the necessary preparation for conducting the run itself. Specifically, steps 1a–1c.

1. Initialization:

   (a) Generate: Create (randomly or otherwise) a *currentPopulation* of decisions.

   (b) Evaluate: Evaluate the performance ("fitness") of each member of the *current-Population* of decisions.

2. Main loop: Do until a stopping condition is met:

   (a) Select: Create a *newPopulation* based on sampling the *currentPopulation,* biased towards decisions with better fitness scores.

   (b) Perturb: Modify members of the new population using genetic operators.

   (c) Replace: Set the *currentPopulation* to the *newPopulation.*

   (d) Evaluate: Assess the performance ("fitness") of each member of the *currentPopulation* of decisions.

3. Report on what was found, then stop.

FIGURE 12.3: High-level pseudocode for the basic GA.

(a) Step 1a, "Initialize the parameters for the model to be solved," involves choosing the model to be solved and its parameters, normally (or at least often) the objective function coefficients, the left-hand-side coefficients, and the right-hand-side values, all of which have been with us since Chapter 1. In terms of an implementation of the basic GA, our approach is typical in placing this task outside of the function that specifies the GA run. The model and its details are determined before the GA procedure is called, and in calling the procedure we give it the model details. Even so, it is worth noting this key step.

(b) Step 1b, "Initialize the run parameters for the GA," manifests some of the distinctive, characteristic aspects of the GA, as distinguished from other forms of EC. The run parameters for our implementation are typical of the basic GA: *popSize* (the population size), r (the mutation rate), *xOver* (the crossover rate), *numGens* (the number of generations in the run), and *elite* (whether elite selection should be used or not). *popSize, r,* and *numGens* have the same meaning in the basic GA as they do in our EP. *xOver* is the probability that a decision in the population will undergo crossover (aka: recombination) during one generation of processing. Our EP did not use recombination; it only used mutation as a genetic operator. It is typical that EP does not and GA does employ recombination. We will explain our crossover method below. Finally, as we will see, it is quite possible with the basic GA for good decisions to appear and be lost from the population by chance. This is not possible in EP as we have formulated it. *elite* is a true-or-false flag that when true has the GA find the best decision in the population each generation and make sure it is present in the population for the next generation. We will discuss some of the subtleties of selection when we get to step 2(a)i.

(c) Step 1c, "Initialize the GA program internal variables," covers any other preparation that needs to be done. Calculating the number of decision variables is the only thing that falls under this heading in our implementation of the basic GA. This completes the preparatory steps for the procedure.

1. Preparation:

   (a) Initialize the parameters for the model to be solved.

   (b) Initialize the run parameters for the GA.

   (c) Initialize the GA program internal variables.

   (d) Initialize the population: Create an initial population of random solutions.

   (e) Evaluation: Evaluate the fitness of each solution in the population.

2. Main loop: Do while not yet done:

   (a) Do until the next generation has been created:

       i. Selection: Select parents for the individuals in the next generation.

       ii. Mutation: Expose parent solutions to mutation.

       iii. Recombination: Expose parent solutions to recombination (crossover).

       iv. Evaluation: Evaluate the fitness of each solution in the population.

3. Postparation (post-processing) and results reporting.

FIGURE 12.4: Elaborated pseudocode for the basic GA.

   (d) In step 1d, "Initialize the population: Create an initial population of random solutions," we create a population of *popSize* decisions, with each decision drawn randomly from the space of possible decisions. It is often the case that every member of this initial population is infeasible.

   (e) This is step 1e, "Evaluate the fitness of each solution in the population." We use the same decision evaluation procedure in all of the model implementations we have been discussing, including in particular the implementations of our three principal test problems. This procedure, given a decision for a model, returns a 3-tuple consisting of: (1) True or False to indicate feasibility of the decision, (2) the objective function value of the decision, and (3) the slack values for the decision (RHS − LHS). Regarding the slacks, if a model has one or more constraints, the (RHS − LHS) value for a decision will be a vector (or list). If the model has no constraints (as implemented), (3) is given the special value of None.

2. Step 2, "Main loop: Do while not yet done," begins the main loop. We use the usual stopping condition of completing *numGens* passes through the body of the loop, each pass constituting one generation.

3. The first thing we do upon entering the main loop is to undertake selection based on the fitness of each member of the current population. This is step 2(a)i, "Select parents for the individuals in the next generation." At this point we sample the current population with replacement to create a new population of size *popSize* consisting of decisions copied from the current population. There are very many selection processes used in practice in GAs. All are designed to take into account the evaluated fitnesses of the decisions in the current population and to favor selection of those with comparatively better fitness values.

Dealing with constraints is an important wrinkle that arises in any GA selection mechanism when the decisions being evaluated are for a constrained optimization model. How are we to compare with respect to fitness solutions that are infeasible? An obvious approach, which is often used, is to compare decisions on their net objective values. A feasible decision will have a net objective value equal to its objective value and an infeasible decision will have a net value equal to its objective value minus a penalty for being infeasible. (We are of course speaking in the context of maximization.) The size of the penalty may depend on the degree of infeasibility.

An issue with the penalty approach is that introducing another parameter (the penalty size) into the algorithm entails having to tune the algorithm to find good values for the parameter. This is substantively complicating. The alternative we have favored in this book is to use what is called *tournament-2 selection*. Here is how it works.

We have a current population, *currentPop*, of size *popSize*. We wish to create a new population, *nextPop*, also of size *popSize*. We wish to do this through sampling of decisions in *currentPop*, biased in favor of decisions with higher fitness values. With tournament-2 selection, we create *nextPop* by putting into it the winners of *popSize* tournaments between two decisions randomly drawn from *currentPop*. If both decisions are feasible, the one with the higher objective value wins the tournament and is placed into *nextPop*. If one decision is feasible and the other is not, the feasible decision wins the tournament. Finally, if both decisions are infeasible the decision with the better infeasibility score wins the tournament. Figure 12.5 on page 191 provides the pseudocode and further detail.

Repeat *popSize* times:

(a) Randomly draw (with replacement) two decisions from *currentPop*. Call them x and y.

    i. If x is feasible and y is infeasible, place x in *nextPop*.

    ii. If y is feasible and x is infeasible, place y in *nextPop*.

    iii. If x is feasible and y is feasible, place into *nextPop* whichever decision has the higher objective function value or, if tied, randomly choose between them.

    iv. If x is infeasible and y is infeasible, place into *nextPop* whichever solution has the higher infeasibility score or, if tied, randomly choose between them.

FIGURE 12.5: Pseudocode for tournament-2 selection of decisions for a constrained optimization model.

Feasibility and objective function value are, as we have seen (item 1e above), returned by the decision evaluation procedure for the model. This leaves to the GA algorithm the task of assigning infeasibility scores to infeasible decisions, which needs to be done when comparing two infeasible decisions. Our policy is to sum the constraint violations (which will be negative in the assumed case of $\leq$ constraints). The decision with the higher violation sum (closer to 0) wins the tournament. This is a sensible procedure. In a real application one should consider alternatives, including scaling or weighting the individual constraint violations in order to make them accurately comparable. Equal weighting, which we use here, is of course a good point of departure.

4. Step 2(a)ii, "Mutation: Expose parent solutions to mutation," is used in essentially all forms of EC, and we used it in our EP. We are employing a simple and straightforward

variety. When a decision is subjected to mutation, we first draw a random number from a Poisson distribution with mean r. This value tells us how many loci in the decision are to be subjected to mutation. The number could be 0, 1, or more. We then select at random (uniformly) that many loci and mutate them. Each locus that is mutated is given a random value (drawn uniformly) from the possible values for that locus. It is possible that the randomly-drawn value is identical to the existing value, so that no net change is effected at that locus.

To illustrate, consider our example GAP model, GAP 1-c5-15-1. There are 15 jobs and 5 agents. A decision is a sequence of 15 integer values, each ranging from 0 to 4. The jth locus of the decision corresponds to the jth and its value, 0–4, indicates which agent is assigned to it. By random choice locus 4 (counting from 0, the fifth from the left) is chosen for mutation.

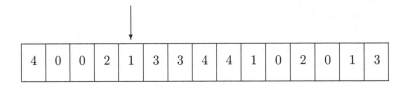

A random number is drawn, say a 3. The new decision has a 3 at locus 4, but is otherwise the same.

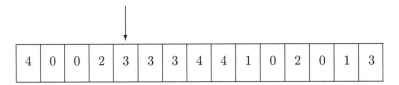

5. Step 2(a)iii, "Recombination: Expose parent solutions to recombination (crossover)," is characteristic of GAs and often absent in other forms of EC, such as EP. Imitating ("inspired by") biological processes (in meiosis), recombination has pairs of decisions exchange "genetic" material with one another. Our implementation uses perhaps the simplest form of recombination, single point crossover. Working on *nextPop* after selection and mutation, crossover matches decisions with neighboring indexes, 0 and 1, 2 and 3, 4 and 5, and so on for the entire population. For each pair so matched, with probability *xOver* (see step 1b) they undergo crossing over (and with probability (1 - *xOver*) they are not subjected to crossing over). For those pairs that do undergo crossing over, we use *single point crossover*. A random integer is drawn in the $[1, n-1]$ interval, where n is the number of loci (or decision variables) in a decision. This is the crossover point. The two decisions often called *chromosomes* in the GA literature, on analogy with biological entities—then swap values in the loci to the right of (higher than) the crossover point. The new decisions are inserted into the population, replacing their parents.

To illustrate, we continue with our GAP example. There are $n = 15$ jobs and 5 agents. A decision is a sequence of 15 integer values, each ranging from 0 to 4. With 15 loci in a decision, there are 14 possible crossover points, which separate adjacent loci. We draw one at random. Let us say it is 3. Then the crossover point is between the third and fourth locus, counting from the left.

Crossover at point 3

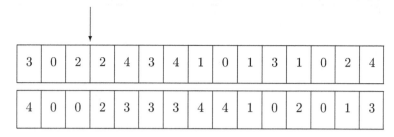

yields two new decisions:

3	0	2	2	3	3	3	4	4	1	0	2	0	1	3

4	0	0	2	4	3	4	1	0	1	3	1	0	2	4

6. Step 2(a)iv, "Evaluation: Evaluate the fitness of each solution in the population," is a repeat (on a new population) of step 1e. This completes one pass through the main loop of the GA.

7. Step 3, "Postparation (post-processing) and results reporting," concludes. Its function is to collect, save, and report data from the run.

12.3.2   Applying the Basic GA Code to a Test Problem

In the interests of brevity, we will discuss just one of our three principal test problems: GAP 1-c5-15-1 (stored in *gap1_c5_15_1.p*). How does our GA code perform on it?

Table 12.5, on page 194, shows the fifty best decision values drawn from the final population (of 200) for a representative test run of the GA code on the GAP 1-c5-15-1 problem. The run also has *popSize* = 200, *xOver* = 0.7, *r* (mutation rate) = 0.8, *numGens* = 6,200, and *elite* = True. Note that the run found the optimal decision we found otherwise. It also found some high quality decisions, but not as many as were found in our EP runs. Notably, the final population of 200 has only 37 feasible decisions; the rest are infeasible. Moreover, the final population has a number of duplications even among the feasible decisions. The ratio of the best objective to the 200th best found objective value is a comparatively large $336/284 = 1.18$.

12.3.3   GA Discussion

GAs in different varieties, including the one we have described here, are widely and routinely used for heuristic optimization. It is safe to say that EAs in practice more commonly and more closely resemble the basic GA than they do the EP we have described. Still, the GA's performance on our test problem is evidently not as good as the EP's. Both found an optimal decision, but the GA produced far fewer decisions of interest.

We should not conclude much by way of generalization from these results. Their primary purpose is to teach the underlying ideas and concepts. Which metaheuristic is best for a particular problem must, with the present state of knowledge, be largely a matter to be resolved empirically. Any new problem should be approached with a tool kit of metaheuristics and a spirit of creative exploration.

Decision ID	Feasible?	Objective Value	Slacks
0	True	336.0	[1.0, 2.0, 0.0, 0.0, 1.0]
1	True	336.0	[1.0, 2.0, 0.0, 0.0, 1.0]
2	True	317.0	[3.0, 0.0, 0.0, 1.0, 25.0]
3	True	315.0	[11.0, 0.0, 0.0, 1.0, 13.0]
4	True	311.0	[10.0, 0.0, 12.0, 0.0, 0.0]
5	True	311.0	[6.0, 10.0, 0.0, 3.0, 0.0]
6	True	311.0	[3.0, 10.0, 0.0, 1.0, 20.0]
7	True	309.0	[11.0, 10.0, 0.0, 1.0, 8.0]
8	True	309.0	[11.0, 10.0, 0.0, 1.0, 8.0]
9	True	308.0	[1.0, 1.0, 16.0, 7.0, 2.0]
10	True	305.0	[9.0, 1.0, 5.0, 1.0, 12.0]
11	True	305.0	[1.0, 6.0, 11.0, 0.0, 8.0]
12	True	304.0	[4.0, 1.0, 0.0, 1.0, 10.0]
13	True	303.0	[11.0, 7.0, 5.0, 3.0, 12.0]
14	True	303.0	[11.0, 7.0, 5.0, 3.0, 12.0]
15	True	303.0	[3.0, 17.0, 16.0, 3.0, 7.0]
16	True	302.0	[11.0, 6.0, 16.0, 2.0, 7.0]
17	True	302.0	[3.0, 6.0, 11.0, 0.0, 15.0]
18	True	300.0	[11.0, 0.0, 16.0, 3.0, 12.0]
19	True	300.0	[3.0, 17.0, 12.0, 2.0, 2.0]
20	True	299.0	[11.0, 0.0, 16.0, 9.0, 0.0]
21	True	299.0	[9.0, 1.0, 5.0, 7.0, 0.0]
22	True	299.0	[1.0, 20.0, 3.0, 0.0, 12.0]
23	True	299.0	[1.0, 20.0, 3.0, 0.0, 12.0]
24	True	299.0	[1.0, 17.0, 5.0, 0.0, 12.0]
25	True	298.0	[19.0, 0.0, 6.0, 8.0, 0.0]
26	True	298.0	[3.0, 17.0, 5.0, 9.0, 2.0]
27	True	297.0	[11.0, 17.0, 5.0, 3.0, 7.0]
28	True	296.0	[3.0, 17.0, 2.0, 9.0, 7.0]
29	True	295.0	[19.0, 6.0, 6.0, 2.0, 7.0]
30	True	294.0	[19.0, 1.0, 2.0, 1.0, 7.0]
31	True	294.0	[11.0, 10.0, 16.0, 3.0, 7.0]
32	True	293.0	[3.0, 1.0, 6.0, 8.0, 15.0]
33	True	293.0	[1.0, 20.0, 7.0, 3.0, 12.0]
34	True	293.0	[1.0, 20.0, 7.0, 3.0, 12.0]
35	True	293.0	[1.0, 8.0, 7.0, 9.0, 12.0]
36	True	284.0	[11.0, 10.0, 2.0, 6.0, 7.0]
37	False	339.0	[-6.0, 9.0, 0.0, -6.0, 3.0]
38	False	330.0	[-7.0, -5.0, 0.0, 6.0, 13.0]
39	False	325.0	[9.0, -16.0, -11.0, 0.0, 13.0]
40	False	325.0	[-6.0, 17.0, 0.0, 0.0, -3.0]
41	False	323.0	[9.0, -11.0, 0.0, -1.0, 15.0]
42	False	322.0	[10.0, 0.0, -10.0, -6.0, 25.0]
43	False	322.0	[1.0, -5.0, 0.0, -8.0, 33.0]
44	False	322.0	[-4.0, -10.0, -14.0, 7.0, 25.0]
45	False	321.0	[-5.0, 7.0, -6.0, 14.0, 3.0]
46	False	319.0	[1.0, 2.0, 2.0, 12.0, -10.0]
47	False	318.0	[3.0, 7.0, -14.0, 1.0, 25.0]
48	False	317.0	[11.0, -16.0, 0.0, 8.0, 13.0]
49	False	317.0	[1.0, -18.0, 0.0, 8.0, 25.0]

TABLE 12.5: GAP 1-c5-15-1: Fifty best decision values from the final population in a representative run of a basic GA.

12.4 For Exploration

1. The following passage occurs on page 188, at the end of the section on EP:

   > The best *popSize* decisions delivered at the end of a run of EP will typically all be feasible. When the model has constraints that are not encoded in its representation, as occurs in our Simple Knapsack 100 and GAP 1-c5-15-1 problems, we will often be interested in high quality infeasible decisions as well. It is also worth considering whether there might be additional ways to capture feasible decisions of high quality. An example might be feasible decisions with objective values above a certain value that have relatively large amounts of slack in their constraints. This is a subject we will explore in some detail in the sequel, especially in Chapter 13.

   Discuss how the EP procedure could be modified to maintain a second population of infeasible decisions, so that at the end of a run the primary population contains the best *popSize* feasible decisions encountered during the run (as it does now) and the secondary population contains the best *popSize* infeasible decisions encountered. Provide pseudocode.

2. Our EP implementation used ranking selection, while our basic GA implementation used tournament-2 selection. Is it possible to use tournament-2 selection with EP instead of ranking selection? What would be the consequences if we tried it? What about using ranking selection with the basic GA, instead of tournament-2 selection?

3. Discuss the difference between ranking and tournament-2 selection in terms of the exploration versus exploitation tradeoff in search and learning, §2.5.1.

4. By going to the literature and/or using your imagination, identify at least three credible forms of selection besides ranking and tournament-2. Discuss their properties and their pros and cons, especially with respect to the exploration versus exploitation tradeoff in search and learning, §2.5.1.

5. Using the open literature describe *uniform crossover* and provide pseudocode for it. How would you design experiments to determine whether uniform crossover is better than the single point crossover we used for the basic GA?

12.5 For More Information

EAs (evolutionary algorithms), including the varieties discussed in this chapter, evolutionary programming (EP), and genetic algorithms (GA), continue to be heavily researched, and have the benefit of many excellent textbooks describing them, e.g., [39, 44, 63, 122, 125, 133]. They are also extensively used in applications. Up to date practices, which constantly change, will be revealed with Internet searches and Web sites that collect these things, e.g., https://en.wikipedia.org/wiki/List_of_genetic_algorithm_applications and http://brainz.org/15-real-world-applications-genetic-algorithms/.

Chapter 13

Identifying and Collecting Decisions of Interest

13.1 Kinds of Decisions of Interest (DoIs)

Recall that given an optimization model we are also presented with, typically, a large space of decisions for the model. There is by definition one decision for each possible setting of the decision variables. Our Simple Knapsack 100 model has 100 binary decision variables, so there are $2^{100} = 1.2677 \times 10^{30}$ decisions associated with the model. Our GAP 1-c5-15-1 model has 15 decision variables, each of which can take on 5 values, yielding $5^{15} = 3.0518 \times 10^{10}$ decisions in all. Finally, in our Philadelphia 2010 Census Tracts model there are 1.6625×10^{19} decisions possible when choosing 10 hubs out of 383 tracts. Our examples have millions and millions of decisions. They are large enough to be good for illustrating the points we wish to make, small enough to be effectively tractable, and towards the low end of complexity for industrial problems.

Considering all these millions of decisions, can we characterize in any way those that are interesting and can we say why they are interesting for decision making? This chapter and the next are aimed at giving a positive answer to this question, and giving it with genuine specificity, although the whole book may be understood as an essay in response to the question.

Optimal decisions, it goes without saying, are certainly interesting. Our focus is on non-optimal decisions that are interesting because they may often be valuable for decision-oriented deliberation based on optimization models. We identify four classes of DoIs (decisions of interest) pertaining to constrained optimization models:

A. **Feasible, high objective function value.** Feasible decisions with high objective function values.[1]

B. **Feasible, high slack value, acceptable on objective value.** Feasible decisions with comparatively larger amounts of slack values on the constraints (but perhaps meeting a threshold requirement on objective function value).

---

[1] We are assuming for the sake of discussion maximization of the objective. High of course is relative to the other decisions for the model.

C. **Infeasible, high slack value.** Infeasible decisions that are nearly feasible.

D. **Infeasible, high objective value, acceptable on slack value.** Infeasible decisions with high objective function values (but meeting a threshold requirement on amount of negative slack).[2]

Points arising:

1. We are assuming throughout that our models are maximization models. This simplifies the discussion without prejudice to minimization models. The transformation is simple: Multiply the objective function by −1 to convert from minimization to maximization, and vice versa.

2. Type A decisions are those that are feasible and have a high objective value, relative to an optimal decision (whether exactly optimal or heuristically optimal). We may be interested in type A decisions for a number of reasons, including a desire to find decisions that will be robust to small changes in the model's parameters, and an interest in exploring ways to reallocate resources assumed available to the model. We will discuss these issues, and more generally what use can be made of information about type A decisions, in the next chapter.

3. Type B decisions are feasible, have a suitably high objective function value, and have large amounts of slack compared to other decisions that are feasible and meet the objective function threshold. Type A and type B decisions, when collected from a number of runs, may overlap. One decision may be in both collections. This happens rather rarely, however, especially with models of the complexities of our examples, or more complexity.

   A decision maker will be interested in type B decisions for many of the reasons type A decisions are interesting. Type B decisions are about discovering opportunities (engaging in candle lighting as we say) to use resources that are in slack for other purposes.

4. If our optimization model is unconstrained—as is our Philadelphia 2010 Census Tracts model in our encoding of it, which forces the constraints to be satisfied for every decision—there will be no constraints to violate and no slack to consider. So there will be no type B decisions when a model is unconstrained. Similarly, there will be no infeasible decisions of types C or D.

5. Type C decisions are infeasible but comparatively close to being feasible. In virtue of being infeasible they are distinct from both type A and type B decisions. They are potentially useful because they will often be nearly feasible, so that candle lighting moves—such as obtaining more of a constrained resource—might price out favorably.

6. Type D decisions are infeasible but comparatively high in objective value, compared that is to other infeasible decisions encountered. In virtue of being infeasible they are of course distinct from both type A and type B decisions. They may overlap with decisions of type C in any collection of these decisions.

7. Besides the usual reasons, type D decisions are interesting because they provide an estimate of how high the objective function may go and at what cost in terms of additional resources to undo the infeasible constraints.

---

[2]For the sake of the discussion we assume ≤ constraints.

If there are four classes of DoIs, could there be more? Certainly. We have characterized types B, C, and D in terms of total slack (positive or negative) on the decisions. There will be many circumstances in which we are more interested in some other function of the slacks. We might, for example, be mainly interested in just two of the constraints and/or we may want to transform the slack values in some nonlinear way (a little matters a little, a little more matters a lot) and collect DoIs under these regimes. It is often the case that some constraints are logical, constitutive of the problem. These constraints do not afford any flexibility or alternative courses of action. Other constraints—based on "data"—may in principle be modifiable by courses of action, such as purchasing more of the resource in short supply. Such considerations may be used to categorize constraints with regard to how we want to assess infeasible decisions.[3] We focus here on types A–D because they are excellent starting types, they may be filtered and combined to yield useful information without imposing a new collection regime, and our experience with them has been favorable.

13.2 The FI2-Pop GA

We have emphasized throughout that metaheuristics are classes or families of heuristics. They are forms that must be filled in to make a specific, operationalized, usable heuristic, and they are forms that may be filled in in indefinitely many ways. This holds for local search as well as population based metaheuristics, including evolutionary algorithms.

The FI2-Pop GA ("feasible-infeasible two-population GA" [100, 110], see also [21]) is a member of the class of population based metaheuristics, is a kind of evolutionary algorithm, and has been shown to be effective at finding good decisions for challenging models. It is also especially well suited for our interest in collecting DoIs, as we shall see.

Recall the high-level pseudocode for the Basic GA, Figure 12.3 on page 189. For ease of comparison, the figure is repeated here as Figure 13.1, page 200. Figure 13.2, also on page 201, holds the high-level pseudocode for the FI2-Pop GA. The key feature of the FI2-Pop GA is that it maintains throughout each run *two* populations of decisions, one holding feasible decisions, the other holding infeasible decisions. Each population is processed separately (and in sequence) using a standard EA (take your pick; we are using a variant on the Basic GA), except that after a new generation is created and evaluated, the feasible decisions are moved to the next generation of feasible decisions and the infeasible decisions are moved to the next generation of infeasible decisions.

During processing in a generation, the fitness of a feasible individual is assessed as its objective function value. The fitness of an infeasible individual is (prototypically) assessed as the sum of its constraint violations—exactly the criterion we have been using for tournament-2 selection. The notion on the infeasible side is that the closer a decision is to being feasible (the smaller its constraint violations) the fitter it is.

During any given generation a population of feasible (infeasible) decisions may contain decisions generated by processing an infeasible (feasible) population. Children, as we say, may migrate away from infeasible parents when they themselves are feasible; and vice versa. Populations may be empty. For example, it can easily happen that no feasible decisions are generated at initialization. The FI2-Pop GA proceeds apace; if there are no decisions to process in a population, then the wand is passed to the other population, which in time may create decisions for the empty population. This is a routine occurrence.

---

[3]Thanks to Fred Murphy bringing this point more explicitly to our attention.

1. Initialization:

   (a) Generate: Create (randomly or otherwise) a *currentPopulation* of decisions.

   (b) Evaluate: Evaluate the performance ("fitness") of each member of the *current-Population* of decisions.

2. Main loop: Do until a stopping condition is met:

   (a) Select: Create a *newPopulation* based on sampling the *currentPopulation,* biased towards decisions with better fitness scores.

   (b) Perturb: Modify members of the new population using genetic operators.

   (c) Replace: Set the *currentPopulation* to the *newPopulation.*

   (d) Evaluate: Assess the performance ("fitness") of each member of the *currentPopulation* of decisions.

3. Report on what was found, then stop.

FIGURE 13.1: Basic genetic algorithm: high-level pseudocode.

A main intuition motivating the FI2-Pop GA is that decisions that are infeasible may well contain valuable "genetic" material (settings of the decision variables) that will be lost, at least for a time, if the decision is eliminated. So, why not set up a second "struggle for existence" among the infeasible decisions? Empirically, we have often found that feasible decisions in the population at the end of a successful run are all or nearly all descended from at least one infeasible ancestor.

Further, consider the matter in terms of an exploration-exploitation tradeoff. It is agreed that heuristic search algorithms face an ineluctable dilemma between exploiting existing knowledge and exploring for more. Greedy hill climbing sits at one extreme: It is entirely greedy. Because no single run is likely to yield a very good decision on other than trivial problems, we normally will undertake multiple runs and use the collective results in deliberation and decision making. Such multiple runs are inherently exploratory. Even with greedy hill climbing we see the practical necessity of exploration, as well as exploitation. Still, greedy hill climbing is not much used in practice, at least not alone.

Purely random search—just creating at random one decision after another—is fully exploratory. It operates without exploitation, without learning from what it has found, and it does not work well at all. A proper balance needs to be struck between exploration and exploitation when it comes to heuristic search (including heuristic optimization). There is, however, no general way known to choose a good balance, aside from experience with different heuristics on specific problems, or kinds of problems. And so, we recommend approaching new problems with a plurality (solution pluralism again!) of approaches. Here that means a variety of metaheuristics we can try on the problem.

Simulated annealing and threshold accepting algorithms are metaheuristics that are motivated in part by the need to find alternatives to greedy hill climbing that are more exploratory, and hopefully perform better. They definitely succeed at this, which is why they are used so much in practice, while greedy hill climbing is not. Population based metaheuristics are motivated (in part) by the perceived need to find principled yet more exploratory alternatives to local search metaheuristics. The intuition is that we can get better decisions for harder problems by being more exploratory (for a given level of computational effort).

1. Initialization:

   (a) Generate: Create (randomly or otherwise) *popSize* decisions.

   (b) Evaluate: Evaluate the performance ("fitness") of each member of the decisions.

   (c) Create populations: Create two populations, *feasiblePop* and *infeasiblePop*, to hold feasible and infeasible decisions respectively.

   (d) Apportion: Place the feasible decisions in the *feasiblePop* and place the infeasible decisions in the *infeasiblePop*.

2. Main loop: Do until a stopping condition is met:

   (a) Process one generation of the *feasiblePop* with a standard EA.

   (b) Add any new infeasible decisions produced to the *infeasiblePop*, and create a new *feasiblePop* consisting of the feasible decisions (if any) produced in the processing of the incoming *feasiblePop*.

   (c) Process one generation of the *infeasiblePop* with a standard EA.

   (d) Add any new feasible decisions produced to the *feasiblePop*, and create a new *infeasiblePop* consisting of the infeasible decisions (if any) produced in the processing of the incoming *infeasiblePop*.

3. Report on what was found, then stop.

FIGURE 13.2: High-level pseudocode for the FI2-Pop GA [100].

One way they may be more exploratory than local search is by having a decision be able to use information from another decision, one that is relatively successful. Another way is to explore beyond the confines of a small neighborhood. That is the purpose of recombination. The FI2-Pop GA metaheuristic is further exploratory because it maintains an infeasible population and applies evolutionary pressure on it to find feasible decisions, which of course must compete in the normal way with other feasible decisions.

More important for our purposes than these theoretical speculations is that the FI2-Pop GA affords collecting *infeasible* decisions of interest. We wish to use the FI2-Pop GA in large portion simply *because* we wish to collect DoIs of types C and D. In other metaheuristcs the infeasible decisions tend to disappear quickly. There is certainly no effort to cultivate interesting ones, but this is just what the FI2-Pop GA does. So, we shall be using it with this purpose in mind.

13.3   Discussion

Points arising:

1. The FI2-Pop GA is a comparatively good way to find infeasible DoIs [93, 94]. The support for this claim has to be almost entirely empirical. In any given situation experience is the best guide to what will work best.

2. The FI2-Pop GA is a metaheuristic, a pattern—of maintaining two populations, one feasible, one infeasible—that can be applied in many ways, certainly to other EAs.

3. Collecting various kinds of DoIs—types A, B, C, D, etc.—is also something that can be done with metaheuristics other than the FI2-Pop GA. Both population based metaheuristcs and local search metaheuristics may be similarly modified to collect DoIs.

4. The GA we use in our FI2-Pop GA code (*FI2PopGA.py*, available on the book's Web site) is close to our Basic GA, but augmented in two ways: (1) We have added elite-n selection, so the user may specify n for putting the n best decisions into the population of the next generation. (2) We have added removal of duplicate decisions, from both populations.

13.4   For Exploration

1. Besides the four kinds we have discussed in this chapter—A, B, C, and D—propose other decisions of interest (DoIs) and explain why or under what circumstances they may be useful in decision making.

2. Suppose that there are 1000 DoIs of type X for a given model and we wish to find them. What is the chance we can find any of them by sampling at random in the decisions for a model with 10^{10} decisions? 2^{100} decisions? What conclusions do you draw from this?

13.5   For More Information

See [21] for an early illustration of many of these ideas. See [100] for a mathematical argument supporting the efficacy of the FI2-Pop GA. See [110] for a detailed investigation of the FI2-Pop GA and its performance on difficult optimization problems.

Part IV

Post-Solution Analysis of Optimization Models

Chapter 14

Decision Sweeping

14.1   Introduction

Recall that decision sweeping with an optimization model consists of:

1. Identifying the decisions of interest (DoIs),

2. Systematically collecting the DoIs, and

3. Using the corpus of collected DoIs in deliberation and decision making.

Chapter 13 discussed identifying DoIs and systematically collecting them. With regard to identification we discussed four main kinds of DoIs: types A (feasible, strong on objective function value), B (feasible, strong on total slack and acceptable on objective value), C (infeasible, nearly feasible), and D (infeasible, strong on objective value). With regard to collecting the DoIs, something that is normally done by heuristic sampling, we presented and discussed the FI2-Pop GA (see Table 13.2 on page 201). The FI2-Pop GA maintains two populations, one consisting of feasible decisions and one containing only infeasible decisions. This affords collection of all four types of DoIs.

The purpose of this chapter is to put these concepts into use, that is, to demonstrate and discuss how using a corpus of collected DoIs can be managed productively in deliberation and decision making with optimization models.

14.2   Decision Sweeping with the GAP 1-c5-15-1 Model

Tables 14.1, 14.2, 14.3, and 14.4 (pages 208–211) present the top 50 A, B, C, and D decisions of interest discovered in a single run of the FI2-Pop GA on the GAP 1-c5-15-1 model. For the sake of sparing the reader the need to mull over the numbers in detail, we note the following general points revealed by these figures.

1. Table 14.1, type A decisions.

   (a) An optimal decision has been found. We know it is optimal because we have solved the model with an exact solver.

   (b) The top 50 type A DoIs are all of high quality. The ratio of the worst to the best is $326/336 = 0.97$, which is to say that the worst of these decisions has an objective value within about 3% of the best.

   (c) The amount of slack among the top 50 type A DoIs varies from 2.0 to 20.0.

   (d) Job 14 (counting from 0, the 15th job) is always assigned to agent 2. Job 3 is also always assigned to agent 2, with one exception, item 29, where it is assigned to agent 4. Job 7 is always assigned to agent 1. The other jobs have more variable assignments, yet tend to be concentrated on one agent.

2. Table 14.2, type B decisions.

   (a) The type B decisions were collected under the constraint of having an objective function value of at least 300, or about 10.7% ($1 - \frac{300}{336} = 0.10714$) off of the optimal value. Subject to this constraint, Table 14.2 presents the top 50 feasible decisions encountered in the run as measured by total slack in the constraints. See the column labeled SSLKs (sum of slacks).

   (b) The worst decisions in Table 14.2 in terms of total slack have a SSLK value of 34.0, which is much larger than the highest slack member of the type A DoIS in Table 14.1.

   (c) The highest SSLKs value is 46.

   (d) The objective function values are all below the values for the type A decisions. The best of the objective values for the type B decisions is 315 (item 38 in Table 14.2), which is 6% off from optimality ($1 - \frac{315}{336} = 0.0625$). That decision, however, does yield a total slack of 34. Intriguingly, that decision beats the objective values in the next 50 best type B decisions (items 50–99, not shown).

   (e) The patterns of assignments of jobs to agents sometimes resemble that of the type A decisions. For example, job 14 is uniformly assigned to agent 2. On the other hand job 3 is uniformly assigned to agent 4, with one exception.

3. Table 14.3, type C decisions.

   (a) Every one of the top 50 type C decisions (infeasible, but nearly feasible) is infeasible by one unit. That is, its SNSLKs (sum of negative slacks) is -1.0.

   (b) One, but only one of the discovered decisions has a higher objective function value than the optimal decision. That is item 0 with an objective value of 342.0. It is infeasible by one unit on constraint 1 (counting from 0).

   (c) The patterns of assignments of jobs to agents sometimes resemble that of the type A decisions. Job 14 is again uniformly assigned to agent 2. On the other hand job 8 is uniformly assigned to agent 1.

4. Table 14.4, type D decisions.

   (a) Every one of the top 50 type D decisions discovered has an objective function value greater than the optimal decision. These 50 range in value from 343 up to 349.

(b) Remarkably, item 1 has an objective value of 348 but a SNSLKs value of only -7, making it a standout among the type D decisions. Item 35, at 343 in objective value and -5 in SNSLKs, is also noteworthy.

The records in Tables 14.1–14.4 constitute a *decision sweep* of the GAP 1-c5-15-1 model. They are, or arise from, a plurality of solutions to the model, obtained by heuristically capturing decisions of interest and calculating the model's value with each of them. Now, with nearly 200 distinct DoIs before us, we turn to the topic of their use in deliberation and decision making.

14.3   Deliberating with the Results of a Decision Sweep

The GAP 1-c5-15-1 model is an artificial model, created to test and compare solvers. There is no real world context it closely represents. That said, there *are* many real world contexts that resemble the GAP sufficiently for present purposes and it is entirely appropriate to entertain hypothetical scenarios for which our model *is* an apt representation. We shall do just that, after offering some comments that are generally applicable for deliberation with decision sweeps of constrained optimization models.

The parameters of a model—the objective coefficients, constraint coefficients, right-hand-side values—are normally given as fixed, static entities. Yet often they are changeable. Additional resources may be available for a price; surplus resources may be redeployed in valuable ways. We use the best available information to arrive at parameter values, knowing at the same time that our knowledge is far from perfect and that these values may change when realized concretely. Constraint right-hand-side values are especially worth considering because of their central role in constraining, in limiting, the objective value that can be reached with feasible decisions. It will often be the case that for a price a right-hand-side value can be increased, raising the question of whether the price on offer is economic. Conversely, it will often be the case that a right-hand-side value, representing an amount of resource, can be decreased and redeployed, with a concomitant cost that plays out as a less favorable (think: reduced, for we are presuming maximization and $\leq$ constraints) objective function value. This raises the question of comparing what is gained by redeployment with what is lost from the objective value.

Recall, Chapter 1, page 15, our general characterization of the matter:

> *What favorable opportunities are there to take action resulting in changes to the assumptions (e.g., parameter values) of the model leading to improved outcomes (net of costs and benefits)?*

These and similar questions are what we whimsically call *candle lighting questions*, with allusion to the motto of the Christopher Society, "It is better to light one candle than to curse the darkness" [86, 87, 88, 89, 102]. In *candle lighting analysis* we use a plurality of solutions, obtained perhaps with decision sweeping, to explore whether and how we might find a better overall decision than that produced by optimizing the model.

We need to make one more general point before getting down to specifics. Given a constrained optimization problem it will nearly always be the case that a superior objective value can be obtained if one or more constraints are *relaxed*. A constraint is relaxed from an existing state if its right-hand-side-value is increased (decreased) in the case it is a $\leq$ ($\geq$) constraint. For the sake of simplicity (and without violence to generality) we will continue to focus on constraints of the $\leq$ variety and optimization models that seek to maximize.

i	Obj	SSLKs	SLKs	Decisions
0	336.0	4.0	(1.0, 2.0, 0.0, 0.0, 1.0)	(1, 1, 3, 2, 0, 4, 0, 1, 0, 3, 3, 3, 0, 4, 2)
1	335.0	5.0	(3.0, 2.0, 0.0, 0.0, 0.0)	(0, 4, 3, 2, 0, 4, 1, 1, 0, 3, 3, 3, 1, 0, 2)
2	335.0	6.0	(1.0, 2.0, 0.0, 0.0, 3.0)	(1, 1, 3, 2, 4, 4, 0, 1, 0, 3, 3, 3, 0, 0, 2)
3	334.0	15.0	(3.0, 7.0, 0.0, 0.0, 5.0)	(4, 1, 3, 2, 0, 4, 0, 1, 0, 3, 3, 3, 1, 0, 2)
4	334.0	6.0	(1.0, 2.0, 0.0, 0.0, 3.0)	(1, 1, 3, 2, 0, 4, 3, 1, 0, 3, 4, 3, 0, 0, 2)
5	334.0	8.0	(1.0, 1.0, 0.0, 1.0, 5.0)	(4, 1, 3, 2, 0, 4, 3, 1, 0, 1, 3, 3, 0, 0, 2)
6	333.0	13.0	(3.0, 2.0, 0.0, 0.0, 8.0)	(1, 1, 3, 2, 0, 4, 0, 1, 0, 3, 3, 3, 4, 0, 2)
7	332.0	7.0	(2.0, 0.0, 0.0, 0.0, 5.0)	(4, 1, 3, 2, 0, 4, 0, 1, 1, 3, 3, 3, 0, 0, 2)
8	332.0	11.0	(1.0, 5.0, 0.0, 0.0, 5.0)	(4, 1, 3, 2, 0, 4, 1, 1, 0, 3, 3, 3, 0, 0, 2)
9	332.0	3.0	(1.0, 2.0, 0.0, 0.0, 0.0)	(1, 4, 3, 2, 0, 4, 3, 1, 0, 3, 1, 3, 0, 0, 2)
10	331.0	11.0	(4.0, 3.0, 0.0, 0.0, 4.0)	(0, 4, 3, 2, 0, 1, 0, 1, 4, 3, 3, 3, 1, 0, 2)
11	331.0	11.0	(3.0, 7.0, 0.0, 0.0, 1.0)	(0, 1, 3, 2, 0, 4, 0, 1, 0, 3, 3, 3, 1, 4, 2)
12	331.0	6.0	(1.0, 2.0, 0.0, 0.0, 3.0)	(1, 1, 3, 2, 0, 4, 3, 1, 0, 3, 3, 4, 0, 0, 2)
13	331.0	4.0	(1.0, 1.0, 0.0, 1.0, 1.0)	(0, 1, 3, 2, 0, 4, 3, 1, 0, 1, 3, 3, 0, 4, 2)
14	330.0	13.0	(3.0, 7.0, 0.0, 0.0, 3.0)	(0, 1, 3, 2, 4, 4, 0, 1, 0, 3, 3, 3, 1, 0, 2)
15	330.0	11.0	(3.0, 1.0, 0.0, 7.0, 0.0)	(4, 1, 3, 2, 0, 4, 0, 1, 0, 1, 3, 3, 4, 0, 2)
16	330.0	12.0	(2.0, 2.0, 0.0, 1.0, 7.0)	(1, 1, 3, 2, 0, 3, 0, 1, 4, 3, 4, 3, 0, 0, 2)
17	330.0	5.0	(1.0, 2.0, 0.0, 2.0, 0.0)	(1, 1, 4, 2, 0, 4, 3, 1, 0, 3, 3, 3, 0, 0, 2)
18	330.0	6.0	(1.0, 1.0, 0.0, 1.0, 3.0)	(0, 1, 3, 2, 4, 4, 3, 1, 0, 1, 3, 3, 0, 0, 2)
19	329.0	13.0	(11.0, 1.0, 0.0, 1.0, 0.0)	(4, 1, 3, 2, 0, 4, 3, 1, 0, 1, 3, 3, 4, 0, 2)
20	329.0	13.0	(3.0, 7.0, 0.0, 0.0, 3.0)	(0, 1, 3, 2, 0, 4, 3, 1, 0, 3, 4, 3, 1, 0, 2)
21	329.0	6.0	(3.0, 3.0, 0.0, 0.0, 0.0)	(0, 4, 3, 2, 1, 4, 0, 1, 0, 3, 3, 3, 1, 0, 2)
22	329.0	7.0	(2.0, 1.0, 0.0, 0.0, 4.0)	(0, 4, 3, 2, 0, 1, 1, 1, 4, 3, 3, 3, 0, 0, 2)
23	329.0	3.0	(2.0, 0.0, 0.0, 0.0, 1.0)	(0, 1, 3, 2, 0, 4, 0, 1, 1, 3, 3, 3, 0, 4, 2)
24	329.0	7.0	(1.0, 5.0, 0.0, 0.0, 1.0)	(0, 1, 3, 2, 0, 4, 1, 1, 0, 3, 3, 3, 0, 4, 2)
25	328.0	20.0	(3.0, 17.0, 0.0, 0.0, 0.0)	(4, 1, 3, 2, 0, 4, 0, 1, 0, 3, 3, 3, 4, 0, 2)
26	328.0	18.0	(3.0, 3.0, 0.0, 0.0, 12.0)	(4, 4, 3, 2, 0, 1, 0, 1, 0, 3, 3, 3, 1, 0, 2)
27	328.0	13.0	(3.0, 1.0, 0.0, 1.0, 8.0)	(0, 1, 3, 2, 0, 4, 3, 1, 0, 1, 3, 3, 4, 0, 2)
28	328.0	17.0	(2.0, 6.0, 0.0, 0.0, 9.0)	(4, 1, 3, 2, 0, 1, 0, 1, 4, 3, 3, 3, 0, 0, 2)
29	328.0	5.0	(2.0, 2.0, 1.0, 0.0, 0.0)	(1, 1, 3, 4, 0, 4, 0, 1, 2, 3, 3, 3, 0, 0, 2)
30	328.0	5.0	(2.0, 0.0, 0.0, 0.0, 3.0)	(0, 1, 3, 2, 4, 4, 0, 1, 1, 3, 3, 3, 0, 0, 2)
31	328.0	9.0	(1.0, 5.0, 0.0, 0.0, 3.0)	(0, 1, 3, 2, 4, 4, 1, 1, 0, 3, 3, 3, 0, 0, 2)
32	327.0	12.0	(12.0, 0.0, 0.0, 0.0, 0.0)	(4, 1, 3, 2, 0, 4, 0, 1, 1, 3, 3, 3, 4, 0, 2)
33	327.0	16.0	(11.0, 5.0, 0.0, 0.0, 0.0)	(4, 1, 3, 2, 0, 4, 1, 1, 0, 3, 3, 3, 4, 0, 2)
34	327.0	10.0	(5.0, 1.0, 0.0, 0.0, 4.0)	(4, 0, 3, 2, 0, 1, 1, 1, 4, 3, 3, 3, 4, 0, 2)
35	327.0	10.0	(3.0, 7.0, 0.0, 0.0, 0.0)	(0, 4, 3, 2, 0, 4, 3, 1, 0, 3, 1, 3, 1, 0, 2)
36	327.0	12.0	(2.0, 2.0, 0.0, 1.0, 7.0)	(1, 1, 3, 2, 0, 3, 0, 1, 4, 3, 3, 4, 0, 0, 2)
37	327.0	5.0	(2.0, 0.0, 0.0, 0.0, 3.0)	(0, 1, 3, 2, 0, 4, 3, 1, 1, 3, 4, 3, 0, 0, 2)
38	327.0	11.0	(1.0, 8.0, 0.0, 2.0, 0.0)	(4, 4, 3, 2, 0, 3, 0, 1, 0, 1, 3, 3, 0, 4, 2)
39	327.0	2.0	(1.0, 1.0, 0.0, 0.0, 0.0)	(0, 4, 3, 2, 1, 4, 1, 1, 0, 3, 3, 3, 0, 0, 2)
40	327.0	2.0	(1.0, 1.0, 0.0, 0.0, 0.0)	(0, 1, 4, 2, 0, 4, 3, 1, 0, 1, 3, 3, 0, 3, 2)
41	326.0	14.0	(11.0, 3.0, 0.0, 0.0, 0.0)	(4, 4, 3, 2, 0, 1, 0, 1, 0, 3, 3, 3, 1, 4, 2)
42	326.0	8.0	(6.0, 0.0, 0.0, 2.0, 0.0)	(4, 1, 0, 2, 0, 4, 3, 1, 1, 3, 3, 3, 4, 0, 2)
43	326.0	12.0	(4.0, 0.0, 0.0, 0.0, 8.0)	(0, 1, 3, 2, 0, 4, 0, 1, 1, 3, 3, 3, 4, 0, 2)
44	326.0	13.0	(3.0, 7.0, 0.0, 0.0, 3.0)	(0, 1, 3, 2, 0, 4, 3, 1, 0, 3, 3, 4, 1, 0, 2)
45	326.0	16.0	(3.0, 5.0, 0.0, 0.0, 8.0)	(0, 1, 3, 2, 0, 4, 1, 1, 0, 3, 3, 3, 4, 0, 2)
46	326.0	14.0	(3.0, 0.0, 0.0, 6.0, 5.0)	(4, 1, 3, 2, 0, 4, 0, 1, 0, 3, 1, 3, 1, 0, 2)
47	326.0	11.0	(2.0, 2.0, 0.0, 3.0, 4.0)	(1, 1, 4, 2, 0, 3, 0, 1, 4, 3, 3, 3, 0, 0, 2)
48	326.0	11.0	(1.0, 9.0, 0.0, 1.0, 0.0)	(1, 4, 3, 2, 4, 3, 0, 1, 0, 3, 4, 3, 0, 0, 2)
49	326.0	12.0	(1.0, 6.0, 0.0, 0.0, 5.0)	(4, 1, 3, 2, 1, 4, 0, 1, 0, 3, 3, 3, 0, 0, 2)

TABLE 14.1: Top 50 type A decisions found in a run of the GAP model with FI2-Pop GA. SSLKs: sum of slack values. SLKs: the slack values. i = item/decision number. Obj = objective value of the decision. SSLKs = sum of the slack values. SLKs = slack values for the constraints. Decisions = associated settings of the decision variables.

i	Obj	SSLKs	SLKs	Decisions
0	303.0	46.0	(3.0, 17.0, 16.0, 3.0, 7.0)	(4, 1, 2, 4, 0, 3, 0, 1, 0, 3, 3, 3, 4, 0, 2)
1	309.0	41.0	(3.0, 7.0, 16.0, 3.0, 12.0)	(4, 1, 2, 4, 0, 3, 0, 1, 0, 3, 3, 3, 1, 0, 2)
2	303.0	41.0	(3.0, 17.0, 6.0, 8.0, 7.0)	(4, 1, 2, 4, 0, 2, 0, 1, 0, 3, 3, 3, 4, 0, 2)
3	309.0	39.0	(3.0, 17.0, 12.0, 0.0, 7.0)	(4, 1, 3, 4, 0, 2, 0, 1, 0, 3, 3, 3, 4, 0, 2)
4	308.0	39.0	(3.0, 2.0, 16.0, 3.0, 15.0)	(1, 1, 2, 4, 0, 3, 0, 1, 0, 3, 3, 3, 4, 0, 2)
5	304.0	39.0	(11.0, 1.0, 16.0, 4.0, 7.0)	(4, 1, 2, 4, 0, 3, 3, 1, 0, 1, 3, 3, 4, 0, 2)
6	304.0	39.0	(3.0, 17.0, 11.0, 1.0, 7.0)	(4, 1, 3, 4, 0, 3, 0, 1, 0, 3, 2, 3, 4, 0, 2)
7	304.0	39.0	(3.0, 7.0, 16.0, 3.0, 10.0)	(0, 1, 2, 4, 0, 3, 3, 1, 0, 3, 4, 3, 1, 0, 2)
8	309.0	38.0	(3.0, 6.0, 22.0, 0.0, 7.0)	(4, 1, 3, 4, 0, 1, 0, 1, 0, 3, 3, 3, 4, 0, 2)
9	308.0	38.0	(11.0, 7.0, 6.0, 2.0, 12.0)	(4, 1, 2, 4, 0, 2, 3, 1, 0, 3, 3, 3, 1, 0, 2)
10	307.0	38.0	(3.0, 0.0, 22.0, 1.0, 12.0)	(4, 1, 3, 4, 0, 3, 0, 1, 0, 3, 1, 3, 1, 0, 2)
11	307.0	38.0	(1.0, 17.0, 6.0, 2.0, 12.0)	(4, 1, 2, 4, 0, 2, 3, 1, 0, 3, 3, 3, 0, 0, 2)
12	307.0	37.0	(11.0, 7.0, 16.0, 3.0, 0.0)	(4, 1, 2, 4, 0, 3, 0, 1, 0, 3, 3, 3, 1, 4, 2)
13	307.0	37.0	(1.0, 6.0, 16.0, 2.0, 12.0)	(4, 1, 2, 4, 0, 1, 3, 1, 0, 3, 3, 3, 0, 0, 2)
14	306.0	37.0	(1.0, 17.0, 16.0, 3.0, 0.0)	(4, 1, 2, 4, 0, 3, 0, 1, 0, 3, 3, 3, 0, 4, 2)
15	305.0	37.0	(3.0, 17.0, 12.0, 0.0, 5.0)	(0, 1, 3, 4, 4, 2, 0, 1, 0, 3, 3, 3, 4, 0, 2)
16	302.0	37.0	(11.0, 5.0, 6.0, 8.0, 7.0)	(4, 1, 2, 4, 0, 2, 1, 1, 0, 3, 3, 3, 4, 0, 2)
17	309.0	36.0	(3.0, 7.0, 6.0, 8.0, 12.0)	(4, 1, 2, 4, 0, 2, 0, 1, 0, 3, 3, 3, 1, 0, 2)
18	307.0	36.0	(11.0, 2.0, 6.0, 2.0, 15.0)	(1, 1, 2, 4, 0, 2, 3, 1, 0, 3, 3, 3, 4, 0, 2)
19	305.0	36.0	(19.0, 7.0, 6.0, 2.0, 2.0)	(4, 1, 2, 4, 4, 2, 3, 1, 0, 3, 3, 3, 1, 0, 2)
20	305.0	36.0	(3.0, 6.0, 22.0, 0.0, 5.0)	(0, 1, 3, 4, 4, 1, 0, 1, 0, 3, 3, 3, 4, 0, 2)
21	304.0	36.0	(11.0, 0.0, 22.0, 1.0, 2.0)	(4, 1, 3, 4, 4, 3, 0, 1, 0, 3, 1, 3, 1, 0, 2)
22	303.0	36.0	(3.0, 17.0, 13.0, 1.0, 2.0)	(4, 1, 3, 4, 0, 3, 0, 2, 0, 3, 3, 4, 1, 0, 2)
23	302.0	36.0	(3.0, 17.0, 8.0, 1.0, 7.0)	(4, 1, 3, 4, 0, 3, 0, 1, 0, 3, 2, 3, 4, 0, 2)
24	300.0	36.0	(11.0, 10.0, 3.0, 0.0, 12.0)	(4, 1, 3, 4, 0, 2, 3, 2, 0, 3, 1, 3, 1, 0, 2)
25	311.0	35.0	(3.0, 1.0, 22.0, 2.0, 7.0)	(4, 1, 3, 4, 0, 3, 0, 1, 0, 1, 3, 3, 4, 0, 2)
26	309.0	35.0	(3.0, 17.0, 3.0, 0.0, 12.0)	(4, 1, 3, 4, 0, 2, 0, 2, 0, 3, 3, 3, 1, 0, 2)
27	308.0	35.0	(11.0, 5.0, 12.0, 0.0, 7.0)	(4, 1, 3, 4, 0, 2, 1, 1, 0, 3, 3, 3, 4, 0, 2)
28	308.0	35.0	(3.0, 2.0, 7.0, 3.0, 20.0)	(1, 1, 2, 4, 0, 3, 0, 2, 0, 3, 3, 3, 1, 0, 2)
29	307.0	35.0	(3.0, 5.0, 12.0, 0.0, 15.0)	(0, 1, 3, 4, 0, 2, 1, 1, 0, 3, 3, 3, 4, 0, 2)
30	306.0	35.0	(3.0, 17.0, 12.0, 0.0, 3.0)	(0, 1, 3, 4, 0, 2, 0, 1, 0, 3, 3, 3, 4, 4, 2)
31	305.0	35.0	(3.0, 0.0, 12.0, 0.0, 20.0)	(0, 1, 3, 4, 0, 2, 3, 1, 0, 3, 1, 3, 1, 0, 2)
32	304.0	35.0	(3.0, 17.0, 2.0, 1.0, 12.0)	(4, 1, 3, 4, 0, 3, 0, 2, 0, 3, 2, 3, 1, 0, 2)
33	302.0	35.0	(3.0, 5.0, 11.0, 1.0, 15.0)	(0, 1, 3, 4, 0, 3, 1, 1, 0, 3, 2, 3, 4, 0, 2)
34	301.0	35.0	(20.0, 0.0, 6.0, 2.0, 7.0)	(4, 1, 2, 4, 0, 2, 3, 1, 1, 3, 3, 3, 4, 0, 2)
35	301.0	35.0	(11.0, 6.0, 2.0, 9.0, 7.0)	(4, 1, 2, 4, 0, 1, 3, 1, 0, 2, 3, 3, 4, 0, 2)
36	300.0	35.0	(9.0, 6.0, 6.0, 2.0, 12.0)	(4, 1, 2, 4, 1, 2, 3, 1, 0, 3, 3, 3, 0, 0, 2)
37	300.0	35.0	(1.0, 17.0, 2.0, 3.0, 12.0)	(4, 1, 2, 4, 0, 3, 3, 1, 0, 3, 2, 3, 0, 0, 2)
38	315.0	34.0	(3.0, 7.0, 12.0, 0.0, 12.0)	(4, 1, 3, 4, 0, 2, 0, 1, 0, 3, 3, 3, 1, 0, 2)
39	312.0	34.0	(4.0, 7.0, 16.0, 3.0, 4.0)	(0, 1, 2, 4, 0, 3, 0, 1, 4, 3, 3, 3, 1, 0, 2)
40	311.0	34.0	(3.0, 10.0, 0.0, 1.0, 20.0)	(4, 1, 3, 2, 0, 3, 0, 1, 0, 3, 1, 3, 4, 0, 2)
41	310.0	34.0	(3.0, 7.0, 11.0, 1.0, 12.0)	(4, 1, 3, 4, 0, 3, 0, 1, 0, 3, 2, 3, 1, 0, 2)
42	309.0	34.0	(3.0, 6.0, 13.0, 0.0, 12.0)	(4, 1, 3, 4, 0, 1, 0, 2, 0, 3, 3, 3, 1, 0, 2)
43	308.0	34.0	(3.0, 7.0, 2.0, 10.0, 12.0)	(4, 1, 2, 4, 0, 3, 0, 1, 0, 2, 3, 3, 1, 0, 2)
44	308.0	34.0	(3.0, 2.0, 6.0, 8.0, 15.0)	(1, 1, 2, 4, 0, 2, 0, 1, 0, 3, 3, 3, 4, 0, 2)
45	306.0	34.0	(19.0, 7.0, 6.0, 2.0, 0.0)	(4, 1, 2, 4, 0, 2, 3, 1, 0, 3, 3, 3, 1, 4, 2)
46	306.0	34.0	(11.0, 7.0, 6.0, 8.0, 2.0)	(4, 1, 2, 4, 4, 2, 0, 1, 0, 3, 3, 3, 1, 0, 2)
47	306.0	34.0	(11.0, 2.0, 2.0, 4.0, 15.0)	(1, 1, 2, 4, 0, 3, 3, 1, 0, 2, 3, 3, 4, 0, 2)
48	305.0	34.0	(11.0, 0.0, 22.0, 1.0, 0.0)	(4, 1, 3, 4, 0, 3, 0, 1, 0, 3, 1, 3, 1, 4, 2)
49	305.0	34.0	(9.0, 17.0, 6.0, 2.0, 0.0)	(4, 1, 2, 4, 0, 2, 3, 1, 0, 3, 3, 3, 0, 4, 2)

TABLE 14.2: Top 50 type B decisions found in a run of the GAP model with FI2-Pop GA, having an objective value 300 or more. i = item/decision number. Obj = objective value of the decision. SSLKs = sum of the slack values. SLKs = slack values for the constraints. Decisions = associated settings of the decision variables.

i	Obj	SNSLKs	SLKs	Decisions
0	342.0	-1.0	(3.0, -1.0, 0.0, 0.0, 0.0)	(1, 4, 3, 2, 0, 4, 0, 1, 0, 3, 3, 3, 1, 0, 2)
1	333.0	-1.0	(3.0, -1.0, 2.0, 0.0, 0.0)	(1, 2, 3, 4, 0, 4, 0, 1, 0, 3, 3, 3, 1, 0, 2)
2	332.0	-1.0	(1.0, 2.0, 0.0, -1.0, 0.0)	(1, 1, 4, 2, 0, 4, 0, 1, 0, 3, 3, 3, 0, 3, 2)
3	331.0	-1.0	(5.0, 3.0, 0.0, 0.0, -1.0)	(4, 0, 3, 2, 4, 1, 0, 1, 4, 3, 3, 3, 1, 0, 2)
4	330.0	-1.0	(3.0, -1.0, 0.0, 1.0, 10.0)	(1, 4, 3, 2, 0, 3, 0, 1, 0, 3, 4, 3, 1, 0, 2)
5	329.0	-1.0	(3.0, -1.0, 0.0, 2.0, 3.0)	(1, 4, 3, 2, 0, 3, 0, 1, 0, 4, 3, 3, 1, 0, 2)
6	329.0	-1.0	(3.0, 1.0, 0.0, 0.0, -1.0)	(4, 0, 3, 2, 4, 1, 1, 1, 4, 3, 3, 3, 0, 0, 2)
7	328.0	-1.0	(6.0, 0.0, 0.0, -1.0, 0.0)	(4, 1, 0, 2, 0, 4, 0, 1, 1, 3, 3, 3, 4, 3, 2)
8	328.0	-1.0	(5.0, 5.0, 0.0, -1.0, 0.0)	(4, 1, 0, 2, 0, 4, 1, 1, 0, 3, 3, 3, 4, 3, 2)
9	327.0	-1.0	(11.0, -1.0, 0.0, 1.0, 0.0)	(1, 4, 3, 2, 4, 3, 0, 1, 0, 3, 4, 3, 1, 0, 2)
10	327.0	-1.0	(4.0, 8.0, 0.0, 2.0, -1.0)	(0, 4, 3, 2, 0, 3, 0, 1, 4, 1, 3, 3, 4, 0, 2)
11	327.0	-1.0	(3.0, -1.0, 0.0, 1.0, 10.0)	(1, 4, 3, 2, 0, 3, 0, 1, 0, 3, 3, 4, 1, 0, 2)
12	327.0	-1.0	(3.0, 7.0, 0.0, -1.0, 0.0)	(0, 1, 4, 2, 0, 4, 0, 1, 0, 3, 3, 3, 1, 3, 2)
13	327.0	-1.0	(1.0, 2.0, 0.0, -1.0, 0.0)	(1, 1, 4, 2, 3, 4, 0, 1, 0, 3, 3, 3, 0, 0, 2)
14	327.0	-1.0	(1.0, 8.0, 0.0, -1.0, 2.0)	(4, 4, 3, 2, 0, 3, 0, 1, 0, 1, 4, 3, 0, 3, 2)
15	326.0	-1.0	(12.0, 7.0, 0.0, 1.0, -1.0)	(4, 1, 3, 2, 0, 3, 0, 1, 4, 3, 4, 3, 1, 0, 2)
16	326.0	-1.0	(3.0, -1.0, 0.0, 3.0, 7.0)	(1, 4, 4, 2, 0, 3, 0, 1, 0, 3, 3, 3, 1, 0, 2)
17	326.0	-1.0	(1.0, -1.0, 0.0, 0.0, 5.0)	(4, 1, 3, 2, 0, 4, 0, 1, 0, 3, 3, 3, 0, 1, 2)
18	325.0	-1.0	(2.0, 17.0, 0.0, 1.0, -1.0)	(4, 1, 3, 2, 0, 3, 0, 1, 4, 3, 4, 3, 0, 0, 2)
19	325.0	-1.0	(2.0, 0.0, 0.0, -1.0, 0.0)	(0, 1, 4, 2, 0, 4, 0, 1, 1, 3, 3, 3, 0, 3, 2)
20	325.0	-1.0	(4.0, 13.0, 0.0, 0.0, -1.0)	(0, 4, 3, 2, 0, 1, 0, 1, 4, 3, 3, 3, 4, 0, 2)
21	325.0	-1.0	(2.0, 6.0, 0.0, 6.0, -1.0)	(4, 1, 3, 2, 0, 1, 0, 1, 4, 3, 4, 3, 0, 0, 2)
22	325.0	-1.0	(10.0, 6.0, 0.0, 0.0, -1.0)	(4, 1, 3, 2, 4, 1, 0, 1, 4, 3, 3, 3, 0, 0, 2)
23	325.0	-1.0	(1.0, 5.0, 0.0, -1.0, 0.0)	(0, 1, 4, 2, 0, 4, 1, 1, 0, 3, 3, 3, 0, 3, 2)
24	324.0	-1.0	(4.0, 1.0, 0.0, 3.0, -1.0)	(0, 4, 1, 2, 0, 3, 0, 1, 4, 3, 3, 3, 4, 0, 2)
25	324.0	-1.0	(10.0, 5.0, 0.0, 1.0, -1.0)	(4, 1, 3, 2, 0, 3, 1, 1, 4, 3, 4, 3, 0, 0, 2)
26	324.0	-1.0	(10.0, 6.0, 0.0, 0.0, -1.0)	(4, 1, 3, 2, 0, 1, 3, 1, 4, 3, 4, 3, 0, 0, 2)
27	324.0	-1.0	(3.0, -1.0, 0.0, 7.0, 0.0)	(1, 4, 3, 2, 0, 3, 0, 1, 0, 3, 4, 4, 1, 0, 2)
28	324.0	-1.0	(12.0, 1.0, 0.0, 0.0, -1.0)	(0, 4, 3, 2, 0, 1, 1, 1, 4, 3, 3, 3, 4, 0, 2)
29	323.0	-1.0	(11.0, -1.0, 0.0, 1.0, 0.0)	(1, 4, 3, 2, 0, 3, 3, 1, 0, 3, 4, 4, 1, 0, 2)
30	323.0	-1.0	(6.0, 0.0, 0.0, -1.0, 0.0)	(4, 1, 0, 2, 3, 4, 0, 1, 1, 3, 3, 3, 4, 0, 2)
31	323.0	-1.0	(1.0, 2.0, 16.0, -1.0, 0.0)	(1, 1, 2, 4, 0, 4, 0, 1, 0, 3, 3, 3, 0, 3, 2)
32	323.0	-1.0	(1.0, 6.0, 0.0, 1.0, -1.0)	(4, 1, 3, 2, 4, 1, 3, 1, 4, 0, 3, 3, 0, 0, 2)
33	323.0	-1.0	(3.0, -1.0, 12.0, 0.0, 7.0)	(1, 4, 3, 4, 0, 2, 0, 1, 0, 3, 3, 3, 1, 0, 2)
34	323.0	-1.0	(12.0, 7.0, 0.0, 1.0, -1.0)	(4, 1, 3, 2, 0, 3, 0, 1, 4, 3, 3, 4, 1, 0, 2)
35	323.0	-1.0	(5.0, 5.0, 0.0, -1.0, 0.0)	(4, 1, 0, 2, 3, 4, 1, 1, 0, 3, 3, 3, 4, 0, 2)
36	323.0	-1.0	(2.0, 5.0, 0.0, 0.0, -1.0)	(4, 1, 3, 2, 4, 0, 1, 1, 4, 3, 3, 3, 0, 0, 2)
37	322.0	-1.0	(2.0, 17.0, 0.0, 1.0, -1.0)	(4, 1, 3, 2, 0, 3, 0, 1, 4, 3, 3, 4, 0, 0, 2)
38	322.0	-1.0	(2.0, 6.0, 0.0, 6.0, -1.0)	(4, 1, 3, 2, 0, 1, 0, 1, 4, 3, 3, 4, 0, 0, 2)
39	322.0	-1.0	(3.0, 7.0, 0.0, -1.0, 0.0)	(0, 1, 4, 2, 3, 4, 0, 1, 0, 3, 3, 3, 1, 0, 2)
40	322.0	-1.0	(1.0, -1.0, 0.0, 0.0, 3.0)	(0, 1, 3, 2, 4, 4, 0, 1, 0, 3, 3, 3, 0, 1, 2)
41	321.0	-1.0	(3.0, 3.0, 0.0, 8.0, -1.0)	(4, 4, 4, 2, 0, 1, 0, 1, 0, 3, 3, 3, 1, 0, 2)
42	321.0	-1.0	(1.0, -1.0, 0.0, 0.0, 3.0)	(0, 1, 3, 2, 0, 4, 3, 1, 0, 3, 4, 3, 0, 1, 2)
43	321.0	-1.0	(3.0, -1.0, 2.0, 1.0, 10.0)	(1, 2, 3, 4, 0, 3, 0, 1, 0, 3, 4, 3, 1, 0, 2)
44	321.0	-1.0	(11.0, -1.0, 0.0, 0.0, 0.0)	(4, 1, 3, 2, 0, 4, 0, 1, 0, 3, 3, 3, 4, 1, 2)
45	321.0	-1.0	(12.0, 2.0, 0.0, 3.0, -1.0)	(1, 1, 4, 2, 0, 3, 0, 1, 4, 3, 3, 3, 4, 0, 2)
46	320.0	-1.0	(1.0, 5.0, 0.0, -1.0, 0.0)	(0, 1, 4, 2, 3, 4, 1, 1, 0, 3, 3, 3, 0, 0, 2)
47	320.0	-1.0	(2.0, -1.0, 0.0, 6.0, 9.0)	(4, 1, 3, 2, 0, 1, 0, 1, 4, 3, 1, 3, 0, 0, 2)
48	320.0	-1.0	(2.0, 0.0, 0.0, -1.0, 0.0)	(0, 1, 4, 2, 3, 4, 0, 1, 1, 3, 3, 3, 0, 0, 2)
49	320.0	-1.0	(3.0, -1.0, 0.0, 0.0, 8.0)	(0, 1, 3, 2, 0, 4, 0, 1, 0, 3, 3, 3, 4, 1, 2)

TABLE 14.3: Top 50 type C decisions found in a run of the GAP model with FI2-Pop GA. i = item/decision number. Obj = objective value of the decision. SNSLKs = sum of the negative (infeasible) slack values. SLKs = slack values for the constraints. Decisions = associated settings of the decision variables.

i	Obj	SNSLKs	SLKs	Decisions
0	349.0	-13.0	(5.0, -1.0, 0.0, -9.0, -3.0)	(1, 0, 3, 2, 0, 4, 0, 1, 4, 3, 3, 3, 1, 3, 2)
1	348.0	-7.0	(-3.0, -1.0, 0.0, 0.0, -3.0)	(1, 0, 3, 2, 0, 4, 0, 1, 4, 3, 3, 3, 1, 0, 2)
2	347.0	-10.0	(5.0, -1.0, 0.0, -6.0, -3.0)	(1, 0, 3, 2, 0, 4, 3, 1, 4, 3, 3, 3, 1, 0, 2)
3	347.0	-14.0	(2.0, -7.0, 0.0, -4.0, -3.0)	(1, 3, 3, 2, 0, 4, 0, 1, 4, 1, 3, 3, 0, 0, 2)
4	347.0	-16.0	(5.0, -13.0, 0.0, 0.0, -3.0)	(1, 0, 3, 2, 0, 4, 1, 1, 4, 3, 3, 3, 1, 0, 2)
5	347.0	-16.0	(-13.0, 9.0, 0.0, 0.0, -3.0)	(1, 0, 3, 2, 0, 4, 0, 1, 4, 3, 3, 3, 0, 0, 2)
6	347.0	-18.0	(-11.0, -3.0, 0.0, -1.0, -3.0)	(1, 0, 0, 2, 0, 4, 1, 1, 4, 3, 3, 3, 0, 3, 2)
7	346.0	-14.0	(-5.0, 9.0, 0.0, -6.0, -3.0)	(1, 0, 3, 2, 0, 4, 3, 1, 4, 3, 3, 3, 0, 0, 2)
8	346.0	-15.0	(12.0, -1.0, 0.0, -11.0, -3.0)	(1, 3, 3, 2, 0, 4, 0, 1, 4, 3, 3, 3, 1, 0, 2)
9	346.0	-17.0	(12.0, -1.0, 0.0, 0.0, -16.0)	(1, 4, 3, 2, 0, 4, 0, 1, 4, 3, 3, 3, 1, 0, 2)
10	346.0	-20.0	(10.0, -7.0, 0.0, -10.0, -3.0)	(1, 3, 3, 2, 0, 4, 3, 1, 4, 1, 3, 0, 0, 0, 2)
11	345.0	-14.0	(2.0, 9.0, 0.0, -11.0, -3.0)	(1, 3, 3, 2, 0, 4, 0, 1, 4, 3, 3, 3, 0, 0, 2)
12	345.0	-14.0	(-4.0, -1.0, 0.0, -9.0, 13.0)	(1, 0, 3, 2, 0, 4, 0, 1, 0, 3, 3, 3, 1, 3, 2)
13	345.0	-16.0	(2.0, 9.0, 0.0, 0.0, -16.0)	(1, 4, 3, 2, 0, 4, 0, 1, 4, 3, 3, 3, 0, 0, 2)
14	345.0	-18.0	(-12.0, 9.0, 0.0, -3.0, -3.0)	(1, 3, 0, 2, 0, 4, 0, 1, 4, 3, 3, 3, 0, 0, 2)
15	345.0	-19.0	(11.0, -17.0, 0.0, -2.0, 0.0)	(1, 4, 3, 2, 0, 4, 0, 1, 0, 1, 3, 3, 1, 3, 2)
16	345.0	-19.0	(10.0, -14.0, 0.0, -2.0, -3.0)	(1, 1, 3, 2, 0, 4, 0, 1, 4, 1, 3, 0, 0, 3, 2)
17	345.0	-19.0	(6.0, -13.0, 0.0, -3.0, -3.0)	(1, 3, 0, 2, 0, 4, 1, 1, 4, 3, 3, 3, 1, 0, 2)
18	345.0	-19.0	(2.0, -7.0, 0.0, -12.0, 0.0)	(1, 4, 3, 2, 0, 4, 0, 1, 3, 1, 3, 3, 0, 0, 2)
19	345.0	-20.0	(-5.0, 9.0, 0.0, 0.0, -15.0)	(1, 0, 3, 2, 0, 4, 0, 1, 4, 3, 3, 3, 0, 4, 2)
20	344.0	-9.0	(1.0, -7.0, 0.0, -2.0, 0.0)	(1, 4, 3, 2, 0, 4, 0, 1, 0, 1, 3, 3, 0, 3, 2)
21	344.0	-13.0	(5.0, -1.0, 0.0, -9.0, -3.0)	(1, 0, 3, 2, 3, 4, 0, 1, 4, 3, 3, 3, 1, 0, 2)
22	344.0	-13.0	(3.0, -7.0, 0.0, -6.0, 0.0)	(1, 4, 3, 2, 0, 4, 0, 1, 0, 1, 3, 3, 0, 0, 2)
23	344.0	-13.0	(-12.0, -1.0, 0.0, 0.0, 13.0)	(1, 0, 3, 2, 0, 4, 0, 1, 0, 3, 3, 3, 1, 0, 2)
24	344.0	-14.0	(-2.0, -1.0, 0.0, -11.0, 0.0)	(1, 4, 0, 2, 0, 4, 0, 1, 3, 3, 3, 3, 1, 0, 2)
25	344.0	-17.0	(10.0, -3.0, 0.0, -11.0, -3.0)	(1, 3, 3, 2, 0, 4, 1, 1, 4, 3, 3, 0, 0, 0, 2)
26	344.0	-17.0	(3.0, -17.0, 0.0, 7.0, 0.0)	(1, 4, 3, 2, 0, 4, 0, 1, 0, 1, 3, 3, 1, 0, 2)
27	344.0	-17.0	(2.0, -14.0, 0.0, 7.0, -3.0)	(1, 1, 3, 2, 0, 4, 0, 1, 4, 1, 3, 0, 0, 0, 2)
28	344.0	-18.0	(3.0, -3.0, 0.0, 0.0, -15.0)	(1, 0, 3, 2, 0, 4, 1, 1, 4, 3, 3, 3, 0, 4, 2)
29	344.0	-18.0	(-3.0, -7.0, 0.0, 7.0, -8.0)	(1, 0, 3, 2, 0, 4, 0, 1, 4, 1, 3, 3, 4, 0, 2)
30	344.0	-18.0	(-5.0, 9.0, 0.0, 0.0, -13.0)	(1, 0, 3, 2, 4, 4, 0, 1, 4, 3, 3, 3, 0, 0, 2)
31	344.0	-19.0	(10.0, -3.0, 0.0, 0.0, -16.0)	(1, 4, 3, 2, 0, 4, 1, 1, 4, 3, 3, 0, 0, 0, 2)
32	344.0	-20.0	(20.0, -8.0, 0.0, -9.0, -3.0)	(1, 1, 3, 2, 0, 4, 0, 1, 4, 3, 3, 3, 1, 3, 2)
33	344.0	-20.0	(12.0, -1.0, 0.0, -19.0, 0.0)	(1, 4, 3, 2, 0, 4, 0, 1, 3, 3, 3, 3, 1, 0, 2)
34	344.0	-20.0	(1.0, -7.0, 0.0, -13.0, 13.0)	(1, 3, 3, 2, 0, 4, 0, 1, 0, 1, 3, 3, 0, 3, 2)
35	343.0	-5.0	(-3.0, -1.0, 0.0, -1.0, 0.0)	(1, 4, 0, 2, 0, 4, 0, 1, 0, 3, 3, 3, 1, 3, 2)
36	343.0	-8.0	(-4.0, 2.0, 0.0, -1.0, -3.0)	(1, 1, 0, 2, 0, 4, 0, 1, 4, 3, 3, 3, 0, 3, 2)
37	343.0	-10.0	(11.0, -1.0, 0.0, -9.0, 0.0)	(1, 4, 3, 2, 0, 4, 0, 1, 0, 3, 3, 3, 1, 3, 2)
38	343.0	-11.0	(12.0, -8.0, 0.0, 0.0, -3.0)	(1, 1, 3, 2, 0, 4, 0, 1, 4, 3, 3, 3, 1, 0, 2)
39	343.0	-11.0	(-4.0, -1.0, 0.0, -6.0, 13.0)	(1, 0, 3, 2, 0, 4, 3, 1, 0, 3, 3, 3, 1, 0, 2)
40	343.0	-12.0	(10.0, 2.0, 0.0, -9.0, -3.0)	(1, 1, 3, 2, 0, 4, 0, 1, 4, 3, 3, 3, 0, 3, 2)
41	343.0	-14.0	(-3.0, 14.0, 0.0, 0.0, -11.0)	(4, 0, 3, 2, 0, 4, 0, 1, 4, 3, 3, 3, 1, 0, 2)
42	343.0	-14.0	(-5.0, -4.0, 0.0, -2.0, -3.0)	(0, 0, 3, 2, 0, 4, 1, 1, 4, 1, 3, 3, 0, 3, 2)
43	343.0	-14.0	(-7.0, -7.0, 0.0, 7.0, 0.0)	(1, 4, 3, 2, 0, 4, 0, 1, 0, 1, 3, 3, 0, 0, 2)
44	343.0	-15.0	(9.0, -7.0, 0.0, -8.0, 0.0)	(1, 4, 3, 2, 0, 4, 3, 1, 0, 1, 3, 3, 0, 3, 2)
45	343.0	-15.0	(3.0, -1.0, -4.0, 0.0, -10.0)	(1, 2, 3, 2, 0, 4, 0, 1, 0, 3, 3, 3, 1, 0, 4)
46	343.0	-15.0	(-3.0, 14.0, 0.0, -9.0, -3.0)	(0, 0, 3, 2, 0, 4, 0, 1, 4, 3, 3, 3, 1, 3, 2)
47	343.0	-16.0	(12.0, 2.0, 0.0, -13.0, -3.0)	(1, 1, 3, 2, 0, 4, 0, 1, 4, 3, 3, 3, 3, 0, 2)
48	343.0	-16.0	(-5.0, 8.0, 0.0, 1.0, -11.0)	(4, 0, 3, 2, 0, 4, 3, 1, 4, 1, 3, 3, 0, 0, 2)
49	343.0	-17.0	(11.0, -17.0, 0.0, 1.0, 0.0)	(1, 4, 3, 2, 0, 4, 3, 1, 0, 1, 3, 3, 1, 0, 2)

TABLE 14.4: Top 50 type D decisions found in a run of the GAP model with FI2-Pop GA, having sum of negative slacks ≥ -20. i = item/decision number. Obj = objective value of the decision. SNSLKs = sum of the negative (infeasible) slack values. SLKs = slack values for the constraints. Decisions = associated settings of the decision variables.

Prototypically, if our constraint is $a_1 x_1 + a_2 x_2 \leq b$ and we increase the value of b, then the constraint is *relaxed* because there are more ways to satisfy it. All of the values of x_1 and x_2 that satisfied the constraint before we increased the value of b will still satisfy it, and so will others. Conversely, if we decrease the value of b, then the constraint is *tightened* and there are fewer ways to satisfy it. In optimization, if we relax a constraint we can hope to do better. That is, we can hope that there is a newly feasible decision with an improved objective value. We are certainly guaranteed to do no worse, for any optimal decision we have will still be feasible if we relax a constraint.

With this in mind, let us revisit the optimal decision, item 0 in Table 14.1:

i	Obj	SSLKs	SLKs	Decisions
0	336.0	4.0	(1.0, 2.0, 0.0, 0.0, 1.0)	(1, 1, 3, 2, 0, 4, 0, 1, 0, 3, 3, 3, 0, 4, 2)

The slacks—SLKs—are the differences between the right-hand-side values of the constraints and the left-hand-side values for the decision at hand. In the case of the optimal decision, the record is telling us that constraint 0 has a slack of 1.0, constraint 1 has a slack of 2.0, constraint 4 has slack of 1.0, and the other two constraints have no slack at all. They are said to be *tight* for this decision (1, 1, 3, 2, 0, 4, 0, 1, 0, 3, 3, 3, 0, 4, 2). Points arising:

1. Constraints typically represent resources that are in limited supply. In the case of a GAP-like problem, the constraints will often represent limitations on time or energy available to the agent processing the jobs. This resource characterization of constraints applies quite generally in constrained optimization, whether GAP-like or not. Universally, increasing the available resources (relaxing constraints) can only help.

2. The record before us tells us that the optimal decision does *not* use all of the resources available to it. Constraints 0, 1, and 4 have positive slack at optimality. This means that for the optimal decision for the model as given we do not need all of what is available. We can choose to implement this decision and redeploy these resources to other purposes without reducing the objective value of 336.

3. The record before us also tells us that constraints 2 and 3 have slacks of 0 for our optimal decision. This means that if we tighten these constraints—reduce their right-hand-side values—even a little, then our optimal decision will no longer be optimal because it will be infeasible.

To repeat: because constraints typically represent resources and resources are typically available to buy or sell, these considerations naturally lead to the question of whether it might be profitable to purchase additional resources, or sell (or redeploy) slack resources. In short we are led to candle lighting analysis, the search for better alternatives to our *problem* than the optimal decision for our *model*. With one more concept in hand we will be in position to move swiftly. That concept is the *challenge value of a decision*.

Definition 1 (Challenge Value) *Given an incumbent (default) decision, the challenge value of a candidate decision is the net benefit of implementing the candidate instead of the incumbent decision.*

What we do to conduct the candle lighting analysis is, in its essentials, to begin with an optimal decision—item 0 of Table 14.1 in our GAP 1-c5-15-1 example—as the incumbent and then consider every other discovered DoI as a candidate to replace the incumbent. A few cases from our GAP 1-c5-15-1 example will be ample for demonstrating how this works.

Suppose we have solved the model and are in possession of the optimal decision (item 0 of Table 14.1). Before we can implement the decision, we are offered a deal. For $2000 we can purchase 1 additional unit of constraint 1. Should we take the deal as offered? Note that the slack on constraint 1 (counting from 0) is 2. At optimality we have 2 units of constraint 1 to spare. Do we want to spend real money buying yet another unit? The answer is easy and perhaps surprising: yes, if the value of increasing the objective function value by 6 units is worth more than $2,000. How do we arrive at this answer? Consider the first record from Table 14.3, showing the type C DoIs, reproduced below.

i	Obj	SNSLKs	SLKs	Decisions
0	342.0	-1.0	(3.0, -1.0, 0.0, 0.0, 0.0)	(1, 4, 3, 2, 0, 4, 0, 1, 0, 3, 3, 3, 1, 0, 2)

It tells us that if we can get 1.0 more units of constraint 1, then the decision (1, 4, 3, 2, 0, 4, 0, 1, 0, 3, 3, 3, 1, 0, 2) will be feasible. Its objective function value is 342, which is 6 units more than the 336 of our current optimum. So if $342 - 336 = 6$ is worth more than $2,000 we would be happy to accept the deal. Actually, it's better than that. The type C DoI before us also has a slack of 3 on constraint 0, so we have an additional source of value to weigh into our deliberations.

As a second case, imagine that for business purposes 336 in the objective is not high enough, nor is 342; we need at least 343 to meet certain goals or exigencies. This is an instance of what we call an outcome reach question; see §1.4.3. The corpora of DoIs can help. Items 35 and 1 in the type D DoIs, in Table 14.4, have objective values of 343 and 348 respectively at comparatively small cost, -5 and -7 in constraint violation. Of course, everything depends upon the availability and cost of acquiring resources that would relax the constraints and make these decisions feasible. A decision with a fairly high sum of constraint violations, e.g., item 0 of the Ds, in Table 14.4, may, depending on circumstances, be quite attractive. The larger issue here is that the corpora of DoIs, particularly the Cs and Ds, can be used to get an idea of what it will take in terms of constraint relations to reach a desired level of objective value. Indeed, the Cs and Ds present actual decisions that reach the levels shown. There may, however, well be other decisions not yet found that would do even better.

In a third case we look at the value of choosing a feasible non-optimal decision. In the optimal decision, constraint 2 has a slack of 0. Worried about availability of the resource associated with constraint 2, or confronted with a shortage on that constraint, a decision maker might want to opt for, or at least have a look at, the best decision available that has a slack of at least 1 on constraint 2. Is there one and if so, what is its objective value? The corpus of type A DoIs in Table 14.1 can answer the question: Item 29 with an objective value of 328 is the best discovered feasible decision that has a slack of at least 1 on constraint 2. Notice that no other decision in Table 14.1 has positive slack on constraint 2. If we go to Table 14.2 we see that nearly all of the Bs have ample slack on constraint 2, at the price of large reductions in the objective value.

As a fourth case, we see that at optimality (item 0 in Table 14.1) job 10 (counting from 0) is assigned to agent 3. Suppose that there arises a reason not to assign job 10 to agent 3 or perhaps something happens so that the assignment cannot be made at all. What is the best alternative? Looking at the As in Table 14.1 we see that item 4 has the decision with the highest objective value in which job 10 is assigned to some agent other than 3, 4 specifically.

Stepping back and reflecting on these cases it should be apparent that well collected corpora of DoIs afford answers and insights for a rich and diverse set of questions, questions that arise naturally and are important in decision making with optimization models.

14.4   Discussion

The chapter has illustrated how, having defined the four types of DoIs (A, B, C, and D), the FI2-Pop GA can be effective at finding ample corpora of each type. Although we have presented only one computational example here, experience indicates that the method we have employed works quite generally and quite well. Further points arising:

1. The data presented in Tables 14.1–14.4, nearly 200 distinct decisions, are from one run of a decision sweep. For real world decision making it is well advised to undertake several decision sweeps, if possible, and pool the results.

2. Extending the run for more generations can also be expected to produce new DoIs. Runs will typically continue to produce improved corpora of DoIs long after they have found an optimal decision. How long to run is a subtle design problem. Ultimately it has to be resolved by experience.

3. Decision sweeping often produces surprising results. This is especially the case with models having integer decision variables. One example we saw was in the type D decisions, where it may be judged surprising that a few of the high objective value decisions discovered have fairly low scores on constraint violation.

14.5   For Exploration

The book's Web site holds four files of DoIs (A, B, C, D) from a run of the FI2-Pop GA on the GAP 1-c5-15-1 model that is different than the run discussed in the body of this chapter. The files are: *gap1_c5_15_1_DoIs_A_20150623.csv*, *gap1_c5_15_1_DoIs_B_20150623.csv*, *gap1_c5_15_1_DoIs_C_20150623.csv*, and *gap1_c5_15_1_DoIs_D_20150623.csv*.

The Web site also holds four files of DoIs from a run of the Simple Knapsack 100 model. The files are: *simpleKS100_DoIs_A_20150624.csv*, *simpleKS100_DoIs_B_20150624.csv*, *simpleKS100_DoIs_C_20150624.csv*, and *simpleKS100_DoIs_D_20150624.csv*.

All of these files are in the convenient CSV (comma separated values) format and can be loaded directly into spreadsheet programs, such as are available in Microsoft Office (Excel), and Open Office. See §14.6 for the URL of the book's Web site.

1. Note that in Table 14.3, showing type C DoIs, every decision has an infeasibility score of -1, meaning that it is nearly feasible: every decision is feasible on all constraints except one, where it is in violation by one unit. Notice as well that only one of the 50 decisions in the figure has an objective value superior to the objective value of the optimal solution. The type D decisions in Table 14.4 all have objective values much larger than that for the optimal solution and are much further from feasibility that the type C decisions.

   (a) Examine the results of a different run of decision sweeping on the model. See *gap1_c5_15_1_DoIs_A_20150623.csv*, *gap1_c5_15_1_DoIs_B_20150623.csv*, *gap1_c5_15_1_DoIs_C_20150623.csv*, and *gap1_c5_15_1_DoIs_D_20150623.csv* on the book's Web site. Does this pattern recur? Discuss what you find and offer an explanation.

(b) Examime the results of decision sweeping on a different model. See *simpleKS100_DoIs_A_20150624.csv, simpleKS100_DoIs_B_20150624.csv, simpleKS100_DoIs_C_20150624.csv,* and *simpleKS100_DoIs_D_20150624.csv* on the book's Web site. Does this pattern recur? Discuss.

(c) What might be done to discover decisions that lie in the gap between type C and D DoIs? These C–D gap decisions would be close to feasibility, but not as close as the type C decisions and otherwise strong on objective value. Discuss one or more general approaches that might work; provide pseudocode.

2. In a study that explores a heuristic for the p-median problem Alp et al. (quoted at greater length in §14.6 below) have the following observation:

> While it is not surprising that near-optimal solutions are quite similar to one another, note that the similarity is not extreme. On average 8 of the 100 locations are different between a pair of solutions at termination. [1, page 40]

How does this finding square with the results in Tables 14.1–14.4? With the results in the book Web site files identified at the beginning of this section? With the results reported earlier for runs of the various metaheuristics we discussed (greedy hill climbing, simulated annealing, threshold accepting algorithms, evolutionary programming)?

3. The evolutionary process we used to produce the DoIs of type A, B, C, and D is a heuristic. It may well miss decisions that are interesting and important. Discuss:

(a) How might you use discovered DoIs as "seeds" to search for other DoIs that are in some sense (which is?) nearby the DoIs actually discovered? Hint: Use them as starting points for local search metaheuristics.

(b) How might you use discovered DoIs with exact solvers to gain insight into your model and perhaps discover new DoIs?

4. We did not give a formal general definition of the challenge value. Instead, we presented it carefully, but informally.

(a) *On the assumption that objective function values and slack values can be straightforwardly represented in monetary quantities,* give a formal definition of challenge value.

(b) Discuss why how much we value an objective function score or an amount of slack may well, in some circumstances, be discontinuous or confined to a limited range.

5. In §14.3 we presented a number of cases in which information gotten from the DoIs could be used to address a realistic business problem. Many more are possible. Find several and write them up. Explain why and how they might arise and how a corpus of discovered DoIs may be used to address the problems at hand.

6. Use Excel or MATLAB or another exact solver of your choice to explore seeding the solver with decisions from the DoIs. Compare random starts or default starts with starting from a suitably chosen DoI. What do you find? Does seeding with DoIs help or hurt or make no difference? In what ways?

7. Repeat the previous exercise but with seeding of other metaheuristics, e.g., greedy hill climbing, simulated annealing, evolutionary programming.

14.6   For More Information

The term decision sweeping is newly minted for this book (to the best of our knowledge). We find the term apt and hope it will come into general use. The underlying concept, which is most important, is hardly new, although it heretofore lacks a proper name. Our term is, as William James said of Pragmatism, "A new name for some old ways of thinking" [80].

The attentive reader of the optimization literature, especially that portion of it directed at practice, will find frequent comments to the effect that finding multiple decisions/solutions to an optimization model has real value. The following passage, quoted in part above, is quite representative.

> While it is not surprising that near-optimal solutions are quite similar to one another, note that the similarity is not extreme. On average 8 of the 100 locations are different between a pair of solutions at termination. Hence, we have a large number of very good solutions to the problem, and this may be more useful than one optimal solution. [1, page 40]

As another example, Jeffrey Camm's article [28] (discussed below with regard to its technical content) is eloquent and insightful on the value of finding "suboptimal" solutions/decisions for a model, that is, decisions that are feasible, of high quality, but not optimal. In short, these are our type A DoIs. The following passage is a quote from the paper.

Why Should You Care About Suboptimal Solutions?

> The first reason you should care about suboptimal solutions is quite simple—a model is a simplification of reality. In almost all cases, however, some details are left out of the model. We use a simplified model for many good reasons. Some examples are the beliefs that (1) collecting the data is not worth the cost it entails, (2) time constraints for the project might preclude getting more detailed data, (3) the more detailed model might not be solvable in a reasonable amount of time, and (4) some factors are just not likely to be the main drivers for the decision at hand. An important point to remember is that the solution to your model that is provably optimal does not consider factors external to the model. . . .

> The second reason you should care about suboptimal solutions is that your optimization model is an investment. If you have taken the time and effort to formulate and solve a model of your problem, why not get the most out of it? Use your model to explore the opportunities. Is the solution to your model unique? If it is unique, how suboptimal is the second-best solution? How do the decision variable values change over the various solutions? In short, you can and should use your model to explore! [28]

We agree entirely! Despite such comments and the continual recurrence of the theme, very little literature exists by way of systematic development of the idea of decision sweeping and of finding multiple solutions or decisions for a given model. Without essaying anything like a comprehensive review of the literature, we do want to note with pleasure two examples from the operations research literature (using exact solution methods) in which the theme has been treated explicitly, innovatively, and carefully.

The first example is "Integrative Population Analysis" introduced by Fred Glover and co-workers. In [62] the authors consider the situation of needing to solve a large number of related constrained optimization models appearing over time. They argue that this need arises naturally and often in practice, and they give examples from new product development, scheduling, and financial services. The Integrative Population Analysis method they develop aims to significantly reduce solution times for new problems as they appear. The approach involves identifying representative problems and finding good ("elite") solutions/decisions for them, which decisions can be used as seeds or starting points for solving new problems. See the paper for details. The motivating problematic they have identified (solving a large number of difficult optimization models appearing over time) is important in practice and their method, Integrative Population Analysis, has much to recommend it. Promising connections with the decision sweeping methods developed here should be apparent: corpora of DoIs can be made available to supply seeds for solving new problems by any method of choice. This is a subject, however, that lies beyond the scope of this introductory book.

The second example we note of using exact solvers to find multiple solutions/decisions for constrained optimization models appears in a recent article by Jeffrey Camm [28]. It describes a simple and elegant method for using any standard exact solver to find the second, third, etc. best decisions for optimization problems with binary integer decision variables. Thus, the method applies straightforwardly to Simple Knapsack problems, p-median problems, and, less straightforwardly, to GAPs. By following the method and repeatedly re-solving the problem (with more and more constraints added), it is possible to sweep the decisions and generate an exact corpus of type A DoIs. Camm's method is limited to feasible DoIs and to binary decision variables. Neither restriction is present for the decision sweeping methods introduced here. Nevertheless, as in the case of Integrative Population Analysis, promising ways to combine the methods appear immediately. For example, when Camm's method applies it can be used to calibrate the results obtained by the heuristic decision sweeping methods developed in this book. As in the case of Integrative Population Analysis, developing these promising ideas is beyond the scope of the book.

Regarding predecessor papers, Branley et al. [21] is an early investigation of using evolutionary computation to find interesting decisions associated with integer programming models. Ann Kuo's Ph.D. thesis [110] uses the FI2-Pop GA and documents its effectiveness at finding DoIs for difficult optimization problems. FI2-Pop GA–based papers that report on effectiveness in finding DoIs include: [93, 94, 96, 97, 98, 100, 101, 102]. For an early statement on candle lighting analysis see [89].

The book's Web site is:

http://pulsar.wharton.upenn.edu/~sok/biz_analytics_rep.

Chapter 15

Parameter Sweeping

15.1 Introduction: Reminders on Solution Pluralism and Parameter Sweeping

Decision sweeping (see especially Chapter 14) is a variety of solution pluralism that applies to models with decision variables, notably constrained optimization models. Parameter sweeping, the subject of this chapter, complements decision sweeping in COModels and also applies to models generally. It is a variety of solution pluralism that is useful for post-solution analysis of all models, whether or not they have decision variables. Given this, it will be helpful at this point to provide some comments and reminders on solution pluralism.

Solution pluralism is a prescriptive approach to deliberating with, indeed to thinking with, models. Given a decision problem and a relevant model, the principle of solution pluralism recommends that we (i) obtain a number of solutions to the model at hand, (ii) assemble this plurality of solutions into an accessible corpus of data, and (iii) mine the corpus for information bearing upon the decision to hand. This is likely to result, according to solution pluralism taken as a prescriptive principle, in better understanding and better decision making. The principle applies quite generally, to any mathematical, computational, or procedural model that might be used to guide decision making.

The general principle of solution pluralism has received widespread assent and much uptake in certain areas of practice. For example,

1. If our model has stochastic outputs, such as a discrete event simulation, then it is imperative to obtain a statistically well-behaved plurality of solutions (by re-running the model multiple times) and to condition our decisions upon statistical properties of the solution corpus (mean, mode, variance, etc.) rather than upon any single outcome.

2. An important form of sensitivity analysis for any model with numeric parameters involves perturbing the parameters by small amounts numerous times, re-solving the model in each case, collecting the solutions, and examining their distribution(s) in order to assess how the model can be expected to behave under small departures from its assumptions. This is called *Monte Carlo analysis* of the model. @RISK is a popular commercial product, an add-in to Excel, for doing this kind of analysis (http://www.palisade.com/risk/).

3. In a *multiobjective model* we seek to optimize in the presence of multiple goals (objectives), as in a portfolio model that seeks to minimize risk and maximize return. Such models will generate a Pareto frontier of solutions, such that for every point (solution) on the frontier, it is impossible to find a solution that improves it on one dimension (objective, goal) without doing worse on at least one other dimension. Standardly, in the presence of a multiobjective model, we seek to find solutions on the Pareto frontier and return these to the decision maker for making decision tradeoffs.

The basic concept and prescriptive force of solution pluralism is perhaps most simply seen in the elementary case of plotting a simple function. The function is a model and we solve it many times, with different parameter values, in order to visualize it. We plot it, so as to understand it better. For example, let our model be the inverse tangent function, atan in MATLAB. Do you know what it looks like? Well, obtain 101 parameter values (between −20 and 20) for the model's parameter, then re-solve the model for each parameter value (here is our plurality or corpus of solutions), and plot the results (use the corpus to extract information about the model). The following MATLAB code does just that.

```
>> x = linspace(-20,20,101);
>> plot(x,atan(x),'Color','black')
>> grid on
>> ylabel('atan(x)')
>> title('Plot of the inverse tangent')
>> xlabel('x')
```

Figure 15.1 displays the resulting plot. Upon examining it you understand much better what the inverse tangent function does (unless of course you were previously well acquainted; but you get the point in any case). This small example, writ large and greatly generalized, is what solution pluralism—and in particular, parameter sweeping—is all about. (It does not, however, *require* visualization.)

In what follows we will mainly be concerned with using computational methods to explore constrained optimization models via solution pluralism, that is, by (i) obtaining a number of distinct solutions for the model, (ii) assembling this plurality into a coherent corpus of solutions, and (iii) mining the resulting corpus for information pertaining to the question at hand.

15.2 Parameter Sweeping: Post-Solution Analysis by Model Re-Solution

Using the inverse tangent function example as a conceptual starting point, we now explore various ways of doing parameter sweeping, which rely upon a schedule of model re-solutions.

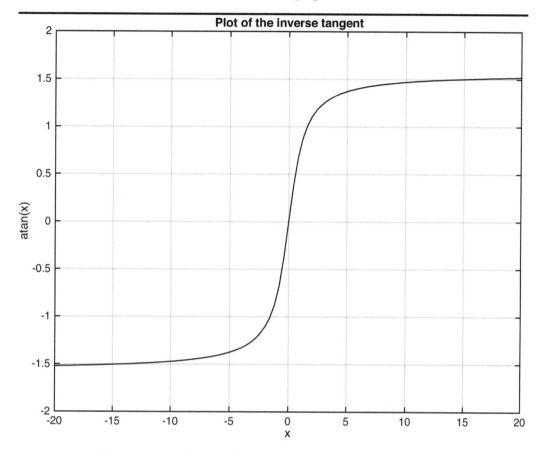

FIGURE 15.1: Plot of the inverse tangent function from -20 to 20.

15.2.1   One Parameter at a Time

Figure 15.2 is a direct analog of Figure 15.1, but for a linear programming model with scores of parameters. What we see plotted is the response of the objective function value at optimality, z^*, to changes in a single parameter, an objective function coefficient, c_3, as it ranges between 10 and 50. The z^* values were obtained by using a linear programming exact solver to re-solve the model for each of the 41 different values of c_3.

The exercise may be repeated for any parameter in the model. An important practical factor, however, is that when there are many parameters, even with fast and cheap re-solution, the amount of insight available is limited, both by effective information overload and by ignoring interactions among parameters.

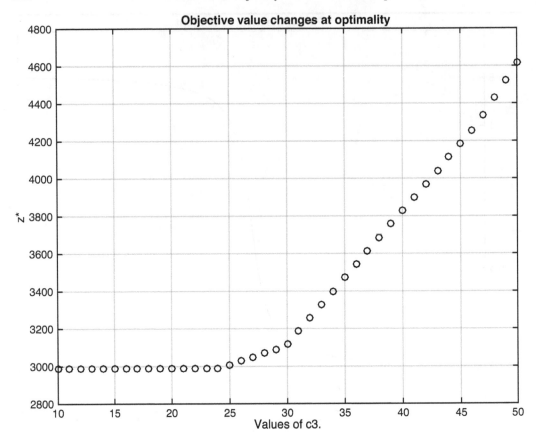

FIGURE 15.2: Example of z^* changes in a linear program as a result of changes in an objective function coefficient.

15.2.2 Two Parameters at a Time

We saw an example of a two parameter sweep in Chapter 1, §1.5, Table 1.2 on page 19, reprinted here as Table 15.1.

A two parameter sweep allows us to visualize interactions between two parameters. With further conditioning more parameters may effectively be examined together. See [149] for an extensive and sophisticated treatment of this sort. Notice as well that the sweep shows areas of similarity, in which the model's output values stay within a circumscribed range, even as both parameters are changing. Finding such stable regions—sometimes referred to as *phases*—is often significant for decision making. In two dimensions we can see the regions. Computational analysis, whether in two or more dimensions, will often be required.

15.2.3 *N* Parameters at a Time

When N is small, say 3 to 6, there are still visualization techniques that can productively convey information. The basic move is to display multiple 2-D and 3-D plots (or tables as in Table 15.1) that show two-way parameter interactions, conditioned upon varying settings of other parameters. Again, see [149] for an illustration. Once N gets much larger these methods no longer suffice and we need to largely give up on visualization and rely on automated processing to discover insights.

	75000	80000	85000	90000	95000
390000	30.4845	31.1727	31.8267	32.45	33.0453
395000	30.3497	31.0362	31.6887	32.3105	32.9046
400000	30.2169	30.9017	31.5527	32.1731	32.7659
405000	30.0861	30.7692	31.4187	32.0377	32.6292
410000	29.9572	30.6387	31.2866	31.9043	32.4945

TABLE 15.1: Parameter sweep results for the Converse formula. Rows: P_a values. Columns: P_b values.

One commonly used method, which some think is constitutive of parameter sweeping (we have a broader conceptualization), is to associate two or more levels (values) for each of the N parameters of interest and then to re-solve the model for each combination of parameter values. Thus, if there are 10 parameters of interest and each parameter is assigned 2 values (a high and a low value, say), then there are $2^{10} = 1024$ distinct combinations of parameter values on which to re-solve the model. This is called a *full factorial* calculation.

The great advantage of having the results of a full factorial calculation, besides simply its thoroughness, is that it is an informationally efficient way to provide data for a regression model in which the parameters serve as predictor variables and the results of the model re-solutions serve as the response variable(s) (see the experimental design literature for details, e.g., [18, 143]). Such a model is an example of a *meta-model*, that is, a model of a model. Meta-models are quite often used as fast shortcuts for large scale simulation models. They have not been much used as surrogates for constrained optimization models, mainly because the existence of constraints tends to limit the accuracy of any summarizing meta-model. Still, there is much worth exploring with the idea.

The obvious disadvantage of pursuing full factorial calculations is that with scale they become intractable. With four levels for each of 20 variables the number of combinations is $4^{20} = 2^{40} \approx 10^{12}$. Remember that these "variables" are actually model parameters. Even our smallish test models are much larger. Our Simple Knapsack 100 model has 201 parameters and our GAP 1-c5-15-1 model has 155 parameters.

15.2.4 Sampling

When the number of parameters to be studied is too large for a full factorial parameter sweep, the obvious move is to sample from the full factorial space. This is in fact what is generally done, in one form or another (see [141] for an introduction and review). A standard practice is to conduct a Monte Carol study by:

1. Identifying the study parameters, $P = \{p_1, p_2, \ldots, p_n\}$.

2. For each study parameter, p_i, identifying a probability distribution from which to draw its values.

   Note, it is standard that a normal distribution is chosen for each parameter, with individual mean and variance values.

3. Setting Y, the store of model results, to empty.

4. Repeating until a stopping condition is reach:

(a) For each of the study parameters obtain an independent draw from its probability distribution, yielding parameter values $X = (x_1, x_2, \ldots, x_n)$.

(b) Solve the model using X to obtain a model result y.

(c) Add y to a master list of model results, Y.

The upshot of this process is to produce, for m repetitions, a collection of m Y values

$$Y = \begin{bmatrix} y^{(1)} \\ y^{(2)} \\ \vdots \\ y^{(m)} \end{bmatrix} \tag{15.1}$$

paired to a collection of m parameter settings

$$M = \begin{bmatrix} x_1^{(1)} & x_2^{(1)} & \cdots & x_n^{(1)} \\ x_1^{(2)} & x_2^{(2)} & \cdots & x_n^{(2)} \\ \vdots & \vdots & \vdots & \vdots \\ x_1^{(m)} & x_2^{(m)} & \cdots & x_n^{(m)} \end{bmatrix} \tag{15.2}$$

where the Y values were calculated by re-solving the model with the randomly drawn X values.

Given M and Y (a plurality of solutions for our model, generated by a form of parameter sweeping) there is much of interest we can learn about the responsiveness, or variability, of the model to different parameters (and combinations of parameters). A simple kind of analysis works as follows. For parameter p_j substitute a constant value (perhaps the expected value of p_j) for the ith column of M. Call this value $x_j^{(c)}$. The new array, M', has the following form:

$$M' = \begin{bmatrix} x_1^{(1)} & \cdots & x_j^{(c)} & \cdots & x_n^{(1)} \\ x_1^{(2)} & \cdots & x_j^{(c)} & \cdots & x_n^{(2)} \\ \vdots & \vdots & \vdots & \vdots & \vdots \\ x_1^{(m)} & \cdots & x_j^{(c)} & \cdots & x_n^{(m)} \end{bmatrix} \tag{15.3}$$

Using M', re-solve the model (m times) to get Y'. Compare the variances of Y and Y', $V(Y)$ and $V(Y')$ and note their difference. If they are close, then there is reason to believe that the model is not very responsive to x_j in the region sampled in M (assuming $x_j^{(c)}$ is itself central in that region). If they are not close, then clearly the model is very responsive to x_j. What counts as close? Use a relative measure: Repeat the exercise for many or all of the parameters and rank them by their resulting $V(Y')$ values. Parameters with comparatively high values are those for which the model is comparatively responsive.

There are any number of useful variants and elaborations of this procedure, and the reader will no doubt be able to think of many of them. In addition, the literature on sensitivity analysis (reviewed, e.g., by [141]) has produced a number of other, rather different, techniques for this sort of analysis. These are, however, beyond the scope of this introductory book.

15.2.5 Active Nonlinear Tests

John Miller has coined the term active nonlinear tests (ANTs) to describe a kind of computational procedure for exploring model behavior [124]. Originally directed at simulation models and at finding possible flaws or anomalies in them ("model busting"), the approach can be used for broader purposes. In a nutshell, the idea is first to characterize formally the anomalous behavior to be sought (e.g., if a probability value is the result calculated by the model, then we might search for values above 1 or below 0), and second to use heuristic search (Miller uses a genetic algorithm) in the space of model parameter values in an attempt to produce the specified anomalous behavior. If, for example, the search produces a valid setting of parameter values for which the model yields a "probability" value of less than 0, then the model may be counted as busted. On the other hand, if a number of such ANTs have been tried without untoward upshot, we gain considerable improvement in confidence in the validity of the model.

The idea certainly can be generalized and applied usefully beyond simulation models and beyond model busting exercises. Beyond simulation, we would argue that any model with parameters is in principle usefully examined with this method. Beyond model busting, we believe that ANTs are credible alternatives to the variance-based methods described in the previous section.

15.3 Parameter Sweeping with Decision Sweeping

The methods described in the previous section, §15.2, have primarily been developed for and applied to models of types other than mathematical programming (or constrained optimization) models. Most of the published discussion of these methods has in mind these other kinds of models (simulations, equational models, etc.; models that is without decision variables). Even so, the methods can be applied in a COModel context, as the example in §15.2.1 demonstrates. We trust to the reader to imagine how the other methods can be used in COModels.

If the methods of §15.2 are to be applied extensively to COModels, two genuine impediments must be overcome. The first is that COModels tend to have a large number of parameters, as we indicated in the discussion above. Parameter sweeping with the methods of §15.2 will be challenging and problematic for *any* model with a large number of parameters. In any such case, the pragmatic alternative of selective attention—studying only a number of parameters small enough to be practicable—will be necessary. It is warranted and is in fact what is done. By reconciling ourselves to reality—not letting the better be enemy of the good—we can intelligently choose which parameters to investigate and thereby circumvent this impediment.

The second impediment arises from the fact that re-solution of COModels, whether by exact or by heuristic solvers, often is prohibitively expensive in time and computing resources. If it takes as little as a minute to solve a COModel (exactly or heuristically) and we need to examine 100,000 parameter settings (a fairly small case, actually), the time required is about 1,667 hours. Small models and simple analyses, as in §15.2.1, can be handled, but this limits the scope of usefulness severely.

As with the first impediment, there is a pragmatic circumvention that will often prove practicable: obtain a corpus of DoIs by a decision sweeping exercise; obtain a collection of PSoIs (parameter settings of interest, the parameter settings to be investigated); and re-evaluate the model for each DoI× PSoI combination and store the results. This in effect gives us the Y and M data discussed in §15.2.4. Putting this explicitly:

1. Use a decision sweeping process to obtain *DoI* as a set of decisions of interest.

2. Specify *PSoI* as a set of parameter settings of interest.

3. For each $d \in DoI$:

    (a) For each $p \in PSoI$:

        i. Set $y(d,p)$ to the result of evaluating the model for decision d and parameter setting p.

We can then construct Y with the $y(d,p)$ as its rows and M with the corresponding (d,p) values as its rows.

15.4 Discussion

Post-solution analysis, particularly sensitivity analysis in both its broad and narrow senses, §1.4.1, is very well entrenched as a necessary aspect of modeling and model-based decision making. A rich literature exists describing methods and cases for models that are not in the most part COModels. These methods can generally be applied to COModels if either they are small and quickly re-solvable or there are effective ways to undertake decision sweeping to obtain appropriate DoIs, as described in §15.3.

The procedure presented in §15.3 loops over a set of DoIs, increasing the computational cost for COModels compared to models without decision variables. Chapters 16 and 17 discuss two methods for filtering sets of DoIs to eliminate their less interesting members.

15.5 For Exploration

1. Models and modeling for decision making loom large in the world of policy analysis. Because decisions are often governmentally or quasi-governmentally based (think: rules from the Environmental Protection Agency, findings from the Intergovernmental Panel on Climate Change (IPCC)), modeling efforts often have associated written policies for model analytics, particularly sensitivity analysis. Exemplars include: [77, 78] from the IPCC, [47] pertaining to environmental impact statements from the European Commission, and [146] from the U.S. Environmental Protection Agency.

    Obtain and read one or more of these publications (or find a similar one on your own). Discuss and assess critically its main directives. What's good? What is perhaps questionable? What is missing?

2. Identify a small or medium-sized model that is interesting to you and that you can implement in Excel or another environment familiar to you. Implement the model and undertake a thorough parameter sweeping analysis. Summarize your findings in a brief report or slide presentation.

3. Implement a COModel in an environment that solves it (exactly, say in Excel or MATLAB, or heuristically, say with simulated annealing or a genetic algorithm). Devise a realistic case story about the model and its implementation, then undertake a thorough parameter sweeping analysis. Summarize your findings in a brief report or slide presentation.

4. You are considering investing in an electric power generation plant. There are three options before you: A is a coal-fired plant, B is a gas-fired plant, and C is a solar PV (photo-voltaic) farm. Your local utility company will write you a 20-year power purchase agreement (PPA) for option C, in which the utility agrees to buy all of the output at a specified price schedule, e.g., $60 a megawatt hour. Electricity produced by options A or B has to be sold on the day-ahead auction market. Using available data both options A and B are, you expect, more profitable to you than option C.

The thing is, today you know with good accuracy what it will cost you to produce a megawatt hour of electricity with the solar farm for the next 20 years. This is because the solar farm does not require fuel that is purchased on the open market, and operations and maintenance are small and well known costs. You also know the price you will be paid for the power the farm produces. On the other hand, you do not know with much certainty what the price of coal or natural gas will be over the next 20 years, nor do you have a lock on the price, which may vary in the market on a daily basis.

So, while options A and B are superior to C in terms of expected profit, they are also much riskier. Discuss how you would model the decision (discuss it at a high level, not with an actual detailed model) and how you would go about using the model(s) to assess risk and uncertainty. Finally, discuss how, once you have a reasonable estimate of risks, you would make the trade-off between the available levels of expected profit and the associated risks and uncertainties.

15.6 For More Information

The relevant literature is extensive. We offer the following short list as a useful introduction.

The book by Morgan and Henrion [127] is something of a classic. It is addressed to the realm of policy making and policy modeling, and it does not discuss COModels. Nevertheless, it has much of value for our context (and much outside our context as well).

Andrea Saltelli and collaborators have produced an important series of works that address sensitivity analysis in the broad sense [139, 140, 141, 142]. They focus on so-called *variance based methods*, one of which we illustrate in §15.2.4. Like Morgan and Henrion [127], this work is addressed to the realm of policy making and policy modeling, and does not discuss COModels.

For illustration of post-solution parameter sweeping analysis to locate phases or regions of stability in parameter space see [149].

Models and modeling for decision making loom large in the world of policy analysis. Because decisions are often governmentally or quasi-governmentally based (think: rules from the Environmental Protection Agency, findings from the Intergovernmental Panel on Climate Change (IPCC)), modeling efforts often have associated written policies for model analytics, particularly sensitivity analysis. Exemplars include: [77, 78] from the IPCC, [47] pertaining to environmental impact statements from the European Commission, and [146] from the U.S. Environmental Protection Agency.

Chapter 16

Multiattribute Utility Modeling

16.1 Introduction

At the end of Chapter 9, in §9.5 on page 140, we commented that

> Our principal aim in this chapter has been to illustrate—with real data on a real location problem—decision sweeping with metaheuristics and to establish in the present case that local search metaheuristics (in particular greedy hill climbing, threshold accepting algorithms, and simulated annealing) are not only able to find high quality decisions, but—what we think is even more important—are able to find quantitatively substantial corpora of high quality decisions.
>
> With this embarrassment of riches, which we shall see again and again, how are we to decide what decision, if any, to implement?

The embarrassment of riches problem has only been exacerbated by what succeeded Chapter 9, in particular our use of the FI2-Pop GA to drive collection of both feasible and infeasible decisions of interest of various kinds. It is time, then, to make good on our promise on page 140 to discuss

> ...a principled and quite general response to the availability of a plurality of solutions: create an index or scoring mechanism—in the form of a mathematical

model—to evaluate and compare the plurality of items, e.g., decisions of interest, on multiple criteria.

To that end, this chapter provides a brief, motivating introduction to decision theory and in particular to *multiattribute utility modeling*, followed by discussion of an example in which we build a multiattribute utility model for a corpus of decisions of interest.

16.2   Single Attribute Utility Modeling

16.2.1   The Basic Framework

To begin, we may think of a decision context as presenting potentially five kinds of factors:

1. Choices: C_i

   These are the options we have, among which we must choose. The C_is are decision variables. We wish to build models that provide a quality score for each of these possible choices. Then, if we accept the scores, there is an obvious and attractive decision rule: pick a choice with the most favorable score.

2. Events: E_j

   Events are actions that may happen independently of our control. We can think of them as produced by Nature, who is ignorant of and indifferent to our interests and welfare, and who does not have interests of her own.[1]

3. Outcomes

   These are choice–event combinations that occur without mutual influence. You might choose either to bring your umbrella today or not. Nature may decide to make it rain today or not, but Nature is not influenced by your decision. While your decision may be influenced by signs of rain or lack thereof, your decision occurs without your knowing for certain whether Nature will make it rain or not. On the other hand, choices may influence, say via a causal path, whether or not an outcome occurs. Nature may decide whether it will rain, but you can decide whether you will get wet if it does rain. (Umbrella? Raincoat? Both?)

   Notationally, we write:

   $$O_k = \langle C_i, E_j \rangle = (E_j | C_i)$$

4. Probabilities

   Outcomes may have probabilities conditioned on, say, the probabilities of various events occurring. We write:

   $$p(O_k) = p(E_j | C_i)$$

   Probabilities may, as appropriate, be objective, based on experience, or subjective, based on individual judgments.

---

[1]So Nature is not an agent, not a player in a game. Decision theory as developed here, and standardly, is in distinction to game theory.

5. Desirabilities

We prefer some outcomes over others. Alternatively, we desire certain outcomes more than others. We write:

$$d(O_k) = d(\langle C_i, E_j \rangle) = d(E_j | C_i)$$

With 1–4 reasonably clear, we now need to determine how we measure desirabilities. First, however, an example.

16.2.2 Example: Bringing Wine

Let us suppose you have been invited to a dinner party and wish to bring a bottle of wine, as is conventionally polite in the context.[2] Your choices, let us say, are three: C_1, bring red; C_2, bring white; C_3, bring rosé. The relevant events are, for the sake of discussion, four in number: E_1, beef is served; E_2, chicken is served; E_3, fish is served; E_4, the meal is vegetarian. This leads to there being the following list of 12 possible outcomes:

1. $O_1 = \langle C_1, E_1 \rangle$

2. $O_2 = \langle C_1, E_2 \rangle$

3. $O_3 = \langle C_1, E_3 \rangle$

4. $O_4 = \langle C_1, E_3 \rangle$

5. $O_5 = \langle C_2, E_1 \rangle$

6. $O_6 = \langle C_2, E_2 \rangle$

7. $O_7 = \langle C_2, E_3 \rangle$

8. $O_8 = \langle C_2, E_4 \rangle$

9. $O_9 = \langle C_3, E_1 \rangle$

10. $O_{10} = \langle C_3, E_2 \rangle$

11. $O_{11} = \langle C_3, E_3 \rangle$

12. $O_{12} = \langle C_3, E_4 \rangle$

Equivalently, and as we shall see usefully, we can place the outcomes in an outcomes table:

	Beef $= E_1$	Chicken $= E_2$	Fish $= E_3$	Veg. $= E_4$
Red $= C_1$	$O_1 = \langle C_1, E_1 \rangle$	$O_2 = \langle C_1, E_2 \rangle$	$O_3 = \langle C_1, E_3 \rangle$	$O_4 = \langle C_1, E_4 \rangle$
White $= C_2$	$O_5 = \langle C_2, E_1 \rangle$	$O_6 = \langle C_2, E_2 \rangle$	$O_7 = \langle C_2, E_3 \rangle$	$O_8 = \langle C_2, E_4 \rangle$
Rosé $= C_3$	$O_9 = \langle C_3, E_1 \rangle$	$O_{10} = \langle C_3, E_2 \rangle$	$O_{11} = \langle C_3, E_3 \rangle$	$O_{12} = \langle C_3, E_4 \rangle$

The outcomes table has an alternative, and entirely equivalent, form:

---

[2]This example is after [81]. We intend no endorsement of consumption of alcohol. Instead, we simply find the example admirably clear.

	Beef $= E_1$	Chicken $= E_2$	Fish $= E_3$	Veg. $= E_4$
Red $= C_1$	$O_1 = (E_1\|C_1)$	$O_2 = (E_2\|C_1)$	$O_3 = (E_3\|C_1)$	$O_4 = (E_4\|C_1)$
White $= C_2$	$O_5 = (E_1\|C_2)$	$O_6 = (E_2\|C_2)$	$O_7 = (E_3\|C_2)$	$O_8 = (E_4\|C_2)$
Rosé $= C_3$	$O_9 = (E_1\|C_3)$	$O_{10} = (E_2\|C_3)$	$O_{11} = (E_3\|C_3)$	$O_{12} = (E_4\|C_3)$

Next, we construct a probabilities table:

	Beef $= E_1$	Chicken $= E_2$	Fish $= E_3$	Veg. $= E_4$
Red $= C_1$	$p(E_1\|C_1)$	$p(E_2\|C_1)$	$p(E_3\|C_1)$	$p(E_4\|C_1)$
White $= C_2$	$p(E_1\|C_2)$	$p(E_2\|C_2)$	$p(E_3\|C_2)$	$p(E_4\|C_2)$
Rosé $= C_3$	$p(E_1\|C_3)$	$p(E_2\|C_3)$	$p(E_3\|C_3)$	$p(E_4\|C_3)$

Note that rows, but not columns need to sum to 1. What do columns sum to? $p(E_i)$ for $i \in \{1, 2, 3, 4\}$.

Next, we have a desirabilities table, which comes in two equivalent forms:

	Beef $= E_1$	Chicken $= E_2$	Fish $= E_3$	Veg. $= E_4$
Red $= C_1$	$d(E_1\|C_1)$	$d(E_2\|C_1)$	$d(E_3\|C_1)$	$d(E_4\|C_1)$
White $= C_2$	$d(E_1\|C_2)$	$d(E_2\|C_2)$	$d(E_3\|C_2)$	$d(E_4\|C_2)$
Rosé $= C_3$	$d(E_1\|C_3)$	$d(E_2\|C_3)$	$d(E_3\|C_3)$	$d(E_4\|C_3)$

and

	Beef $= E_1$	Chicken $= E_2$	Fish $= E_3$	Veg. $= E_4$
Red $= C_1$	$d(O_1)$	$d(O_2)$	$d(O_3)$	$d(O_4)$
White $= C_2$	$d(O_5)$	$d(O_6)$	$d(O_7)$	$d(O_8)$
Rosé $= C_3$	$d(O_9)$	$d(O_{10})$	$d(O_{11})$	$d(O_{12})$

Granting that probabilities may be estimated or gathered from evidence and/or subjective assessment, what about desirabilities? How do we get numerical values—which we definitely need—for them? The full story is a long one and the present context is a brief. Here is a very short version of the story, to be elaborated somewhat below.

We follow the excellent treatment by Richard Jeffrey [81] in using "desirability" as an a-theoretic or pre-theoretic term to indicate preference or value (the more the better). To give actual numbers that are warranted, however, we want to appeal to a theoretical base. That base exists and is called *utility theory* (among other things). Operationally, what we do is to ask the decision makers for their judgments regarding the desirability numbers, $d(O_1), \ldots, d(O_{12})$ in the present case. The questioning, in essence, goes like this. "Of the 12 outcomes, which one do you like the best? OK, let's give that a value of 100. Now, which one do you like the least? Fine, that one gets a value of 0. Now for each of the remaining outcomes please tell me how you would score its value, remembering that the number should be between 0 and 100." Once the decision maker has complied we have our desirability numbers.

Utility theory tells us that if the decision maker's preferences or desirabilities have four fundamental properties, then there will be a function, called a *utility function*, that coherently describes the decision maker's preferences in the sense that for any two lotteries (probabilistic distribution of outcomes), if the decision maker prefers one to the other, then the value of the utility function on those outcomes will be such that the expected value of the more preferred lottery is higher than the expected value of the less preferred lottery (and if the decision maker is indifferent, then the utility values will be equal). So, the theoretical result is saying that under the right conditions a mathematical function exists that correctly represents the decision maker's preferences. The "right conditions" may be described with four fundamental assumptions of utility theory, as follows:

1. With sufficient calculation an individual faced with two prospects (possible outcomes) P_1 and P_2, will be able to decide whether he or she prefers prospect P_1 to P_2, P_2 to P_1, or whether he or she likes each equally well.

   ("You can decide.")

2. If P_1 is regarded at least as well as P_2 and P_2 at least as well as P_3, then P_1 is regarded at least as well as P_3.

   (transitivity of preference).

3. If P_1 is preferred to P_2, which is preferred to P_3, then there is a probabilistic mixture (called a *lottery*) of P_1 and P_3 which is preferred to P_2, and there is a mixture of P_1 and P_3 over which P_2 is preferred.

   (continuity)

4. Suppose the individual prefers P_1 to P_2, and P_3 is some other prospect. Then the individual prefers any probabilistic mixture of P_1 and P_3 to the same mixture of P_2 and P_3.

   ("independence of irrelevant alternatives")

The practical upshot of this is that we ask the decision maker questions for the purpose of getting an estimate of this function. In asking we normally (and should) do some checking to ascertain that the assumptions of utility theory are met, and then we use the answers from the decision maker to estimate the presumed utility function.

There is very much theory behind all of this, which the reader may approach through the references we give at the end of the chapter. For practical purposes, however, it is more important to be broadly aware of the theory and then proceed to use the robust and simple methods for applying it that have been developed. The latter is what we are introducing in this chapter.

Assuming now that we have obtained numbers properly for our probabilities and our desirabilities tables, we can proceed to use the information. What we do is to multiply the tables together, item by item, and collect the row totals. Table 16.1 is an example of just such a *probabilities-desirabilities* table, tailored here for the wine problem. The row totals are *expected utilities* for the choices. Put otherwise, we calculate the expected (average, probability-weighted) desirability for each choice. Our decision rule is then to take the choice with the highest expected utility/desirability. Points arising:

1. Of course, this is just a model and as such it should be explored with post-solution analysis, but we have to stop somewhere.

	Beef $= E_1$	Chicken $= E_2$	Fish $= E_3$	Veg. $= E_4$	Row Totals (Expected Utilities)
Red $= C_1$	$p(O_1) \times$ $d(O_1)$	$p(O_2) \times$ $d(O_2)$	$p(O_3) \times$ $d(O_3)$	$p(O_4) \times$ $d(O_4)$	$\sum_{i=1}^{4} p(O_i) \times d(O_i)$
White $= C_2$	$p(O_5) \times$ $d(O_5)$	$p(O_6) \times$ $d(O_6)$	$p(O_7) \times$ $d(O_7)$	$p(O_8) \times$ $d(O_8)$	$\sum_{i=5}^{8} p(O_i) \times d(O_i)$
Rosé $= C_3$	$p(O_9) \times$ $d(O_9)$	$p(O_{10}) \times$ $d(O_{10})$	$p(O_{11}) \times$ $d(O_{11})$	$p(O_{12}) \times$ $d(O_{12})$	$\sum_{i=9}^{12} p(O_i) \times d(O_i)$

TABLE 16.1: Probabilities-desirabilities table for the wine decision.

2. The procedure is sensible and intuitive even without the underlying theory. This is, in our view, a strong reason to use it.

3. Figure 16.1 presents a *decision tree* representation for a different decision problem, whether to plug a parking meter or not. Broadly speaking, decision trees and probabilities-desirabilities tables are equivalent. We find the tables to be simpler and more quickly understandable. Trees are more often used in practice because they afford various kinds of analyses (e.g., expected value of sample information) that would be more awkward to conduct with a tabular representation. Because our attention is directed at multiattribute models, we eschew detailed discussion of decision tree analysis, however interesting and valuable it is.

16.3   Multiattribute Utility Models

The outcomes in the wine example of the previous section were simple: a wine is brought and a main course is served. The desirability of an outcome was assessed holistically, without further analysis. Often, however, things are not so simple and we need to, or at least wish to, evaluate multiple outcomes on several dimensions or attributes. In comparing apartments to rent, for example, we consider price, quality, distance to work, and likely other attributes as well. Rarely, if ever, do we find a single outcome (e.g., apartment) that is as good as or better than the alternatives on every single attribute of interest. When this happens, as it normally does, we need to make tradeoffs among the attributes in order to arrive at an accurate overall score (desirability or utility) for a possible choice. Multiattribute utility modeling, which we now discuss, has as its purpose and aim the construction of mathematical models for making these tradeoffs.

Recalling the previous section, note that

1. In the analysis we rely upon having the outcome desirabilities/utilities, $d(O_k)$s, as numbers. If our outcomes are multiattribute and we can find a desirability/utility function that produces a utility number from the multiple attributes, then the analysis may proceed exactly as in the previous section, exactly as in the single attribute case. And this is exactly what we will do.

2. If in a probabilities-desirabilities table, e.g., Table 16.1, the probabilities are all either 0 or 1 (so the situation is deterministic) the problem reduces to the trivial one of picking the choice with the highest desirability/utility. The deterministic case is just

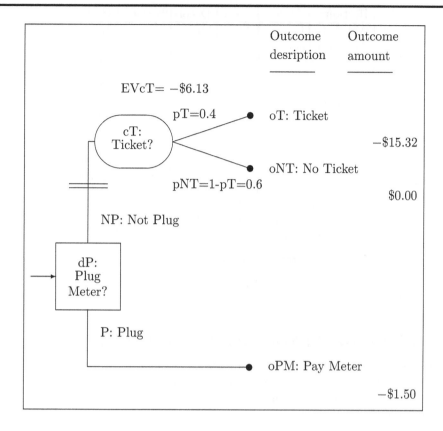

FIGURE 16.1: Decision tree representation of a decision problem.

a special case of the probabilistic, except that it is not trivial when multiple attributes are in play. In the interests of clarity and simplicity we henceforth focus the discussion on the deterministic case. The reader should be able to make the generalization, which is straightforward.

16.3.1 Multiattribute Example: Picking a Restaurant

Consider now another take on the problem of choosing a restaurant. Table 16.2 shows scores for a number of restaurants on each of four dimensions, viz., the attributes food quality, decor quality, service quality, and cost.

RestaurantName	Food	Decor	Service	Cost
Abbey Grill	14	16	16	$26.00
Academy Café	15	16	15	$23.00
Adobe Café	16	12	15	$18.00
Alaina's Fine Food	21	15	20	$19.00
Alberto's Newtown Squire	19	20	19	$31.00
Al Dar Bistro	16	14	14	$20.00
Alexandar's Table at Chaucer's	17	14	16	$19.00
Alexander's Café	17	17	16	$26.00
Alfio's	14	12	16	$22.00
Alisa Café	24	13	21	$31.00
Al Khimah	21	20	21	$24.00
Allies American Grill	16	14	17	$21.00

TABLE 16.2: Restaurants with attributes.

16.3.2 The SMARTER Model Building Methodology

We now proceed to build a multiattribute utility model for comparing these restaurants. In doing so we step through an established methodology, called SMARTER [43], which comports well with our policy of being unremittingly practical. SMARTER consists of nine steps, as follows.

16.3.2.1 Step 1: Purpose and Decision Makers

Identify the purpose of the decision exercise and who the decision makers are. In our restaurant example, you the reader are the decision maker and the purpose is, hypothetically, to pick a restaurant. In more formal, consequential settings, this step should be carefully considered, with all stakeholders identified and informed of the purpose and process.

16.3.2.2 Step 2: Value Tree

Elicit from the decision makers, or obtain in some other way, a list of the attributes to be considered in conjunction with the decision on the available prospective outcomes. A value tree is a hierarchical arrangement of attributes. Discussion of value trees is beyond our scope given our purposes, but see [84, 148] for more information. The most important point for this step is to find a consensus list of attributes that covers what is important and that is non-redundant. On the latter, our model risks serious distortion if we count a source of value more than once.

In our present example, the list of attributes (already stated) is food quality, decor quality, service quality, and cost.

16.3.2.3 Step 3: Objects of Evaluation

Identify what is to be evaluated, that is, what are the prospective outcomes we wish to consider? Normally, the list would fall out of step 1, but it is wise to check again that the list to hand is correct and complete.

In our present example it is the restaurants listed in Table 16.2, and possibly future others, that are to be evaluated.

16.3.2.4 Step 4: Objects-by-Attributes Table

Step 4 is to create a table on the order of Table 16.2 to record scores for object-attribute (table entry) combinations. These scores should be based on physical or judgmental (subjective) measurements, as they are in Table 16.2. This is the logical next step after step 2 (creating the list of attributes; columns in the table) and step 3 (identifying entities to be evaluated; rows in the table).

16.3.2.5 Step 5: Dominated Options

Eliminate any dominated options. An option (row in the objects-by-attributes table) is dominated if its scores are equal to or worse than the scores for another option.

Given that the overall score of any dominated option will be worse than that for any option dominating it, this step may be skipped without any real loss.

16.3.2.6 Step 6: Single-Dimension Utilities

Create a utility function for each of the attributes. Here is how. For each attribute:

1. Find its best score and its worst score in the objects-by-attributes table. Adjust as appropriate to accommodate possible new objects/choices. The best score will be given a utility value of 100, the worst a value of 0.

   In Table 16.2 the scores on the food attribute range from 14 (worst) to 24 (best). We can let 12 have a value of 0 and 26 a value of 100. Decor ranges from 12 (worst) to 20 (best). We'll use 10 as the worst and 22 as the best. Service ranges from 14 (worst) to 21 (best). We'll use 12 to 23. Finally, cost ranges from 31 (worst) to 18 (best). We'll use 33 to 16.

2. Choose a class of single-dimensional utility functions.

   The number of functions is in principle infinite. We have to be practical, so we recognize that the overall scores will not be terribly sensitive to nuances of single-dimensional utility functions. (This can, of course, be tested during post-solution analysis.) The practical advice, then, is to ask first whether the utilities/desirabilities are monotonic. They are not monotonic if sometimes they increase and sometimes they decrease, depending on their objects-by-attributes table scores.

   We will assume the usual case, viz., that the single-dimensional utility functions are monotonic. The analyst should verify this by consulting the decision makers with appropriate questions. If non-monotonic functions are needed, they can be supplied, but the analysis is less simple than what we are discussing here. See for details [43, 84, 148].

   Assuming monotonicity, then, let

$$u_j(x_{ij}) = 100 \times \frac{x_{ij} - x_j^-}{x_j^+ - x_j^-} \tag{16.1}$$

   where:

   - x_j^+ (x_j^-) is the best (worst) score on attribute j.

     For example, from Table 16.2, if the attribute is food (j=food), then x_j^+=26 and x_j^-=12.

- x_{ij} is the score of object/choice i on attribute j.

  For example, from Table 16.2, if the object is Al Dar Bistro (i=Al Dar Bistro) and the attribute is cost (j=cost), then the score is \$20 and so x_{ij}=\$20.

- $u_j(x_{ij})$ is the utility (desirability) on attribute j of the score on attribute j of object/choice i.

  For example, from Table 16.2, if the object is Al Dar Bistro (i=Al Dar Bistro) and the attribute is cost (j=cost), then the object is x_{ij} and its utility u_j on i is score

$$u_j(x_{ij}) = 100 \times \frac{\$20 - \$31}{\$18 - \$31} = 84.6 \qquad (16.2)$$

Finally, we need to check and verify that the attributes are *conditionally monotonic*. At this point we have already ascertained (tentatively, always tentatively) that the attributes are monotonic. Now we ask whether they are still monotonic *conditioned on the values of the other attributes*. That is, for example, is lower cost always better no matter what the value of, say, decor is? Is there ever a case in which, say, the decision maker prefers lower cost when decor is poor but higher cost when decor is good?

There are legitimate cases of conditional non-monotonicity, but they are rare. They cannot be accommodated easily or gracefully with the additive utility model we are building, so other mathematical forms will normally be needed. Again, see [43, 84, 148] for details. We shall proceed on the assumption that our attributes have been found to be conditionally monotonic.

3. Using the single-dimensional utility functions, obtain utility scores for each attribute of each object/choice under consideration.

   In other words, plug the scores into the formulas just developed and record the resulting single-dimensional utility values.

16.3.2.7   Step 7: Do Part I of Swing Weighting

Steps 7 and 8 are about finding weights—conventionally, non-negative numbers adding to 1—for combining the single-dimensional utility scores into an overall score for an object. In step 7 we first find a ranking of the attributes such that the higher-ranked ("more important") attributes will have larger weights.

Our method for finding a ranking of the attributes (and hence their weights) is called swing weighting. It is quite simple. If there are n attributes we ask the decision maker(s) to consider n objects/choices, constructed as follows: object i, $i \in \{1, \dots, n\}$ has a utility score of 100 on attribute i and 0 on every other attribute. For example, adverting to Table 16.2, object 1 has a food score of 26, a decor score of 10, a service score of 12, and a cost of \$33. Object 2 has a food score of 12, a decor score of 22, service score of 12, and a cost of \$33. And so on for objects 3 and 4.

Now we ask the decision maker: Which of the four objects do you like best? If the answer is object 3, then service is the most important attribute and will have the highest weight. Next we ask: Which of the three remaining objects do you like best? If the answer is object 4, then cost is the second most important attribute. And so until all attributes have been ranked.

16.3.2.8   Step 8: Obtain the Rank Weights

Swing weighting has a Part II, but this is not it. We present a simpler and easier method, one that has been validated in practice and is advocated in [43]. We hasten to add that we see nothing wrong with Part II of swing weighting. In fact it potentially is more accurate. Its only downside is that it requires additional, somewhat difficult judgments from decision makers. Rank weights avoid going back to decision makers at all (after Part I) and calculate weights based on the rankings only. There is a formula for doing this:

$$w_k = \frac{1}{n} \sum_{j=k}^{n} \frac{1}{j} \qquad (16.3)$$

where w_k is the weight on the kth attribute, n is the number of attributes, and the highest ranking attribute has a k value of 1. (That is, the best to worst rank order is 1, 2, 3, ..., n.)

For example, with four attributes as in our restaurant example the (rank, weight) pairs are: (1, 0.5208), (2, 0.2708), (3, 0.1458), and (4, 0.0625).

16.3.2.9 Step 9: Calculate the Choice Utilities and Decide

Our additive model is

$$U(x_i) = \sum_{j=1}^{n} w_j u_j(x_{ij}) \qquad (16.4)$$

where $i \in$ objects or choices, $j \in$ the n attributes, x_i is choice i, w_j is the weight on attribute j, u_i is the utility function on attribute j, and x_{ij} is the score of object/choice i on attribute j.

By this point we have all the information needed on the right-hand side of expression (16.4), so we calculate $U(x_i)$ for each restaurant, in the example, and decide based on the rule of choosing the one (or a one) with a maximum utility score.

16.4 Discussion

We have given an exposition of the SMARTER method for constructing multiattribute utility models. The method originated with [43], which can be consulted for further details and elaborations. SMARTER is a general technique for building multiattribute utility models and as such has no unique purchase for post-solution analysis of constrained optimization models. That use is, we think, rather straightforward at this point and so we develop it in the exercises (next section).

We close by quoting with approval a passage from [43], which introduced SMARTER:

> *The strategy of heroic approximation.* Two beliefs motivated SMART [the original, but flawed method] and motivate SMARTS [SMARTER but using swing weighting Part II in step 8], SMARTER, and this paper. One is that simpler tools are easier to use and so more likely to be useful. The second is that the key to appropriate selection of methods is concern about the trade-off between modeling error and elicitation error. ... we believe that more nearly direct assessments of the desired quantities are easier and less likely to produce elicitation errors. ...
>
> We call that view the strategy of heroic approximation. Users of that strategy do not identify formally justifiable judgments and then figure out how to elicit them. Rather they identify the simplest possible judgments that have any hope

of meeting the underlying requirements of multiattribute utility measurements, and try to determine whether they will lead to substantially suboptimal choices *in the problem at hand.* If not, they try to avoid elicitation errors by using those methods. [43, page 310]

16.5   For Exploration

1. Use Excel or another spreadsheet program (or indeed any programming environment of your choice) to calculate the utilities of the restaurants in Table 16.2. (You will first have to use your own judgments to complete steps 7 and 8 of the SMARTER method.) Discuss your findings. How sensitive are the results?

2. Using data from rating services such as Yelp and Zagat's expand the restaurant objects-and-attributes table in Table 16.2 and build a SMARTER utility model for the new data. Explore with the model. Are there clusters of very similar restaurants? Are there discernible "holes" in the offerings, indicating perhaps commercial opportunities for new products?

3. Recall from Chapter 14: The book's Web site holds four files of DoIs (A, B, C, D) from a run of the FI2-Pop GA on the GAP 1-c5-15-1 model that is different than the run discussed in the body of Chapter 14. The files are:

   - *gap1_c5_15_1_DoIs_A_20150623.csv*
   - *gap1_c5_15_1_DoIs_B_20150623.csv*
   - *gap1_c5_15_1_DoIs_C_20150623.csv*
   - *gap1_c5_15_1_DoIs_D_20150623.csv*

   The Web site also holds four files of DoIs from a run of the Simple Knapsack 100 model. The files are: *simpleKS100_DoIs_A_20150624.csv, simpleKS100_DoIs_B_20150624.csv, simpleKS100_DoIs_C_20150624.csv,* and *simpleKS100_DoIs_D_20150624.csv.*

   All of these files are in the convenient CSV (comma separated values) format and can be loaded directly into spreadsheet programs, such as are available in Microsoft Office (Excel), and Open Office.

   Using the objective function values and the slack values from these files and using the SMARTER method, build multiattribute utility models to assist in post-solution analysis of the GAP 1-c5-15-1 model and of the Simple Knapsack 100 model.

4. Discuss how you could use SMARTER models to undertake post-solution analysis of simultaneous changes to multiple objective function coefficients.

16.6   For More Information

The literature on utility modeling is vast. Certain classics, however, stand out. Our short list includes: [131] and [16] for clear and comprehensive introductions to decision analysis with single attributes; [84] as the standard for the theory of multiattribute utility modeling; [148] for a behavioral and application-driven general overview of decision analysis; [43] for the SMARTS and SMARTER methods discussed in this chapter; [81] for a very careful, thoughtful, and clear development of utility theory by a first-rate philosopher; and [38] for bracing evidence on the usefulness of simple linear models.

The book's Web site is:

`http://pulsar.wharton.upenn.edu/~sok/biz_analytics_rep.`

Chapter 17

Data Envelopment Analysis

17.1   Introduction

Data envelopment analysis (DEA) is an optimization based approach to measuring the comparative efficiencies of a number of entities that produce outputs from inputs. In the basic setup, which we explore here, there are n entities—"decision making units" (DMUs)—to be compared with regard to their performance in converting one or more input factors to one or more output factors. We assume that we have the factor data for all of the DMUs.

The core intuition in DEA is that we can assess the *relative* efficiency of a specific DMU, o, as follows. If we can produce o's levels of output factors with reduced levels of input factors, then o is not fully efficient; otherwise it is. DEA assesses this in a relative or comparative manner. (How else could you do it, practically speaking?) If a mixture of DMUs can produce at least o's levels of outputs, requiring lower amounts of inputs, then o is relatively inefficient. Contrariwise, if there is no way to obtain o's levels of output with fewer resources than o is using, by combining any or all of the other DMUs, then o must be judged to be relatively efficient.

DEA embodies a clever and insightful way of translating this core intuition into a linear programming formulation. To see how this works, we begin with a specific example, then abstract it.

There are $n = 4$ decision making units (DMUs), $out = 2$ output measures, and $in = 3$ input factors. The DMUs might be, say, 4 different manufacturing facilities. The output measures might be profit and growth rate, and the input factors might be hours of operation per week, number of full time equivalent employees, and operating expenses. The essential data we require consists of two tables or matrices: X is the table/matrix of inputs, with dimensions $in = 3$ by $n = 4$. In tabular, labeled form it is:

	u_1	u_2	u_3	u_4
operating hours	84	96	100	96
FTE's	14	16	20	18
operating expenses	880	900	1,000	1,100

Expressing this mathematically we have:

$$X = \begin{bmatrix} 84 & 96 & 100 & 96 \\ 14 & 16 & 20 & 18 \\ 880 & 900 & 1000 & 1100 \end{bmatrix} \tag{17.1}$$

For outputs we correspondingly have

	u_1	u_2	u_3	u_4
profit	4,200,000	4,300,000	5,000,000	4,800,000
growth rate	9.2	8.8	8.0	9.0

and

$$Y = \begin{bmatrix} 4200 & 4300 & 5000 & 4800 \\ 9.2 & 8.8 & 8.0 & 9.0 \end{bmatrix} \tag{17.2}$$

(after dividing the profit values by 1,000 in the interest of numerical and display convenience).

We can now specify a DEA linear programming formulation for this problem, *in the case in which we wish to measure the relative efficiency of the second DMU* (represented in the second columns of X and Y), so that $o = 2$.

Parameters:

X: As defined above, X_{rj} = consumption of resource r by DMU j, $r = 1, \ldots, n = 3$, $j = 1, \ldots, n$.

Y: As defined above, Y_{ij} = production of output i by DMU j, $i = 1, \ldots, out = 2$, $j = 1, \ldots, n$.

(Implicitly) o, the index of the target DMU, = 2.

Variables:

w_j: weight assigned to DMU j's inputs and outputs, $j = 1, \ldots, n$.

E: Efficiency factor for input use.

Model:

Minimize E

Subject to:

w_1	$+$	w_2	$+$	w_3	$+$	w_4	$=$	1
$4200w_1$	$+$	$4300w_2$	$+$	$5000w_3$	$+$	$4800w_4$	$\geq$	4300
$9.2w_1$	$+$	$8.8w_2$	$+$	$8.0w_3$	$+$	$9.0w_4$	$\geq$	8.8
$84w_1$	$+$	$96w_2$	$+$	$100w_3$	$+$	$96w_4$	$\leq$	96E
$14w_1$	$+$	$16w_2$	$+$	$20w_3$	$+$	$18w_4$	$\leq$	16E
$880w_1$	$+$	$900w_2$	$+$	$1000w_3$	$+$	$1100w_4$	$\leq$	900E

$$0 \leq w_j, E, \quad j = 1, \ldots, n$$

Let us see how this works. The equality constraint on the sum of the weight parameters

$$w_1 + w_2 + w_3 + w_4 = 1 \tag{17.3}$$

along with their non-negativity constraints

$$w_j \geq 0, \quad j = 1, \ldots, n \tag{17.4}$$

ensure that the weights specify a proper mixture of DMUs. These are decision variables. We seek to find a setting for them that satisfies all of the constraints and that minimizes $E \geq 0$. With the output constraints

$4200w_1$	$+$	$4300w_2$	$+$	$5000w_3$	$+$	$4800w_4$	$\geq$	4300
$9.2w_1$	$+$	$8.8w_2$	$+$	$8.0w_3$	$+$	$9.0w_4$	$\geq$	8.8

we specify that the total output of the mixture must equal or exceed the output of DMU 2, whose outputs are represented as the right-hand-side values of these constraints.

Finally, we seek to find a mixture (setting of the w_i's) that uses a minimal amount of resources, compared to DMU 2.

$84w_1$	$+$	$96w_2$	$+$	$100w_3$	$+$	$96w_4$	$\leq$	$96E$
$14w_1$	$+$	$16w_2$	$+$	$20w_3$	$+$	$18w_4$	$\leq$	$16E$
$880w_1$	$+$	$900w_2$	$+$	$1000w_3$	$+$	$1100w_4$	$\leq$	$900E$

Given a mixture, a setting of the w_i's, its resource consumption is captured by the left-hand side of the $\leq$ constraints shown just above. Minimizing E, if we find a feasible setting of the w_i's at which $E < 1$, then the target DMU, $o = 2$ in the present example, is inefficient relative to the production available to the mixture. On the other hand if E at its minimum feasible setting equals 1, then the target DMU is relatively efficient because there is no mixture that does better. Note that 1 is also the upper bound on E, because if we set $w_o = 1$ and $w_i = 0$ for $i \neq o$, then $E = 1$ is feasible and optimal. So the question the LP is to answer for us is whether there is a setting of the w_i's that is feasible for which the optimal value of E is between 0 and 1.

Our model, as presented above, is correct, although it is not presented in a standard form for most solvers. Transforming it into such a standard form will be useful.

Minimize

$$z = 0w_1 + 0w_2 + 0w_3 + 0w_4 + 1E$$

Subject to:

w_1	$+$	w_2	$+$	w_3	$+$	w_4	$+$	$0E$	$=$	1
$-4200w_1$	$+$	$-4300w_2$	$+$	$-5000w_3$	$+$	$-4800w_4$	$+$	$0E$	$\leq$	-4300
$-9.2w_1$	$+$	$-8.8w_2$	$+$	$-8.0w_3$	$+$	$-9.0w_4$	$+$	$0E$	$\leq$	-8.8
$84w_1$	$+$	$96w_2$	$+$	$100w_3$	$+$	$96w_4$	$+$	$-96E$	$\leq$	0
$14w_1$	$+$	$16w_2$	$+$	$20w_3$	$+$	$18w_4$	$+$	$-16E$	$\leq$	0
$880w_1$	$+$	$900w_2$	$+$	$1000w_3$	$+$	$1100w_4$	$+$	$-900E$	$\leq$	0

$$0 \leq w_j, E, \quad j = 1, \ldots, n$$

FIGURE 17.1: Formulation of the example as a DEA LP of type BCC.

Using X and Y as defined above (with dimensions r rows and n columns for X, and m rows and n columns for Y), we can abstract this version to a matrix-based equivalent formulation. Let:

$$f = \begin{bmatrix} 0 \\ 0 \\ 0 \\ 0 \\ 1 \end{bmatrix} \qquad (17.5)$$

More generally, we can write:

$$f = \begin{bmatrix} \mathbf{0}(n,1) \\ 1 \end{bmatrix} \qquad (17.6)$$

where $\mathbf{0}(n,1)$ is a column vector (n by 1 matrix) having n rows, all its elements equal to 0. Our vector x of decision variables is

$$x = \begin{bmatrix} w_1 \\ w_2 \\ w_3 \\ w_4 \\ E \end{bmatrix} \qquad (17.7)$$

equivalently as

$$x = \begin{bmatrix} w \\ E \end{bmatrix} \qquad (17.8)$$

with w defined generally as

$$w = \begin{bmatrix} w_1 \\ w_2 \\ \vdots \\ w_n \end{bmatrix} \qquad (17.9)$$

which is notation we use to indicate a construction identical to expression (17.7) when $n = 4$.

Next we define

$$Aeq = [\ 1 \quad 1 \quad 1 \quad 1 \quad 0\ ] = [\ \mathbf{1}(1,n) \quad 0\ ] \qquad (17.10)$$

and

$$beq = 1 \qquad (17.11)$$

The expression $\mathbf{1}(1,n)$ stands for a row vector of length n consisting of all 1's. Finally, we define

$$Aineq = \begin{bmatrix} \begin{bmatrix} -Y \\ X \end{bmatrix} \begin{bmatrix} \mathbf{0}(m,1) \\ -X(:,o) \end{bmatrix} \end{bmatrix} \qquad (17.12)$$

and

$$bineq = \begin{bmatrix} -Y(:,o) \\ \mathbf{0}(r,1) \end{bmatrix} \qquad (17.13)$$

The expression $X(:,o)$ indicates column o of the matrix X. Thus, in our example, with $o = 2$,

$$X(:,2) = \begin{bmatrix} 96 \\ 16 \\ 900 \end{bmatrix} \qquad (17.14)$$

The meaning of $Y(:,o)$, and indeed $Z(:,o)$ for any matrix Z, has the same determination: It is vector constituting the *o*th column of the associated matrix. Recall that m is the number of rows in Y and r the number of rows in X, and the notation and expression of our DEA abstraction as a linear program is complete. We may formulate the general LP for BCC type DEA linear programs as in Figure 17.2.

$$\text{Minimize} \quad f'x$$

Subject to:

$$Aeq \cdot x = beq \tag{17.15}$$

$$Aineq \cdot x \leq bineq \tag{17.16}$$

$$x \geq [\mathbf{0}(n+1,1)] \tag{17.17}$$

FIGURE 17.2: General formulation of a BCC type DEA linear program (based on characterizations of the elements, as explained in the text).

17.2 Implementation

With this general formulation to hand, its conversion to a computational function is immediate. Figure 17.3 provides such a function for MATLAB. Notice how directly the mathematical formulations above can be expressed in a general MATLAB function.

17.3 Demonstration of DEA Concept

The principle of solution pluralism, as we have discussed, enjoins us in conjunction with a constrained optimization model to

1. Identify decisions of interests (DoIs),

2. Obtain samples from the DoIs in sufficient quantities for the purposes to hand, and

3. Use the discovered DoIs in deliberations pertaining to post-solution analysis.

We have seen a number of examples of the principle of solution pluralism being put to good use. Our purpose now is to illustrate and explore how DEA can contribute to post-solution analysis of constrained optimization models by extracting information from a plurality of DoIs.

We take for our example the GAP 1-c5-15-1 test problem we have been using throughout the book. Recall that it has five agents, each of whom we consider as providing a resource that is in short supply. The right-hand-side values for the five agents are 36, 34, 38,27, and 33. Using evolutionary computation, we obtained 100 feasible decisions of interest (FoIs). The objective function values and slack values for the five agents are given in the file *feas1500dea.csv*, which is available at the book's Web site. Further, the file *gap1.mat* (a MATLAB data file) contains left-hand-side data, objective data, and slack data for these 100 FoIs in MATLAB array/matrix format. It too is available on the book's Web site.

Tables 17.1 and 17.2 present results from using the MATLAB function `deabcc` shown in Figure 17.3 to solve the DEA BCC model for 100 feasible decisions of interest (FoIs). Decision #1, with a z (objective function) value of 336 is an optimal decision. It has a value of 1.000 in the column headed E, indicating that it (as it should) has an efficiency rating of 1. The $X(j)$ columns show the left-hand-side values of the constraints on the capacities

```
function [x,fval,exitflag,output,lambda] = deabcc( X, Y, o )
%deabcc Constructs and executes a data envelopment analysis
% (DEA) model of the BCC type on DMU (decision making unit) o,
% given inputs X and outputs Y. X and Y must have the
% same number of columns, here called n. X(i,j) is the
% input quantity of type i for DMU j, Y(i,j) the output
% quantity of type i for DMU j. After constructing
% an appropriate LP, the function runs an LP using
% linprog and returns the full complement of return values:
% [x,fval,exitflag,output,lambda].
[infactors,n] = size(X);
[outfactors,~] = size(Y);
f = [zeros(n,1); 1];
Aeq = [ones(1, n) 0]; % E is in the last position.
% w(i)'s in the first 4 positions.
beq = 1;
Aineq = [[-Y; X] [zeros(outfactors,1); -X(:,o)]] ;
bineq = [-Y(:,o); zeros(infactors,1)];
lb = zeros(n+1,1);
% Now call linprog:
[x,fval,exitflag,output,lambda] = ...
    linprog(f,Aineq,bineq,Aeq,beq,lb);
end
```

FIGURE 17.3: MATLAB implementation of the general formulation of a BCC type DEA linear program.

of the agents, while the $S(j)$ columns show the associated slack values. $X(j) + S(j) = b(j)$ where $b(j)$ is the right-hand-side value of the associated ($\leq$) constraint. In looking at the data, the $S(i)$ values are often easier to read and interpret than the $X(j)$ values, but the two contain the same information.

Points arising from Tables 17.1 and 17.2:

1. Decision #2 has an E rating of 0.9984 and so is not efficient. It is not, however, dominated by decision #1, or any other decision. Thus, decision #2 is Pareto efficient, but not DEA efficient. It weakly dominates decision #1 on the slacks but has a lower z value. Decision #3 is efficient and weakly dominates decision #2 on the slacks, but has itself a lower z value.

2. No decision with a z value of 332 or 331 is efficient.

3. Looking at decisions 25 through 31 we see that two of them, #26 and #27, are efficient, while the rest are not. None of the five inefficient decisions are dominated by either of the efficient decisions; they are in fact all Pareto efficient.

Further observations could be made. The important generalization to notice, however, is that of the 100 FoIs, 10 are efficient according to the DEA BCC calculations and the remaining are all or nearly all Pareto efficient. This raises the question of whether, and if so how, identifying the DEA efficient decisions adds value beyond simply knowing the Pareto efficient decisions (among the FoIs). The DEA efficient decisions are Pareto efficient, while not all of the Pareto efficient decisions are DEA efficient. How are the DEA efficient decisions potentially useful, beyond the fact that they are Pareto optimal?

To answer this question we will look at a specific example and then be able to draw generalizations. Consider decision #28, which is not DEA efficient but is Pareto efficient. When we solve the LP associated with #28, E^*, the optimal value of E equals, as we see in Table 17.1, 0.9815. From the LP output we can discover the optimal values of the weights, w_j's, found by the LP. The non-zero weights (summing to 1) are:

Decision #	weight	z
14	0.0411	330
15	0.4428	330
27	0.2461	328
56	0.1123	325
78	0.0408	324
89	0.1169	324

$$Y \cdot w \geq Y(:, o) \tag{17.18}$$

We in fact have $Y \cdot w = 328 = Y(28)$, so the optimal mixture of decisions found by the LP produces an objective function value equal to that of decision 28. Further, again using the outputs of the LP, we find that

$$X \cdot w = \begin{bmatrix} 32.3910 \\ 32.3910 \\ 37.2987 \\ 25.5202 \\ 24.5386 \end{bmatrix} < X(28) = \begin{bmatrix} 33 \\ 33 \\ 38 \\ 26 \\ 25 \end{bmatrix} \tag{17.19}$$

What is true in this particular case will be true in general: the DEA optimization will judge a decision (relatively) inefficient if it can find a weighted mixture (weights in $(0, 1]$, that is $0 < w_j \leq 1$ for all j) of efficient decisions producing equal or better output and requiring less input for at least one resource and no more input from any resource.

Consider now the case in which the DEA inefficient decision (think: #28) is preferred to any of the other available decisions (in this case, to any of decisions #1,..., 27, 29,..., 100). This can happen when the DEA inefficient decision is Pareto efficient.

For any single instance, of course, it is not possible to undertake a weighted average of several decisions. If the problem is to be solved multiple times, however, DEA is telling us that we can *probabilistically mix* the DEA efficient decisions having non-zero weights and achieve in expectation the output of the inefficient decision while using fewer inputs than it requires.

Note further that even if the optimization decision is made only once, treating the DEA weights like a mixed strategy in game theory, we can choose a particular decision probabilistically, according to the weights, and *in expectation* achieve a superior performance, albeit with variance. This suggests an interesting policy for any organization that undertakes optimizations frequently:

1. For each optimization problem, characterize the FoIs and obtain by sampling a (heuristically) sound corpus of them.

2. Evaluate the corpus with respect to DEA efficiency.

3. As part of post-solution analysis, examine the corpus for attractive alternatives to any (exactly or heuristically) optimal decisions available.

4. Consider replacing any such attractive alternatives that are DEA inefficient with the corresponding mixture of DEA efficient decisions.

17.4 Discussion

DEA is interesting in our context for at least two reasons. First, it is a linear programming technique that is widely used for business analytics. Second, as we have shown in this chapter, DEA can be applied very usefully in our primary context of solution pluralism: Having generated a large number of decisions for purposes of post-solution analysis, DEA can be used to identify the most efficient decisions in the corpus. Further points arising:

1. Our discussion has been limited to feasible decisions of interest, yet the lessons and techniques are applicable to a corpus of infeasible decisions of interest (IoIs).

2. The *policy* of focusing on DEA efficient decisions during post-solution analysis is heuristically sound. Searching among the IoIs during post-solution analysis, one looks for attractive decisions that can be made feasible with investment to relax one or more constraints. Focusing on the DEA efficient decisions is a reasonable heuristic for minimizing the investment required to obtain additional output.

17.5 For Exploration

1. Referring to Table 17.1, page 251, we see that decision #1, which is optimal for the original problem, is (relatively) efficient, as we would expect it to be. Decision #3 is also efficient, while decision #2 is not. Discuss: Under what conditions might a decision maker prefer decision #2 to #1? #2 to #3? #3 to #2? #3 to #1?

2. Referring to Table 17.1, page 251, we see that decisions #25–31 all yield an objective function value of 328, yet only decisions #26 and #27 are efficient in the sense of the BCC DEA. What about Pareto efficiency? Considering only these 7 decisions, which are Pareto optimal?

17.6 For More Information

The BCC version(s) of DEA models were introduced in [7]. [30] and [34] contain a great deal of information about the concepts and uses of DEA.

#	z	X(1)	X(2)	X(3)	X(4)	X(5)	E	S(1)	S(2)	S(3)	S(4)	S(5)
1	336.0	35.0	32.0	38.0	27.0	32.0	1.0000	1.0	2.0	0.0	0.0	1.0
2	335.0	35.0	32.0	38.0	27.0	30.0	0.9984	1.0	2.0	0.0	0.0	3.0
3	334.0	33.0	27.0	38.0	27.0	28.0	1.0000	3.0	7.0	0.0	0.0	5.0
4	334.0	35.0	33.0	38.0	26.0	28.0	1.0000	1.0	1.0	0.0	1.0	5.0
5	334.0	35.0	32.0	38.0	27.0	30.0	0.9872	1.0	2.0	0.0	0.0	3.0
6	333.0	33.0	32.0	38.0	27.0	25.0	1.0000	3.0	2.0	0.0	0.0	8.0
7	332.0	34.0	34.0	38.0	27.0	28.0	0.9778	2.0	0.0	0.0	0.0	5.0
8	332.0	35.0	29.0	38.0	27.0	28.0	0.9823	1.0	5.0	0.0	0.0	5.0
9	332.0	35.0	32.0	38.0	27.0	33.0	0.9634	1.0	2.0	0.0	0.0	0.0
10	331.0	33.0	27.0	38.0	27.0	32.0	0.9810	3.0	7.0	0.0	0.0	1.0
11	331.0	35.0	33.0	38.0	26.0	32.0	0.9608	1.0	1.0	0.0	1.0	1.0
12	331.0	32.0	31.0	38.0	27.0	29.0	0.9785	4.0	3.0	0.0	0.0	4.0
13	330.0	34.0	32.0	38.0	26.0	26.0	0.9830	2.0	2.0	0.0	1.0	7.0
14	330.0	33.0	33.0	38.0	20.0	33.0	1.0000	3.0	1.0	0.0	7.0	0.0
15	330.0	34.0	33.0	38.0	25.0	24.0	1.0000	2.0	1.0	0.0	2.0	9.0
16	330.0	33.0	27.0	38.0	27.0	30.0	0.9775	3.0	7.0	0.0	0.0	3.0
17	330.0	35.0	33.0	38.0	26.0	30.0	0.9585	1.0	1.0	0.0	1.0	3.0
18	329.0	25.0	33.0	38.0	26.0	33.0	1.0000	11.0	1.0	0.0	1.0	0.0
19	329.0	33.0	27.0	38.0	27.0	30.0	0.9743	3.0	7.0	0.0	0.0	3.0
20	329.0	34.0	32.0	38.0	25.0	33.0	0.9578	2.0	2.0	0.0	2.0	0.0
21	329.0	35.0	29.0	38.0	27.0	32.0	0.9507	1.0	5.0	0.0	0.0	1.0
22	329.0	33.0	31.0	38.0	27.0	33.0	0.9540	3.0	3.0	0.0	0.0	0.0
23	329.0	34.0	34.0	38.0	27.0	32.0	0.9422	2.0	0.0	0.0	0.0	1.0
24	329.0	34.0	33.0	38.0	27.0	29.0	0.9522	2.0	1.0	0.0	0.0	4.0
25	328.0	34.0	28.0	38.0	27.0	24.0	0.9972	2.0	6.0	0.0	0.0	9.0
26	328.0	33.0	17.0	38.0	27.0	33.0	1.0000	3.0	17.0	0.0	0.0	0.0
27	328.0	33.0	31.0	38.0	27.0	21.0	1.0000	3.0	3.0	0.0	0.0	12.0
28	328.0	33.0	33.0	38.0	26.0	25.0	0.9815	3.0	1.0	0.0	1.0	8.0
29	328.0	34.0	32.0	38.0	26.0	29.0	0.9593	2.0	2.0	0.0	1.0	4.0
30	328.0	34.0	34.0	38.0	27.0	30.0	0.9404	2.0	0.0	0.0	0.0	3.0
31	328.0	35.0	29.0	38.0	27.0	30.0	0.9560	1.0	5.0	0.0	0.0	3.0
32	327.0	35.0	26.0	38.0	25.0	33.0	0.9752	1.0	8.0	0.0	2.0	0.0
33	327.0	34.0	32.0	38.0	26.0	26.0	0.9713	2.0	2.0	0.0	1.0	7.0
34	327.0	34.0	33.0	38.0	25.0	28.0	0.9658	2.0	1.0	0.0	2.0	5.0
35	327.0	24.0	34.0	38.0	27.0	33.0	1.0000	12.0	0.0	0.0	0.0	0.0
36	327.0	25.0	29.0	38.0	27.0	33.0	1.0000	11.0	5.0	0.0	0.0	0.0
37	327.0	33.0	27.0	38.0	27.0	33.0	0.9569	3.0	7.0	0.0	0.0	0.0
38	327.0	35.0	33.0	38.0	27.0	33.0	0.9240	1.0	1.0	0.0	0.0	0.0
39	327.0	31.0	33.0	38.0	27.0	29.0	0.9571	5.0	1.0	0.0	0.0	4.0
40	327.0	34.0	34.0	38.0	27.0	30.0	0.9357	2.0	0.0	0.0	0.0	3.0
41	327.0	35.0	33.0	38.0	27.0	33.0	0.9240	1.0	1.0	0.0	0.0	0.0
42	326.0	25.0	31.0	38.0	27.0	33.0	0.9933	11.0	3.0	0.0	0.0	0.0
43	326.0	35.0	33.0	38.0	27.0	21.0	0.9930	1.0	1.0	0.0	0.0	12.0
44	326.0	34.0	33.0	38.0	25.0	26.0	0.9733	2.0	1.0	0.0	2.0	7.0
45	326.0	35.0	26.0	38.0	25.0	31.0	0.9807	1.0	8.0	0.0	2.0	2.0
46	326.0	32.0	34.0	38.0	27.0	25.0	0.9686	4.0	0.0	0.0	0.0	8.0
47	326.0	33.0	34.0	38.0	21.0	28.0	1.0000	3.0	0.0	0.0	6.0	5.0
48	326.0	35.0	28.0	38.0	27.0	28.0	0.9646	1.0	6.0	0.0	0.0	5.0
49	326.0	33.0	29.0	38.0	27.0	25.0	0.9839	3.0	5.0	0.0	0.0	8.0
50	326.0	35.0	25.0	38.0	26.0	33.0	0.9682	1.0	9.0	0.0	1.0	0.0

TABLE 17.1: Objective function values (z), constraint left-hand-side values (X(j)), DEA BCC efficiency values (E), and slack values (S(j)) for fifty feasible decisions of interest (numbering 1–50) for a GAP with five agents.

#	z	X(1)	X(2)	X(3)	X(4)	X(5)	E	S(1)	S(2)	S(3)	S(4)	S(5)
51	326.0	35.0	32.0	32.0	27.0	33.0	1.0000	1.0	2.0	6.0	0.0	0.0
52	326.0	33.0	27.0	38.0	27.0	30.0	0.9647	3.0	7.0	0.0	0.0	3.0
53	326.0	34.0	32.0	38.0	24.0	29.0	0.9707	2.0	2.0	0.0	3.0	4.0
54	326.0	30.0	33.0	38.0	26.0	29.0	0.9665	6.0	1.0	0.0	1.0	4.0
55	326.0	30.0	34.0	38.0	25.0	33.0	0.9616	6.0	0.0	0.0	2.0	0.0
56	325.0	24.0	33.0	38.0	25.0	29.0	1.0000	12.0	1.0	0.0	2.0	4.0
57	325.0	33.0	20.0	38.0	26.0	31.0	1.0000	3.0	14.0	0.0	1.0	2.0
58	325.0	35.0	26.0	38.0	25.0	31.0	0.9792	1.0	8.0	0.0	2.0	2.0
59	325.0	35.0	21.0	38.0	27.0	33.0	0.9775	1.0	13.0	0.0	0.0	0.0
60	325.0	33.0	31.0	38.0	27.0	25.0	0.9722	3.0	3.0	0.0	0.0	8.0
61	325.0	24.0	32.0	38.0	26.0	31.0	1.0000	12.0	2.0	0.0	1.0	2.0
62	325.0	25.0	34.0	38.0	27.0	28.0	1.0000	11.0	0.0	0.0	0.0	5.0
63	325.0	25.0	31.0	38.0	27.0	31.0	0.9954	11.0	3.0	0.0	0.0	2.0
64	325.0	33.0	31.0	38.0	21.0	31.0	1.0000	3.0	3.0	0.0	6.0	2.0
65	325.0	31.0	32.0	38.0	27.0	31.0	0.9458	5.0	2.0	0.0	0.0	2.0
66	325.0	32.0	27.0	38.0	26.0	26.0	0.9985	4.0	7.0	0.0	1.0	7.0
67	325.0	31.0	31.0	38.0	25.0	33.0	0.9690	5.0	3.0	0.0	2.0	0.0
68	325.0	35.0	32.0	36.0	25.0	33.0	0.9697	1.0	2.0	2.0	2.0	0.0
69	325.0	33.0	34.0	38.0	26.0	33.0	0.9361	3.0	0.0	0.0	1.0	0.0
70	325.0	31.0	34.0	38.0	27.0	33.0	0.9372	5.0	0.0	0.0	0.0	0.0
71	325.0	31.0	32.0	38.0	24.0	33.0	0.9747	5.0	2.0	0.0	3.0	0.0
72	325.0	32.0	31.0	38.0	27.0	30.0	0.9493	4.0	3.0	0.0	0.0	3.0
73	325.0	35.0	32.0	32.0	27.0	33.0	1.0000	1.0	2.0	6.0	0.0	0.0
74	325.0	35.0	32.0	32.0	27.0	33.0	1.0000	1.0	2.0	6.0	0.0	0.0
75	325.0	34.0	34.0	38.0	27.0	33.0	0.9219	2.0	0.0	0.0	0.0	0.0
76	324.0	35.0	33.0	38.0	25.0	20.0	1.0000	1.0	1.0	0.0	2.0	13.0
77	324.0	35.0	32.0	38.0	26.0	22.0	0.9862	1.0	2.0	0.0	1.0	11.0
78	324.0	32.0	33.0	38.0	25.0	21.0	1.0000	4.0	1.0	0.0	2.0	12.0
79	324.0	35.0	24.0	38.0	27.0	28.0	0.9907	1.0	10.0	0.0	0.0	5.0
80	324.0	33.0	26.0	38.0	25.0	26.0	1.0000	3.0	8.0	0.0	2.0	7.0
81	324.0	25.0	32.0	38.0	26.0	31.0	0.9912	11.0	2.0	0.0	1.0	2.0
82	324.0	33.0	25.0	38.0	26.0	28.0	0.9894	3.0	9.0	0.0	1.0	5.0
83	324.0	32.0	31.0	38.0	27.0	33.0	0.9443	4.0	3.0	0.0	0.0	0.0
84	324.0	25.0	31.0	38.0	27.0	31.0	0.9954	11.0	3.0	0.0	0.0	2.0
85	324.0	33.0	34.0	38.0	27.0	20.0	1.0000	3.0	0.0	0.0	0.0	13.0
86	324.0	32.0	31.0	38.0	27.0	30.0	0.9485	4.0	3.0	0.0	0.0	3.0
87	324.0	33.0	31.0	38.0	27.0	23.0	0.9859	3.0	3.0	0.0	0.0	10.0
88	324.0	33.0	27.0	36.0	25.0	31.0	0.9982	3.0	7.0	2.0	2.0	2.0
89	324.0	33.0	32.0	32.0	27.0	28.0	1.0000	3.0	2.0	6.0	0.0	5.0
90	324.0	35.0	21.0	38.0	27.0	31.0	0.9868	1.0	13.0	0.0	0.0	2.0
91	324.0	27.0	33.0	38.0	27.0	33.0	0.9659	9.0	1.0	0.0	0.0	0.0
92	324.0	30.0	32.0	38.0	24.0	31.0	0.9806	6.0	2.0	0.0	3.0	2.0
93	324.0	32.0	27.0	38.0	25.0	33.0	0.9811	4.0	7.0	0.0	2.0	0.0
94	324.0	34.0	31.0	38.0	26.0	33.0	0.9432	2.0	3.0	0.0	1.0	0.0
95	324.0	34.0	28.0	38.0	27.0	26.0	0.9778	2.0	6.0	0.0	0.0	7.0
96	324.0	33.0	32.0	38.0	27.0	33.0	0.9350	3.0	2.0	0.0	0.0	0.0
97	324.0	34.0	32.0	38.0	25.0	33.0	0.9485	2.0	2.0	0.0	2.0	0.0
98	324.0	33.0	32.0	38.0	25.0	31.0	0.9555	3.0	2.0	0.0	2.0	2.0
99	324.0	31.0	32.0	38.0	26.0	31.0	0.9552	5.0	2.0	0.0	1.0	2.0
100	324.0	34.0	33.0	38.0	27.0	32.0	0.9261	2.0	1.0	0.0	0.0	1.0

TABLE 17.2: Objective function values (z), constraint left-hand-side values $(X(j))$, DEA BCC efficiency values (E), and slack values $(S(j))$ for fifty feasible decisions of interest (numbering 51–100) for a GAP with five agents.

Chapter 18

Redistricting: A Case Study in Zone Design

18.1   Introduction

Zone design, also known as *territory design,* is the problem of grouping small geographic areas, called *basic areas* or *areal units*, into larger geographic clusters, called *zones* or *territories* or *districts*, in such a way that the resulting clusters are optimal, or at least acceptable, according to the relevant planning and design criteria.

For example, a geographical region may be partitioned into a large number of subregions and it may be desired to aggregate each of the n subregions into k sales districts, each overseen by a single manager. The value function involved might be the total travel time expected for the agents; we would like to form k sales districts (territories, zones) while minimizing the amount of travel required to service them. There may also be a number of constraints on our designs, such as a requirement of rough equality in likely sales among the k zones.

Democracies routinely face zone design problems when they undertake *redistricting*, the revising of the boundaries of legislative districts in response to population changes. Redistricting is an especially important activity in the United States, which requires it after every decennial census in order to maintain rough population equality among legislative electoral districts. After every census, the seats in the House of Representatives are *reapportioned,* that is, with a fixed number of seats in the House (currently 435), each state is given a number of seats proportional to its percentage of the total population in the country. More populous states are assigned more seats than less populous states, with the constraint that every state gets at least one representative in the House.

Once reapportionment is completed by the Bureau of the Census, the individual states conduct redistricting for all legislative districts using their individually-specified decision procedures. The process is highly contentious and consequential, as captured in the title of a recent book about it, *Redistricting: The Most Political Activity in America* [24]. Redistricting that significantly favors, or harms, one group over another is so common that it has a frequently invoked name, *gerrymandering*, after Massachusetts Governor Elbridge Gerry's plan in 1812.

Multiple forms of gerrymandering have been identified. The problem of how best to do redistricting from the perspective of social welfare is exceedingly, even deliciously, complex and contested. It is an important open social problem for our time. In this chapter we will focus primarily on a simplified, stylized version of the redistricting problem that

- Accords with the bulk of the traditional modeling for it in operations research,

- Will illuminate zone design problems in other areas, outside of politics and government, and

- Serves as a good starting point for discussion of more realistic approaches to the problem.

18.2   The Basic Redistricting Formulation

The basic redistricting formulation proposes a constrained optimization representation of the problem, in which a measure of *compactness* is optimized (minimized or maximized, depending upon the measure), subject to constraints on *population equality* and *contiguity*. We will briefly discuss each of these elements before presenting algorithmic approaches to solving the problem. See [67, 128, 155] for relevant introductions and overviews of the traditional operations research literature on the basic redistricting problem. We proceed within the U.S. context. The lessons generalize.

The requirement for approximate population equality among the legislative districts in a given plan was imposed in the United States by a series of court decisions beginning in the 1960s. See [24, 155] for historical reviews. Absolute population equality is, of course, generally impossible and even undesirable because other values, such as compactness, will conflict and in any event the census population figures are snapshots of continually changing distributions. Redistricting exercises normally set a threshold for deviations from population equality and make it a hard constraint. Given a total population count of p and k districts to be formed, then at strict population equality every district has $\frac{p}{k}$ members. Typical thresholds are 1% for Congressional districting and 5% for all other districts. With a hard threshold of $t \times 100$ percent, each of the k districts will have a population $p(k)$ within a range, $(1-t) \times \frac{p}{k} \leq p(k) \leq (1+t) \times \frac{p}{k}$.

A district is contiguous if for every pair of locations in the district it is possible to travel from one to the other without leaving the district. This seemingly simple and clear idea turns out to be somewhat problematic in practice. What about islands and bodies of water, such as rivers and bays? Then there is point contiguity, when two parts of a district meet at a single point. Consider the U.S. states of Utah, Colorado, Arizona, and New Mexico, which meet at a single geographic point. Could Arizona and Colorado form a contiguous district? What about Utah and New Mexico? Opinions differ but in fact political districting normally counts point contiguity as permitted. In any event, at the federal level in the United States there is no requirement for contiguity. The requirements in place are there because of individual state laws and tradition. Monmonier [126] gives examples of non-contiguous electoral districts that have been maintained with equanimity, and notes that there may well be cases in which non-contiguous districting is in the public's interest. An example might be an elongated district along a highway on either side of which we find agricultural communities. It might make good sense to unite the hinterlands on both sides in a common district, while keeping the corridor united in a different district. The point applies at least as well for non-legislative zone design problems. We will stick to a broad concept of contiguity here.

The third basic requirement for districting (legislative redistricting, zone design, etc.) is that the districts be compact. Intuitively, a compact district is one that resembles a circle or square or other regular geometric object. Social scientists have devised more than two dozen measures of compactness, all of which have problems [126, 162]. There is as yet no generally agreed upon definition of compactness for zone design. U.S. federal law does not mandate compactness, but many of the states do. The U.S. Supreme Court has been known to reject redistricting plans because of failure of compactness. This is without a definition of compactness, which has long been compared to pornography: you can't define it, but you know it when you see it. Figure 18.1 is a case in point. Districts 5 and 7 are noticeably gerrymandered and "unnatural." Searching the Web with any standard search engine will turn up worse cases and very many of them, at all levels of political organization and throughout the United States.

Critics have alleged that the Supreme Court rejects non-compact districts only when they are tailored to elect minorities and leaves in place equally non-compact districts designed to protect other interests. Further, whether and when non-compactness is a bad thing is itself disputed, with some arguing that there may often be excellent reasons to eschew compactness and even contiguity. These important issues are beyond our scope in this book; we leave them to the reader for exploration.

The matter of compactness is very much up the air (at least in the U.S. electoral system). Because we have in mind the more general zone design problem, which is less fraught and is very interesting on its own, we shall work with simple concepts of compactness that are easily computable with available data. We start with that in the next section.

18.3 Representing and Formulating the Problem

It will be helpful for the discussion to proceed in the context of a specific example. We shall use the case of redistricting for the City of Philadelphia City Council after the 2010 census. For background information see [32, 33, 65, 66, 67, 105]. In brief, we wish to allocate 66 wards into 10 councilmanic districts. The number of possibilities is a Sterling number of the second kind, or $\approx 2.7 \times 10^{59}$.[1] This is a surprisingly large search space and it grows very quickly with the number of areal units. The problem's complexity certainly suggests an approach based on heuristics and so it is well suited as an example in this book, which emphasizes the use of heuristics (especially metaheuristics) for business analytics. We will find, however, good uses of exact as well as heuristic solvers.

The necessary files, or data sources, for dealing with the simple version of the problem we are now considering are available at the book's Web site, in the *teamfred/* directory. They are

1. *unitPopulation.csv*

2. *unitXYCoordinates.csv*

3. *warddistancetablelabeled.txt*

   (This is a tab-delimited file, created by *makewarddistancetable.py,* which is also in the directory.)

4. *unitAdjacencyMatrix.csv*

---

[1] Equal to 272946434461341722642441137907806754803339290349146989479015 according to MATLAB.

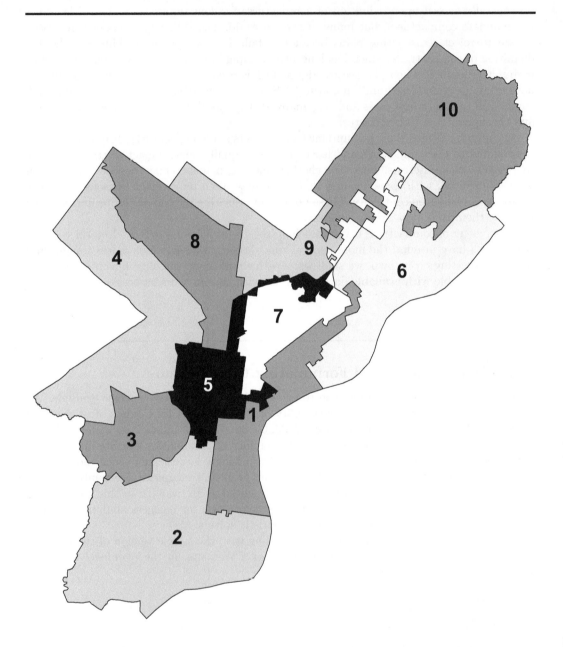

FIGURE 18.1: Philadelphia City Council districts after the 2000 census.

Each contains a table of data in comma separated values (CSV) format (except the third table). In the first file, *unitPopulation.csv*, the data table has 66 rows and 2 columns. The first column holds the IDs of the 66 Philadelphia wards, numbered 1–66. The second column holds the populations of the wards, as determined by the 2010 U.S. census. The second table, *unitXYCoordinates.csv*, is a 66×3 table in which the first column holds the ward (basic unit) IDs (1–66), the second column holds the X coordinate value for the ward's centroid, and the third column holds the Y value. The coordinate values are measured in distances in miles from a reference point outside of Philadelphia.

The third table, *warddistancetablelabeled.txt*, is a 66×66 table in which the entries are the distances in miles from the center of the row district (1–66) to the center of the column district (1–66). The table is symmetric and has zeros on the diagonal. It is derived from the second table, *unitXYCoordinates.csv*. Its rows and columns are headed with ward IDs in the form `w1, w2,..., w66`.

The fourth table, *unitAdjacencyMatrix.csv*, is also a 66×66 table. Its entries are equal to 0 or 1. An entry is 1 if the ward of its row index (1–66) is directly adjacent to the ward of its column index (1–66). This table has 1s on its diagonal; wards are treated as adjacent to themselves. These three data tables and the files holding them are all we need to proceed with a treatment of our basic zone design problem.

We represent a solution to the problem as a vector of length 66, with entries that are integers between 1 and 10. The value of an entry at an index indicates the district to which the ward at that index is assigned.

We score the compactness of a solution using a *centroid* measure. For each of the 10 districts in a given solution, we determine the centroid from the centroids of its constituent wards (the wards are our basic units). These centroids are given in the *unitXYCoordinates.csv* file. Given the centroids ($X - Y$ coordinates) for the wards within a given district, we find the coordinates of the centroid of the district by averaging the X and Y coordinates of its member wards. The compactness score for the district is the sum of the squared distances from the district's centroid to the centroids of its member wards. The compactness of a districting plan is the sum of the compactness scores of its ten districts. Determined in this way, we wish to minimize compactness and make it our objective function.

We treat population equality as a constraint. Given a solution and a corresponding districting plan we require that every district in the plan have a population within 5% of the equal-population number. From the *unitPopulation.csv* file we find that the total population of Philadelphia in the census is 1,528,217. Dividing that number into 10 districts, we have an ideal population per district of 152,822. Allowing a 5% deviation at max, we count solutions having a district with fewer than 145,180 or more than 160,462 people as infeasible.

We can determine the compactness of the plan corresponding to a given solution as follows. For each district we create an adjacency matrix for its constituents, $\mathbf{D}$. We do this by extracting information from the data table (matrix) created from *unitAdjacencyMatrix.csv*. We then determine the number of rows (or columns) in the matrix, that is, the number of wards (basic units), w, for the district in question. Finally, we compute $\mathbf{D}^{w-1}$ and pick an arbitrary row or column. If it contains no zeros, the district is contiguous; otherwise it is not. A plan is contiguous if all of its districts are contiguous. This is a fairly expensive step computationally, but it is what a group calling itself Team Fred used successfully in its project [32, 33, 65, 66, 67, 105]. Macmillan [120] describes two methods that can be expected to be speedier, but they are less general.

In short, we may state our optimization model as follows in English. We seek solutions that minimize compactness (as scored with the centroid method described above), subject to feasibility on population (every district in a plan must have a population within 5% of the ideal value), and contiguity (as determined from the ward adjacency values in *unitAdjacencyMatrix.csv*).

Having stated the problem clearly, how might we solve it? We turn to that beginning in the next section.

18.4 Initial Forays for Discovering Good Districting Plans

A good districting plan, we are assuming, is strong on compactness (our objective function), and both is population balanced and has contiguous districts. For starters we might try a manual approach to districting. To do this, one typically uses visualization software—such as a GIS (geographic information system)—interactively to create contiguous districts. There are normally a very large number of ways to form districting plans, so the trick is to use well-informed judgment to find plans that are reasonably good with regard to population balance and compactness. Figure 18.2, for example, presents a map of Philadelphia and its 66 wards. It would be possible, if tedious, to assemble from it manually one or more districting plans that maintain contiguity. The figure does not contain information on population (or for that matter really anything else). Population information is contained in the *unitPopulation.csv* file, mentioned previously. One could proceed manually, as it were, and search for a good plan. We do not recommend this except when absolutely necessary. We trust the reader will agree, so we turn now to automated methods.

Our first automated method follows [67] in using a p-median or facility location formulation of the districting problem. Without contiguity or population constraints, we begin with the model formulation in Figure 18.3.

Points arising:

1. $i \in I$ indexes the set of basic units, Philadelphia wards in our present example.

2. j is an alias for i, allowing us to use $d(i,j)$ as the distance from unit (ward) i to unit (ward) j.

3. The decision variables are the $X(i,j)$s and the $Y(j)$s. We want $Y(j) = 1$ if and only if ward j is the center or *hub* (our preferred term) of one of the districts. Further, we want $X(i,j) = 1$ only if ward j is a hub ward. This is achieved via constraints (18.3).

4. Constraints (18.2) in conjunction with (18.5) guarantee that every basic unit (ward in our example) is assigned to exactly one district.

5. Constraint (18.4) guarantees that exactly K districts will be created in any optimal solution.

6. The objective function minimizes the sum of the distances from each areal unit to its assigned district hub areal unit. As such it is a measure of compactness. In this formulation of the districting problem, then, we seek to find a solution that minimizes this compactness measure subject to constraints on the number of districts and unique assignments.

If we solve the p-median model of Figure 18.3 we obtain the contiguous, compact plan shown in Figure 18.4. We obtained this solution using the GAMS formulation of Figure 18.5.[2] GAMS is a modern algebraic modeling system. It is not within the scope of this book to teach GAMS in any detail. The reader, with a little work, should be able to see

---

[2]In the *philly2010/GAMS/* on the book's Web site, the reader will find the files: *pmedian.gms* (the GAMS code shown in Figure 18.5), *warddistancetablelabeled.txt* (the table of ward–ward distances; see

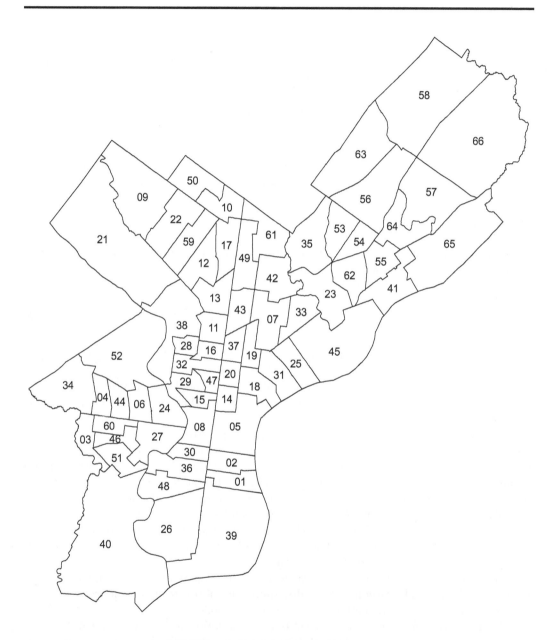

FIGURE 18.2: Philadelphia's 66 wards.

$$\text{Minimize } z = \sum_{i \in I} \sum_{j \in I} d(i,j) X(i,j) \tag{18.1}$$

Subject to:

$$\sum_{j \in I} X(i,j) = 1, \quad \forall i \in I \tag{18.2}$$

$$X(i,j) - Y(j) \le 0, \quad \forall i \in I, \forall j \in I \tag{18.3}$$

$$\sum_{j \in I} Y(j) = K \tag{18.4}$$

$$X(i,j), Y(j) \in \{0,1\}, \quad \forall i \in I, \forall j \in I \tag{18.5}$$

FIGURE 18.3: Districting problem formulated as a p-median problem.

the correspondence between Figures 18.3 and 18.5. With little effort it should be possible to duplicate our results.

Although the districting plan of Figure 18.4 is visually pleasing, it is not population balanced. To achieve that, we need to augment the p-median model of Figure 18.3 to add population balance constraints:

$$\text{POPLOW} \le \sum_{i \in I} P_i X(i,j) \le \text{POPHIGH}, \quad \forall j \in J \tag{18.6}$$

P_i is the population of areal unit (ward) i. We will set POPLOW to 95% of the equal population value, and POPHIGH to 105% of it.

Figure 18.6 presents the GAMS encoding of the augmented model. When we solve the model to optimality and display it in a GIS, Figure 18.7 is what we get. Notice that district 1 is not contiguous.

Our p-median model with population constraints is fairly easy to solve as these things go. Unfortunately it fails to find an optimal solution with contiguity, so we need to add contiguity constraints. The problem this presents is that there will be very many constraints required for a mathematical programming formulation. In consequence, the formulation will not scale very well. One thing we can, and will, do is to use a mathematical programming formulation that saves on constraints—and so should scale better—by fixing the district centers before optimization. What we will do is to use the centers found by the p-median model with population constraints. They are: w2, w19, w22, w32, w48, w49, w53, w57, w60, and w62 (that is, ward 2, ward 19, and so on).[3]

With these available to us we can use a model, in Figure 18.8, based upon Shirabe's clever formulation [144]. Although it requires quite a lot of constraints to enforce contiguity, it represents a significant reduction from previous methods.

The model is easily explained. I is a set of the areal units, here the 66 wards of Philadelphia, w1, w2, ..., w66. K, a subset of I, is the set of given district hubs, obtained from

---

philly2010/READMEandFileInfo/REDME.txt on the book's Web site for how this file was created), *pmediancenters.txt* (an output file that lists the 10 wards found as values of $Y(j)$ that serve as district centers), and *pmedianassignments.txt* (an output file that maps each ward to one of the 10 ward centers).

[3]Alternatively, we could use the centers found by the p-median model without population constraints: w1, w15, w19, w22, w28, w48, w49, w57, w60, and w62. We leave this as an exercise for the reader.

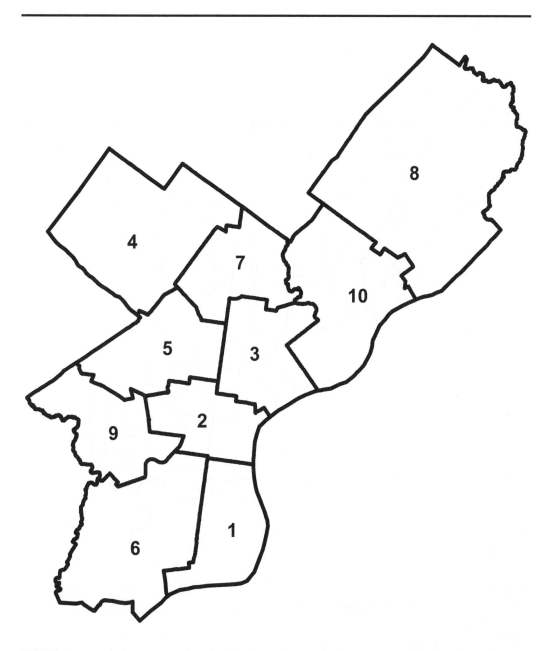

FIGURE 18.4: A districting plan for Philadelphia using the *p*-median model of Figure 18.3.

```
Sets
i wards / w1*w66 / ;

alias(i,j)
Table d(i,j) distance miles
$include "warddistancetablelabeled.txt" ;

Scalar k number of districts to design /10/ ;

Variables
   x(i,j)  indicates whether ward i is assigned or not to district center
           ward j
   y(j)  indicates whether ward j is a district center ward or not
   z     total distance in miles;

Binary Variable x ;
Binary Variable y ;

Equations
   cost define objective function
   uniqueassignment each ward is assigned to exactly one center
   uniquecenters   each district has one center
   numwards        the number of districts to find ;

cost .. z =e= sum((i,j), (d(i,j) * x(i,j))) ;

uniqueassignment(i) .. sum(j, x(i,j)) =e= 1;

uniquecenters(i,j) .. x(i,j) - y(j) =l= 0 ;

numwards .. sum(j, y(j)) =e= k ;

Model pmedian /all/ ;

Solve pmedian using mip minimizing z ;

*  Export results to files:
file output /'pmediancenters.txt'/ ;
put output ;
loop(j,
if(y.l(j) = 1.0,
put j.tl/;););
putclose;
file xijoutput /'pmedianassignments.txt'/;
put xijoutput ;
loop((i,j),
if (x.l(i,j) = 1.0,
put i.tl, j.tl/;););
putclose;
```

FIGURE 18.5: GAMS listing for *p*-median model of Figure 18.3.

```
Sets
i wards / w1*w66 / ;
alias(i,j)
Table d(i,j) distance miles
$include "warddistancetablelabeled.txt" ;
Scalar k number of districts to design /10/ ;
Scalar poplow minimum acceptable population (0.95 * 1528217 * 0. 10) /145181/ ;
Scalar pophigh maximum accpetable population (1.05 * 1528217) *0.10) /160462/ ;
Parameters
pop(i) population of ward i in 2010 census
$include "population.txt" ;
Variables
x(i,j)  indicates whether ward i is assigned or not to district center ward j
y(j)  indicates whether ward j is a district center ward or not
z       total distance in miles;
Binary Variable x ;
Binary Variable y ;
Equations
cost define objective function
uniqueassignment each ward is assigned to exactly one center
uniquecenters constraint that each district has one center
numwards        the number of districts to find
populationlow minimum population requirement
populationhigh maximum population requirement ;

cost .. z =e= sum((i,j), (d(i,j) * x(i,j))) ;

uniqueassignment(i) .. sum(j, x(i,j)) =e= 1;

uniquecenters(i,j) .. x(i,j) - y(j) =l= 0 ;

numwards .. sum(j, y(j)) =e= k ;

populationlow(j)  .. (sum(i, (pop(i) * x(i,j)))) =g= poplow * y(j)  ;

populationhigh(j) .. sum(i, (pop(i) * x(i,j))) =l= pophigh;

Model pmedian /all/ ;
Solve pmedian using mip minimizing z ;
*  Export results to files:
[Removed for display.]
```

FIGURE 18.6: GAMS listing for *p*-median model of Figure 18.3 with population constraints of expression (18.6) added.

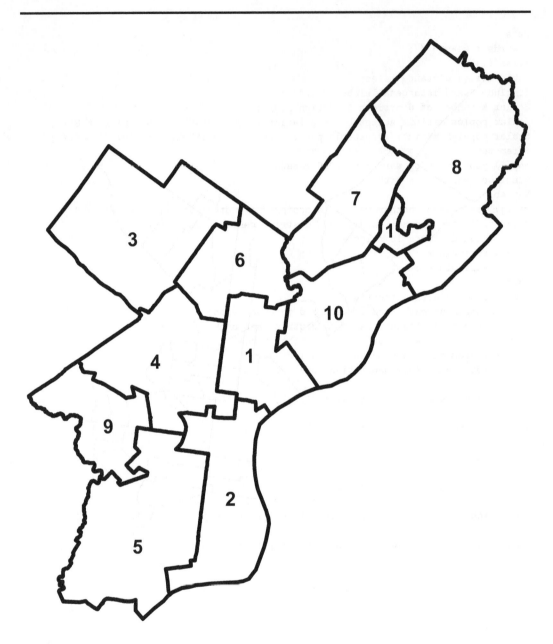

FIGURE 18.7: A districting plan for Philadelphia using the p-median model with population constraints of $\pm 5\%$. District 1 is not contiguous.

$$\text{Minimize } z = \sum_{i \in I} \sum_{k \in K} d(i,k) x(i,k) \tag{18.7}$$

Subject to:

$$\sum_{i \in I} p(i) x(i,k) \geq POPLOW, \quad \forall k \in K \tag{18.8}$$

$$\sum_{i \in I} p(i) x(i,k) \leq POPHIGH, \quad \forall k \in K \tag{18.9}$$

$$x(k,k) = 1, \quad \forall k \in K \tag{18.10}$$

$$\sum_{k \in K} x(i,k) = 1, \quad \forall i \in I - K \tag{18.11}$$

$$y(k,i,j) = 0, \quad \forall k \in K, \forall i \in I, \forall j \in K \tag{18.12}$$

$$y(i,j,k) = 0, \quad \forall i \in I, \forall j \in K, \forall k \in K, \forall j \neq k \tag{18.13}$$

$$\sum_{j|a(i,j)} y(i,j,k) - \sum_{j|a(j,i)} y(j,i,k) = x(i,k), \quad \forall k \in K, \forall i \in I - K \tag{18.14}$$

$$x(i,k) \in \{0,1\}, \quad \forall i \in I, \forall k \in K \tag{18.15}$$

$$y(i,j,k) \geq 0, \quad \forall i \in I, \forall j \in I, \forall k \in K \tag{18.16}$$

FIGURE 18.8: Districting problem formulated as a p-median problem with population and contiguity constraints, and known centers. After [144, page 1059].

---

outside the model (in our case, from the p-median model above), here the wards w2, w19, w22, w32, w48, w49, w53, w57, w60, and w6. Regarding model parameters,

1. The distance table, d, has elements $d(i,k)$ whose values are the distances from areal unit i to areal unit k, where $k \in K$ and $i \in I$.

2. The populations table, p, has elements $p(i)$ whose values are the populations of the areal units i, where $i \in I$.

There are two kinds of decision variables:

1. $x(i,k)$ is intended to be 1 if areal unit i is assigned to the hub k, which is also an areal unit. In the case to hand, if ward i is assigned to district k, then $x(i,k) = 1$; otherwise $x(i,k) = 0$ (under our intended interpretation, to be enforced by the model).

2. $y(i,j,k)$ is intended to be the nonnegative flow from unit i to unit j when both i and j are assigned to district k. Contiguity will be enforced by the constraints. The flow is a modeling concept, introduced explicitly for the purpose of enforcing contiguity. It is not intended to represent anything tangible in the real world. This is a common "modeling trick" in integer programming and indeed it often constitutes a savvy move. We shall see how it works when we discuss the constraints.

3. *POPLOW* is the minimum population permitted for a valid district. In our case this is 95% of the total population of Philadelphia, divided by 10, the number of districts.

4. *POPHIGH* is the maximum population permitted for a valid district. In our case this is 105% of the total population of Philadelphia, divided by 10.

Now to the model's equations and expressions.

1. Expression (18.7) is the *cost* equation and contains our objective function. For a given solution, the objective function sums the values of the distances from each areal unit to the areal unit assigned to it in the solution and given as the hub of its district. In short, we sum up the $d(i, k)$ values whenever $x(i, k) = 1$ in the solution. This is a value we wish to minimize.

2. Expressions (18.8) and (18.9) are the *populationlow* and *populationhigh* equations respectively. They constitute the constraints on the district populations.

3. Expression (18.10) is the *centerselfassignment* equation. It constitutes the constraint for each areal unit $k \in K$ serving as a district hub that it is assigned to itself in any feasible solution.

4. Expression (18.11) is the *exactlyone* equation. It constitutes the constraint for each areal unit $i \in I - K$ (the areal units not serving as district hubs) that it is assigned to exactly one district hub $k \in K$ in any feasible solution.

5. Expression (18.12) is the *nohubout* equation. It requires for feasibility that the flow from any hub is 0. Hubs receive flow as indicated by $y(i, j, k)$ values but they are never a source of flow.

6. Expression (18.13) is the *noothercenter* equation. It requires no flow (realized with $y(i, j, k)$) from a non-hub node to any hub node except to the one it is assigned with $x(i, k)$.

7. Expression (18.14) is the *flowbalance* equation. Its job is to enforce contiguity. To see how this works, consider a simple example with four nodes or areal units: a, b, n, and h. Assume for the sake of the example that the adjacent pairs are (a, b), (a, n), and (b, h), and assume that adjacency is symmetric: (b, a), (n, a), and (h, b). Let $y(i, j, k)$ represent the directed flow from i to j in the context of district k. Suppose that a, b, and h are assigned to district k, so we have $x(a, k) = x(b, k) = x(h, k) = 1$ and $x(n, k) = 0$ because n is not assigned to hub node k, $x(n, k') = 1$ for another district with hub k' and in virtue of the constraints of expression (18.11). Expression (18.14) then requires:

$$[y(a, b, k) + y(a, n, k)] - [y(b, a, k) + y(n, a, k)] = 1 \qquad (18.17)$$

$$[y(b, a, k) + y(b, h, k)] - [y(a, b, k) + y(h, b, k)] = 1 \qquad (18.18)$$

$$[y(n, a, k)] - [y(a, n, k)] = 0 \qquad (18.19)$$

$$[y(h, b, k)] - [y(b, h, k)] = 1 \qquad (18.20)$$

If we add these four equations we reach the result of

$$0 = 3$$

Assuming, as we have that none of the nodes is a hub, this illustrates the general point that expression (18.14) cannot be satisfied.

Now, let us assume that indeed node h is the hub k. This causes expression (18.20) to drop out, according to the mandate of expression (18.14). When we add the three remaining equations we reach the satisfying result:

$$[y(b, h, k)] - [y(h, b, k)] = 2 \qquad (18.21)$$

But from expression (18.12) we require that $[y(h, b, k)] = 0$. Having found that $y(b, h, k) = 2$, we can proceed to solve for the remaining values (although there is no unique solution, we only need to find one).

The example, we submit, has quite general applicability. Extrapolating in a way we hope is more or less obvious to the reader, what this example teaches us is how the model constraints in question do in fact enforce contiguity. We do note that one key possibility has been left out of the example, the possibility that one of the nodes, other than h, is in fact a hub for a different district. This would block some flow and quite possibly lead to a specious feasibility assessment. The constraints of expression (18.13) prevent this from happening.

8. Expression (18.15) requires the $x(i, j)$ variables to be 0–1 binary variables.

9. Expression (18.16) requires the $y(i, j, k)$ variables to be nonnegative. Note that they are not required to be integers.

Figures 18.9 and 18.10 present the model formulation in GAMS. Given our description and a basic understanding of GAMS, the reader should be able to take in the correspondences. Figure 18.11 shows a map of the optimal solution found by the model.

18.5   Solving a Related Solution Pluralism Problem

The optimal solution shown in Figure 18.11 is only optimal for the specific model actually optimized. The model we did optimize, shown in Figure 18.8 on page 265, is itself only a heuristic for the originating real world problem, for we arrived at the ten hub wards given to it by using the model of Figure 18.3 (augmented with the population balance constraints, expression (18.6), pages 260 and 260, respectively). This model neglects contiguity constraints and in fact produced a noncontiguous district at its optimum. These considerations alone would suggest that we define a set of solutions of interest, perhaps based on near neighbors of the ten hub wards, and seek to solve an associated solution pluralism problem.

More fundamental reasons beckon for embracing solution pluralism. The objective function in the model with contiguity constraints (Figure 18.8, page 265) implements one notion of compactness. There are many others. Perhaps some of these will give more satisfactory results. Much more important is that nearly every redistricting plan will have attributes of interest that are not represented in any of the three optimization models we have discussed.

This is certainly the case for Philadelphia, where neighborhood integrity, socioeconomic mixtures, communities of interest (such as LGBTx communities), and ethnic balance and distribution are intensely monitored and contested in the political realm. For example, an important driver behind the gerrymandering of district 7 from the 2000 census (Figure 18.1, page 256) appears to have been the intention of preventing the election of an Hispanic representative. As it happened, an Hispanic won anyway in the 2008 election. Moreover, she was on the City Council and up for re-election, anticipated to be difficult, during the deliberations on redistricting in 2012. Complicating all this is the fact that for the first

```
Sets
        i basic units (wards) / w1*w66 / ;
Sets
      kk(i) basic units serving as district centers
$include "districtcenters.txt" ;
alias(i,ii)
alias(i,k)
alias(i,j)
Sets
        a(i,ii) adjacent wards excludes self-adjacency
$include "gamsadjacency.txt"    ;
Table d(i,ii) distance miles
$include "warddistancetablelabeled.txt" ;

Parameter numdistricts /10/ ;
Parameters
pop(i) population of unit i
$include "population.txt" ;
totpop total population;
totpop = sum(i, pop(i));
Parameter alpha permitted population variance as a percentage /0.05/;
Parameter poplow minimum acceptable population;
poplow = (1 - alpha) * totpop / numdistricts ;
Parameter pophigh maximum acceptable populaton;
pophigh = (1 + alpha) * totpop / numdistricts ;

Variables

   x(i,ii)  indicates whether unit i is assigned to district center unit ii
   y(i,ii,k) flow from i to ii when both assigned to k
   z      total distance in miles;

Binary Variable x ;
Positive Variable y;
```

FIGURE 18.9: GAMS code for districting with population and contiguity constraints and known centers, part 1.

Equations

```
    cost define objective function
    exactlyone each unit is assigned to exactly one district
    populationhigh maximum population requirement
    populationlow minimum population requirement
    flowbalance if you are 1 you are connected
    nohubout no flow out of hubs
    noothercenter no flow to other than your assigned center
    centerselfassignment center units assigned to themselves
    ;

cost .. z =e= sum(i, sum(k$(kk(k)), (d(i,k) * x(i,k)))) ;

populationhigh(k)$(kk(k)) .. sum(i,pop(i)*x(i,k)) =l= pophigh;

populationlow(k)$(kk(k)) .. sum(i, pop(i)*x(i,k)) =g= poplow;

exactlyone(i)$(not kk(i)) .. sum(k$(kk(k)), x(i,k)) =e= 1 ;

flowbalance(k,i)$(kk(k) and not kk(i)) ..
  sum(ii$a(i,ii), y(i,ii,k)) - sum(ii$a(ii,i), y(ii,i,k)) =e= x(i,k) ;

centerselfassignment(k)$kk(k) .. x(k,k) =e= 1 ;

nohubout(i,j,k)$(kk(k) and kk(j)) .. y(k,i,j) =e= 0;

noothercenter(i,j,k)$(kk(k) and kk(j)) .. y(i,j,k)$(ord(j) ne ord(k)) =e= 0;

Model shirabe1 /all/ ;

Solve shirabe1 using mip minimizing z ;

file xijoutput /'contiguityassignments.txt'/;
put xijoutput ;
loop((i,j),
if (x.l(i,j) = 1.0,
put i.tl, j.tl/;););
putclose;
```

FIGURE 18.10: GAMS code for districting with population and contiguity constraints and known centers, part 2.

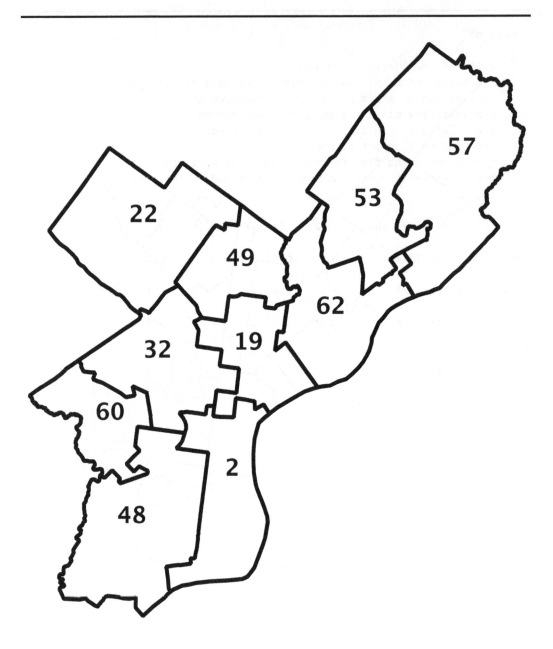

FIGURE 18.11: A districting plan for Philadelphia that is population balanced (±5%), contiguous, and minimizes distance to district centers. Districts are labeled with the IDs of their ward centers.

census since 1950 Philadelphia grew in population, by about 59,000 people, and the fact that during the 2000-2010 period Philadelphia's Hispanic population grew by approximately the same number.

None of these conditions are represented at all in any of the three optimization models we have discussed. What might be done to make the models more relevant? Broadly, there are two approaches. The first is to revise the models so as to reflect the most important attributes of the problem and then solve the revised model(s). A number of difficulties immediately present themselves. We want to preserve neighborhood integrity? Fine, but where are the neighborhoods? The census yields some information regarding location of population by ethnicity, but it is hardly enough for these purposes. The fact is there is no official or even generally accepted delineation of the neighborhoods in Philadelphia and if there were they would surely overlap. These considerations merely hint at the difficulty of articulating a full complement of attributes of interest for the problem, of obtaining the necessary data, of formulating an optimization model, and of solving it to optimality. This is not to say it should not be contemplated and tried if found likely to succeed, even with great approximation. Rather, it is to say that the impediments are so severe that exploring other approaches is well advised.

Team Fred explored the solution pluralism approach for the Philadelphia districting problem. Its basic approach was to use a population based metaheuristic, specifically a form of evolutionary computation, to heuristically optimize a measure of compactness, subject to population balance and contiguity constraints. This model—the EA model—was a procedural analog of the model of Figure 18.8, page 265. The two models were, however, quite different in detail. Most importantly, the EA model was formalized as a procedure rather than represented mathematically and solved heuristically with an EA rather than exactly.

The Team Fred approach to solving the solution pluralism problem was in two phases. Throughout, the 66 wards served as the areal units to be distributed. Smaller units could have been used (there are about 1,600 ward divisions, where people vote, in Philadelphia) and were by other parties, such as City Council itself. The object of the first phase was to obtain a population of 50 contiguous, but not necessarily feasible, solutions to the districting problem. These formed the starting population for the evolutionary algorithm whose running constituted phase two. Regarding phase one, given even a single plan maintaining contiguity, local search can be used to generate a plurality of new contiguous plans. This is in fact what Team Fred did in its Philadelphia districting work. Using a single solution from an exact optimization solver of an approximation to the problem, local search was used to generate a population of 50 solutions. The solutions obtained were contiguous, but infeasible because they violated the population balance constraints.

Phase two consisted of multiple runs of an EA starting with a population of 50 solutions obtained from phase 1. Throughout the multiple runs, whenever a feasible solution was found it was recorded in a log file. At the end of the experiment, 116 distinct feasible solutions had been found, most of which scored at least reasonably well on compactness. This group of 116 solutions constituted Team Fred's response to the solution pluralism problem.

When individuals knowledgeable of Philadelphia, informally chosen, were presented with the 116 solutions, quite a number of the solutions were judged to be of good quality overall, including such attributes of interest as neighborhood preservation. Figures 18.12 and 18.13 are two such examples of good quality plans discovered by the EA. The ward numbering is arbitrary and does not correspond to the actual numbers present in the map for 2000, Figure 18.1, page 256. It is notable that ward 5 in Figures 18.12 and 18.13 does well by the Hispanic community of Philadelphia.

All 116 solutions are available on the Web at `http://opim.wharton.upenn.edu/~sok/` `phillydistricts/`. See [32, 33, 65, 66, 67, 105] for more information on the Team Fred project.

18.6 Discussion

There is a third way to discover and produce districting plans besides the two ways we have discussed here, viz. exact optimization of a mathematical program and using meta-heuristics to discover a plurality of solutions for a solution pluralism formulation. Because the third way is what is most commonly done in practice we could call it the 0th way. In any event it amounts to mucking around, groping manually with a computerized data base and GIS software to discover favorable plans. It is largely what legislators and their assignees do privately in order to find plans that will be favorable to their interests. We might expect that something very similar is routinely done for non-legislative zone design problems, both in the public and the private sectors.

This *GIS groping* approach is both private and unsystematic. Depending on circumstances and views, privacy may or may not be a good thing. Non-partisans interested in redistricting generally find it to be against the public's interest. Political operatives have a different perspective. Regardless of perspective, that GIS groping is unsystematic has to count as a liability. Even the most partisan of gerrymanderers will want to have a thoroughgoing exploration of the possibilities. Very likely, this cannot be done without recourse to a well-designed solution pluralism approach.

Team Fred in its presentation for the Informs Wagner prize [65] offered a fresh perspective on zone design, especially its reapportionment variety. Zone design, observed Team Fred, is not an optimization problem and should not be viewed that way. Instead, zone design is a negotiation game in which it is inevitable that power and special interests will come into play. There is no way around it, yet we might hope to protect the public's or the organization's overall interest by circumscribing the game. We could do this by characterizing a set of solutions of interest and requiring that any plan chosen be among them. This is, in effect, what is done today for redistricting. The courts have limited the permitted solutions to those that have population balance and that meet requirements of the Voting Rights Act. Various state laws impose additional restrictions, which may or may not actually be honored in practice. Team Fred's suggestion amounts to extending existing requirements to include generally recognized desiderata such as compactness, responsiveness (if a party gets a majority of the votes it gets a majority of the seats), symmetry (if one party gets X% of the seats with Y% of the votes, then so would the other party). These would be recognized with tolerances and used to characterize the solutions of interest. Then there would be an open invitation to produce conforming plans. Any conforming plan would be considered and only conforming plans could be adopted. Negotiation would be effectively constrained to the solutions of interest.

Putting aside this proposal, it is apparent that zone design is a rich and rewarding subject for analytics, and that solution pluralism problems are apt for it. They arise naturally and serve to make optimization models, albeit only approximately correct models, valuable contributors to deliberation about zone design.

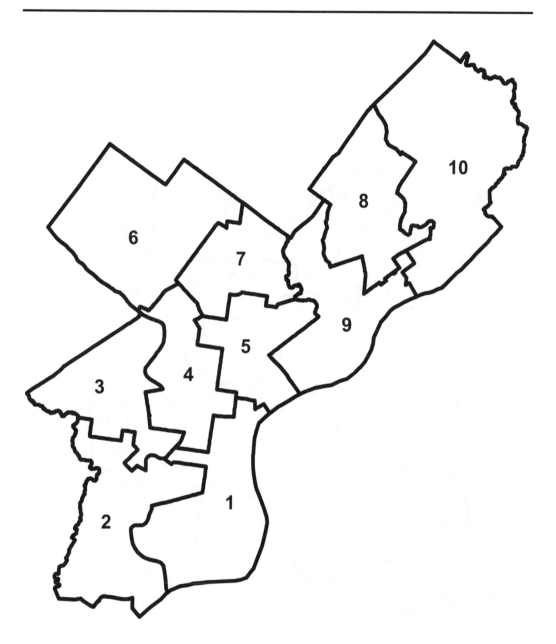

FIGURE 18.12: Districting plan discovered by the evolutionary algorithm. Team Fred's favorite because of neighborhood integrity.

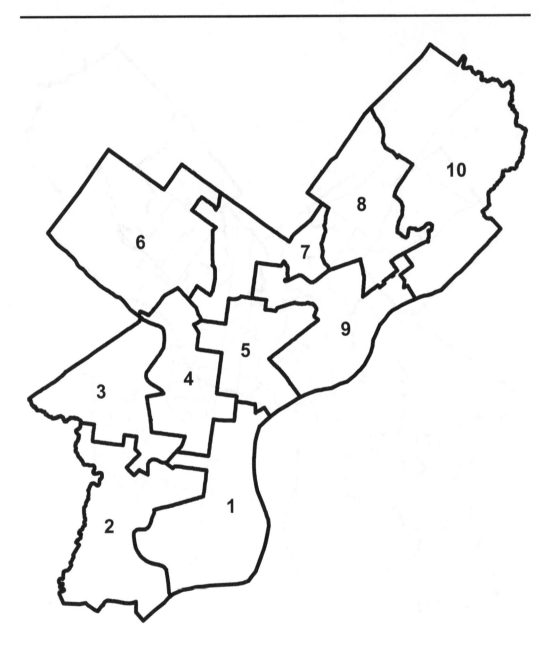

FIGURE 18.13: Districting plan discovered by the evolutionary algorithm. Strong on protecting neighborhoods in the "river wards" (in districts 1, 5, and 9 in the map).

18.7   For Exploration

1. Each of the mathematical programming optimization models discussed in this chapter has the objective of minimizing the sum of distances of the areal units to a hub. Arguably, the sum of distances weighted by population gives us a preferred measure of compactness. Reformulate each of the mathematical programming models to minimize the sum of the population weighted distances.

2. Commercial firms typically divide their areas of business coverage into sales districts. Discuss:

    (a) What are the design criteria commonly used?

    (b) Can you suggest ideal design criteria?

    (c) Sales districts need not be balanced with regard to their potential for sales. Discuss sales incentives in this context. What is common practice? What, in your view, would be a preferred incentive policy?

3. There have been political districting designs that have not been contiguous. Is this always a bad thing, somehow unfair or flawed in some other way? Why or why not? Give examples. Does your assessment change for other contexts of zone design, such as sales districts and service districts (e.g., police, fire, and ambulance services)?

4. Very often there are political districting designs that are far from being compact on any reasonable geographic measure. Is this always a bad thing, somehow unfair or flawed in some other way? Why or why not? Give examples. How does your assessment change (if it does) for other contexts of zone design, such as sales districts and service districts (e.g., police, fire, and ambulance services)?

5. Before the 1960s in the United States, it was common for there to be very substantial (even 10:1) population imbalances between political districts. The Supreme Court put an end to this through a number of decisions on cases appearing before it. Is population imbalance always a bad thing, somehow unfair or flawed in some other way? Are there fair ways to circumvent it? Why or why not? Give examples. How does your assessment change (if it does) for other contexts of zone design, such as sales districts and service districts (e.g., police, fire, and ambulance services)?

6. Suppose you could identify a subgroup of the population in a zone design situation, so that you had as data the population counts of this group for each areal unit. Two commonly used forms of gerrymandering are packing and cracking [126]. In packing, you cram as many of the target population as you can into the smallest number of districts, concentrating this population in as few districts as possible so that its influence is limited to just these districts. In cracking, you spread the target population across as many districts as possible with the aim of minimizing its influence in every district.

    (a) Formulate cracking and packing models as mathematical programs.

    (b) Formulate cracking and packing models procedurally, for solution with meta-heuristics.

    (c) Discuss what one might do in defense of fairness if presented with packed or cracked designs.

7. Team Fred used an EA to find a plurality of feasible solutions to the Philadelphia districting problem. Discuss how a local search heuristic might be used instead. Present high-level pseudocode as part of your analysis.

8. Discuss how to obtain a plurality of solutions from the p-median model, Figure 18.3, page 260. Why might we want to do this? What would it be useful for?

9. Formulate a model—we'll call it a *p-hub* model—that identifies p maximally separated areal units. Formulate

    (a) A mathematical programming model,

    (b) A procedural model suitable for solution with a metaheuristic and able to find a plurality of good solutions.

10. Critically assess and discuss Team Fred's suggestion for organizing the game of redistricting.

18.8   For More Information

The book's Web site is

http://pulsar.wharton.upenn.edu/~sok/biz_analytics_rep.

There, in the *teamfred/* and *philly2010/* directories, you will find the files we mention in conjunction with the chapter. The data for Figure 18.4 were produced with the GAMS model contained in *pmedian.gms*. The data for the plan in Figure 18.7 were produced by the GAMS model contained in *pmedianpopulation.gms*. The data for Figure 18.11 were produced with the GAMS model contained in *shirabe1.gms*, which implements the model in Figure 18.8, page 265. All three files, along with supporting data files, can be found in the *philly2010* directory of the book's Web site.

Redistricting in the United States is considerably more complicated and varied than we have indicated. See *Redistricting: The Most Political Activity in America* by Bullock [24] and Monmonier's *Bushmanders & Bullwinkles: How Politicians Manipulate Electronic Maps and Census Data to Win Elections* [126] for recent overviews. Also valuable are *Elbridge Gerry's Salamander: The Electoral Consequences of the Reapportionment Revolution* [35] and *The Reapportionment Puzzle* [27], the latter being something of a memoir by a participant in the 1980 California redistricting. The more mathematical treatment in [72] is older yet remains useful both for zone design and for evaluation of voting systems.

Two historically important and still interesting algorithmic treatments of districting and zone design are "On the Prevention of Gerrymandering" [147] and "A Procedure for Nonpartisan Districting: Development of Computer Techniques" [153].

See "School Redistricting: Embedding GIS Tools with Integer Programming" [29] for an exemplary modern treatment of zone design with mathematical programming.

There are a number of papers on design, including redistricting, using metaheuristics. These include "Simulated Annealing: An Improved Computer Model for Political Redistricting" [23], "Redistricting in a GIS Environment: An Optimisation Algorithm Using Switching Points" [120], and "Applying Genetic Algorithms to Zone Design" [6].

See "The Computational Complexity of Automated Redistricting: Is Automation the Answer?" [2] for a good discussion of how zone design and redistricting are computationally intractable problems.

For information on and work product of Team Fred, see [32, 33, 65, 66, 67, 105] and `http://opim.wharton.upenn.edu/~sok/phillydistricts/`.

There are a number of public service Web sites relevant to legislative redistricting. We recommend especially the Smart Voter site from the League of Women Voters, `http://www.smartvoter.org/`. Philadelphia is especially rich in well documented, high quality data sources. The Committee of Seventy's site, `www.seventy.org`, is aimed at public policy matters pertaining to Philadelphia. Open Data Philly `http://www.opendataphilly.org/` covers all aspects of life in Philadelphia. The City of Philadelphia has made a strong commitment to making its data publicly available and posts at CityMaps for Philadelphia `http://citymaps.phila.gov/portal/`. PASDA, Pennsylvania Spatial Data Access at `http://www.pasda.psu.edu/default.asp`, covers the Commonwealth of Pennsylvania. For those without a specific interest in Philadelphia or Pennsylvania, these sources constitute a rich and easily accessible library of real world data on which algorithms and methods of business analytics may be tested.

Part V

Conclusion

Chapter 19

Conclusion

19.1   Looking Back

As announced at the outset, our treatment of the material in the book has been introductory. We principally introduce:

- Constrained optimization models, focusing on combinatorial (or integer programming) models, that are or are very like models used extensively in practice and for business analytics.

- Exact solvers for these models (discussed lightly).

- Heuristic solvers, including both local search and population based metaheuristics (discussed in some depth).

- Post-solution analysis, including sensitivity analysis, and deliberation with models for the kinds of models we introduce.

Also as indicated at the outset, our treatment of post-solution analysis is advanced in the sense that we present concepts and methods that can be recommended for actual use but that have to be seen as not elementary. They do not usually appear in elementary presentations of constrained optimization. Our view is that discussion of post-solution analysis is, or should be, central to any serious discussion of constrained optimization modeling, and indeed to any discussion of business analytics. So we have done that in this book, aiming to make the material accessible to those who are being introduced to constrained optimization, that is, to those for whom the remainder of the book is addressed.

19.2   Revisiting Post-Solution Analysis

Given our emphasis on post-solution analysis and deliberation with models, it is appropriate now to revisit our initial discussion and comment on what has and has not been taught in the intervening material. Recall that Chapter 1 contains an extended discussion of the categories in a framework for post-solution analysis. That framework is summarized in

1. Sensitivity

   Are there small changes in parameter values that will result in large changes in outcomes according to the model? If so, what are they and which parameters have the largest sensitivities, i.e., which parameters if changed by small amounts result in large changes in model outcomes? Which parameters have small sensitivities?

2. Policy

   Are there policy reasons, not represented in the model, for deviating from either choosing a preferred (e.g., optimal) solution identified by the model or for modifying one or more parameter value? If so, what changes are indicated and what are the consequences of making them? Policy questions are about factoring into our decision making qualitative factors or aspects of the situation that are not represented in the model formulation.

3. Outcome reach

   Given an outcome predicted or prescribed by the model, and a specific desired alternative outcome, how do the assumptions of the model need to be changed in order to reach the desired outcome(s)? Is it feasible to reach the outcome(s) desired? If so, what is the cheapest or most effective way to do this? Outcome reach questions arise on the degradation side as well. By opting to accept a degraded outcome we may free up resources that can be used elsewhere.

4. Opportunity

   What favorable opportunities are there to take action resulting in changes to the assumptions (e.g., parameter values) of the model leading to improved outcomes (net of costs and benefits)?

5. Robustness

   Which solutions or policy options of the model perform comparatively well across the full range of ambient uncertainty for the model?

6. Explanation

   Why X? Given a predicted or prescribed outcome of the model, X, why does the model favor it? Why not Y? Given a possible outcome of the model, Y, why does the model not favor it over X?

7. Resilience

   Which decisions associated with the model are the most dynamically robust? That is, which decisions best afford deferral of decision to the future when more is known and better decisions can be made?

FIGURE 19.1: Seven categories of representative questions in post-solution analysis.

Figure 1.5, reprinted here as Figure 19.1 on page 282. The key techniques or methods we developed for post-solution analysis involved solution pluralism (obtaining multiple solutions for models as an aid to analysis), which we divided into decision sweeping and parameter sweeping.

We now revisit the seven categories in Figure 19.1 for post-solution analysis. This will serve to draw together in a coherent rope the various threads we have presented and discussed. It will also serve to describe something of the state of the art, indicating what is and what is not known and available for post-solution analysis.

1. Sensitivity

   It is always possible to do sensitivity analysis by undertaking multiple re-solutions of a model,[1] varying parameters with each run. A main problem with this approach is that the search space easily becomes too large to investigate meaningfully, because the cost in computational resources becomes prohibitive.

   Decision sweeping with heuristics is the book's principal response to the problem. When we solve a constrained optimization model with metaheuristics, which is something that is very often naturally preferred for the sorts of models in Part II, it happens that a large number of good decisions are found in the process. This is especially the case with evolutionary programming and the FI2-Pop GA in particular. These good decisions can be collected during the runs as DoIs (decisions of interest), and then used for sensitivity analysis. The additional cost of collecting these decisions, beyond running the heuristic solver itself, is minimal.

2. Policy

   Sensitivity questions are mainly about how changes in parameter values affect model results. Policy questions as we formulated the concept are mainly about the consequences of changing choice of decision from one that is optimal or heuristically optimal to one that is not, but that has a property in accord with a specific policy (unlike the optimal decision).

   Decision sweeping with heuristics is again the book's principal response to the problem. If we can find among the feasible DoIs of higher quality (objective value) decisions that do meet the policy condition and that are otherwise attractive, then decision sweeping has contributed to answering the question. See Chapter 14 for examples.

3. Outcome reach

   There are two aspects to this question. The feasible aspect is: What does it take in terms of parameter settings (especially, but not exclusively, right-hand-side constraint parameter changes) to be able to find a feasible decision that has a much better objective value than the optimum for the present model? The infeasible aspect is: How much can the right-hand-side constraint values be tightened without reducing the resulting optimum value below a given level of quality? Again, decision sweeping is available as a useful tool. Collecting infeasible decisions of interest during runs of a metaheuristc may well yield decisions that fit the bill (providing they exist!). See Chapter 14.

4. Opportunity

   With outcome reach questions the questioner has particular target values in mind. We would frame opportunity questions as exploratory versions of outcome reach questions.

---

[1] Or re-evaluations in the case of a model that is not a constrained optimization model.

Instead of asking Can we achieve a certain target? we are asking Are there any valuable and obtainable targets in evidence? The questions are closely related and may be addressed with similar decision sweeping approaches.

5. Robustness

   Combining decision sweeping with parameter sweeping, as described in Chapter 15, affords identification of decisions that perform relatively well (in terms of objective value) and that are feasible with probability exceeding a specified level. Details are beyond the scope of this book, but see [92, 110] for demonstration studies.

6. Explanation

   This is very much an open, unsolved problem in modeling. We submit, however, that decision sweeping affords a small step forward. Recall the summary description of these questions (see Figure 19.1):

   > Why X? Given a predicted or prescribed outcome of the model, X, why does the model favor it? Why not Y? Given a possible outcome of the model, Y, why does the model not favor it over X?

   Note the similarity with policy questions. Corpora obtained by decision sweeping may well contain decisions that have the X property as well as decisions that have the Y property. The two sets of decisions can then be compared, e.g., on constraint slacks and violations. A small step, but something.

7. Resilience

   This is even more of an open, unsolved problem in modeling than is explanation. Quite evidently answering resilience questions depends upon a great deal of information about possible options and actions that is not represented in the model. The small point we would make is that to the extent that decision sweeping presents good alternative decisions it contributes to solving the problem. In looking for decisions that are attractive and that afford flexible actions, one needs a list of attractive decisions.

19.3   Looking Forward

There are obvious directions to go in augmenting the material in the book:

- Enlarge the kinds of constrained optimization models considered.

- Explore exact solution methods in more depth.

- Explore metaheuristic solution techniques in more depth.

- Extend the post-solution analysis techniques discussed in the book.

To these we should add: investigate more modeling cases in depth, explore possibilities for using visualization for post-solution analysis, augment solution pluralism to encompass representational pluralism (multiple ways of modeling or representing the target system; thanks to Steven Bankes, personal communication, for the suggestion), and additional methods, such as dimension reduction techniques, beyond those in Chapters 16 and 17, for filtering large corpora of solutions.

These post-introductory topics are each and all exciting possibilities, intellectually interesting, and of great practical utility. We close by briefly discussing two less obvious, much more speculative directions for development of model analytics.

19.3.1 Uncertainty

It is instructive to consider models as ranged along a dimension of uncertainty. Some models are quite certain in that their parameter values are well and precisely known, and the models themselves have achieved excellent track records. There are many such examples in the realm of COModels. Industrial plant operations, scheduling, and even typesetting a manuscript for publication are activities that often (although hardly always) are undertaken with the benefit of optimization outcomes that are so reliable that they are immediately and automatically accepted upon production.

The prototypical use case presumed in this book is an intermediate one, typical of business applications of COModels in which some parameter values are at bottom decisions, e.g., a constraint value on the amount of labor available can be changed by a managerial directive. Further, as the models are built and used, the degree of uncertainty is moderate. The model structure may be well established and parameter values known with a fair degree of precision. The model then operates in a context of middling uncertainty. This is fertile ground for post-solution analysis. It can provide genuine assistance to decision makers.

Policy contexts, such as military or diplomatic or healthcare planning for 20 years from the present or climate change modeling (and very much else), are typically at the high end of uncertainty with respect to their model parameters and indeed to the very structures of their models. Model validation becomes tenuous in the extreme when planning arrangements for a future that is not only unknown but likely to include factors previously non-existent. Given the genuine necessity of modeling in these extremely uncertain contexts, it is not surprising that thoughtful participant-observers have had valuable insights into how to conduct modeling in these situations. Many of these insights will, we believe, prove useful in the development of model analytics for COModels. Conversely, the comparatively more certain and structured contexts often present in business analytics contexts can yield example cases that readily afford demonstration and testing of model analytics contexts of all degrees of uncertainty.

Thoughtful comment and workable techniques for dealing with uncertainty in policy modeling have been available for some time (see, e.g., [127], which reviews work dating back to the 1950s). The work of Steven Bankes, extending over the last 25 years, is an especially apt example of insightful thought on policy modeling that has much to offer, and exchange with, business analytics models of middling uncertainty, as well as the larger business analytics enterprise.

Bankes is writing from a perspective that prototypically assumes large scale simulation modeling for policy planning. The models are complex and the context is fraught with uncertainty. In an early work, Bankes distinguishes two kinds of models, or purposes in modeling.

> [There are] two very different uses for computer models. I call these two uses consolidative and exploratory modeling.

> Building a model by consolidating known facts into a single package and then using it as a surrogate for the actual system, which I call *consolidative modeling,* is in many ways the standard approach to model development and use. Where successful, it is a powerful technique for understanding the behavior of complex systems. Unfortunately, the consolidative approach is not always possible.

When insufficient knowledge or unresolvable uncertainties preclude building a surrogate for the target system, modelers must make guesses at details and mechanisms. While the resulting model cannot be taken as a reliable image of the target system, it does provide a computational experiment that reveals how the world would behave if the various guesses were correct. *Exploratory modeling* is the use of series of such computational experiments to explore the implications of varying assumptions and hypotheses. [8, page 435]

The distinction—described here in terms of consolidative versus exploratory modeling—is an important one. We misunderstand modeling and fail to benefit from its full potential if we fail to recognize that in the face of significant uncertainty consolidative modeling cannot achieve much, but exploratory modeling can. Ten years later Bankes (with co-authors) restates the distinction with some new terminology (the term predictive modeling replaces the term consolidative modeling) and provides additional rationale for undertaking exploratory modeling. Because his description is so clear and succinct (and because the reader might find a new prose style refreshing, coming as it does at the end of a long book), we quote from the article extensively.

[W]e can distinguish two very different images of modeling that can be observed in the thinking of computational scientists In the first image, a good computer model is a veridical model of the system, a mirror of the world sufficiently correct that we can learn about the world by peering into it. [10, page 378]

This is all well and good so far as it goes.

Such models can be validated by comparing model outputs to data from the actual system or physical experiments [can be] conducted to check model performance. [10, page 378]

The problem is that predictive models are limited in scope.

However, many problems remain difficult or intractable to solve through predictive modeling. Predictive modeling can only be applied to closed systems that are well enough understood that the uncertainties are manageable and linear enough that whatever uncertainty does exist does not overwhelm the knowledge that we do have about the system. . . .

Unfortunately, many very interesting scientific and policy problems have properties that do not really allow the mirror-world approach. Problems that combine significant complexity with uncertainty can make the classical model-building strategy of representing accurately the important details and neglecting or simplifying others difficult to employ. For deeply uncertain problems, no matter how detailed a model is constructed, we cannot be confident that the model's predictions about the behavior of the real system can be relied on. [10, page 378]

What can be done?

A very different framework for thinking about computer modeling is rooted in the phrase *computational experiment*. . . .

The second approach to modeling that uses models as platforms for computational experiments has been called "exploratory modeling" . . . . The differences between exploratory and predictive modeling are large and not widely appreciated. When building a model intended to predict outcomes, there is a strong motivation to limit the explicit uncertainties, sometimes to the detriment of the scientific value of the result. When building laboratory equipment, on the other hand, adding additional knobs and switches serves to increase the utility of the instrument. [10, page 379]

And how does exploratory modeling work?

> Over the past decade, we and our colleagues have developed and used exploratory modeling-based methods to address numerous problems, including modeling cognition, global climate policy, electronic commerce, technology diffusion, and product planning. Although individual studies have employed diverse methodological tricks, we can synopsize these techniques into four fundamental methodological principles: (a) conceive and employ ensembles of alternatives rather than single models, policies, or scenarios; (b) use principles of invariance or robustness to devise recommended conclusions or policies; (c) use adaptive methods to meet the challenge of deep uncertainty; and (d) iteratively seek improved conclusions through interactive mechanisms that combine human and machine capabilities. [10, page 380]

The attentive reader will find lines of connection leaping out between these passages and ideas bruited in this book. And here we must stop for the present.

For more information: [9] is a recent exploration of these ideas in the context of the complexity sciences. In addition, the paper has a number of intriguing, and surely useful, suggestions for applicable techniques to use in the context of exploratory modeling. A short list of other work well worth reading is [11, 12, 115, 116].

19.3.2 Argumentation

Once we recognize that a model has associated with it a nontrivial amount of uncertainty, a gap is opened between what the model tells us and how we should act on its advice. It is no longer automatic that we should take the model's recommendation or version of the facts at face value. There is a gap that has to be bridged with reasons and evidence. The model outputs are germane to our decisions, but they are no longer dispositive.

Solution pluralism, decision sweeping, parameter sweeping, exploratory modeling, and their ilk are ways of extracting information from uncertain models in order to find compelling, or at least persuasive, reasons and evidence for taking action. There remains the task of marshaling the evidence and assessing the reasons. As it happens, it is the job of logic and the science of argumentation to do just this.

An *argument* in this context is a collection of one or more statements taken as premises and a statement taken as the conclusion. Good arguments, roughly, are arguments in which the premises collectively provide strong reasons for thinking that the conclusion is true. Logic and argumentation theory study arguments and argumentation (the process of constructing arguments). The work is both descriptive and evaluative or prescriptive.

The speculation we wish to float is that when models are complex and uncertain, and when the decision context is fraught and consequential, it will prove valuable to augment the decision making process with explicit representation of the reasons—the argument—that bridges the model and the decision to be taken.

For more information: The terms logic, argumentation theory, informal logic, and related terms are where to begin a general search on this huge literature, which dates back at least to Aristotle. For starters in informal logic and argumentation theory one would do well with Toulmin [145], Walton [152], and Dung [42]. For hints on how arguments might be presented graphically and visually see [90].

Appendix A

Resources

A.1   Resources on the Web

1. The book's Web site is `http://pulsar.wharton.upenn.edu/~sok/biz_analytics_rep`. Some teaching materials can also be downloaded from `http://www.mysmu.edu/faculty/hclau/teach.html`.

2. The OR-Library, source for many benchmark problems, is at `http://www.brunel.ac.uk/~mastjjb/jeb/info.html`.

3. IBM ILOG CPLEX Optimization Studio. The IBM CPLEX Optimization Studio Web page is at `http://www-01.ibm.com/support/knowledgecenter/SSSA5P/welcome`.

   See also the IBM Academic Initiative Licensing scheme at: `https://www-304.ibm.com/ibm/university/academic/pub/page/academic_initiative`.

4. Python. Its home page is at `http://www.python.org`. See `http://www.scipy.org/` for the SciPy Stack, including Python, NumPy, SciPy library, Matplotlib, pandas, SymPy, IPython, nose, and many other packages.

   The Enthought distribution is at `http://www.enthought.com`. It has a free Canopy distribution and development environment.

   Anaconda from Continuum Analytics is at: `https://store.continuum.io/cshop/anaconda/`. It offers a Python distribution, many packages, directed at data analytics, and exploratory scientific computing, and is also free.

5. GAMS. The home page for GAMS is `http://www.gams.com/`.

6. INFORMS. Its home page is `http://www.informs.org`. INFORMS is a major professional society with a primary interest in constrained optimization models and business analytics. Its conferences and publications will be of interest to readers of this book.

Bibliography

[1] Osman Alp, Erhan Erkut, and Zvi Drezner. An efficient genetic algorithm for the p-median problem. *Annals of Operations Research*, 122:21–42, 2003.

[2] M. Altman. The computational complexity of automated redistricting: Is automation the answer? *Rutgers Computer and Technology Law Journal*, 23(1):81–142, 1997.

[3] D. L. Applegate, R. E. Bixby, V. Chvátal, and W. J. Cook. On the solution of travelling salesman problems. *Documenta Mathematica*, 3:645–656, 1998.

[4] D. L. Applegate, R. E. Bixby, V. Chvátal, and W. J. Cook. *The Traveling Salesman Problem: A Computational Study.* Princeton University Press, Princeton, NJ, 2006.

[5] Robert L. Axtell and Steven O. Kimbrough. The high cost of stability in two-sided matching: How much social welfare should be sacrificed in the pursuit of stability? In *Proceedings of the 2008 World Congress on Social Simulation (WCSS-08)*, 2008. http://mann.clermont.cemagref.fr/wcss/.

[6] Fernando Bação, Victor Lobo, and Marco Painho. Applying genetic algorithms to zone design. *Soft Computing*, 9:341–348, 2005. DOI 10.1007/s00500-004-0413-4.

[7] R. D. Banker, A. Charnes, and W. W. Cooper. Some models for estimating technical and scale inefficiencies in data envelopment analysis. *Management Science*, 30:1078–1092, 1984.

[8] Steven Bankes. Exploratory modeling for policy analysis. *Operations Research*, 41(3):435–449, May–June 1993.

[9] Steven Bankes. The use of complexity for policy exploration. In Peter Allen, Steve Maguire, and Bill McKelvey, editors, *The SAGE Handbook of Complexity and Management*, pages 570–589. SAGE Publications, Inc., Thousand Oaks, CA, 2011.

[10] Steven Bankes, Robert Lempert, and Steven Popper. Making computational social science effective. *Social Science Computer Review*, 20(4):377–388, Winter 2002.

[11] Steven C. Bankes. Tools and techniques for developing policies for complex and uncertain systems. *Proceedings of the National Academy of Sciences*, 99:7263–7266, 2002.

[12] Steven C. Bankes. Robust policy analysis for complex open systems. *Emergence*, 7(1), 2005.

[13] John J. Bartholdi, III. The knapsack problem. In Dilip Chhajed and Timothy J. Lowe, editors, *Building Intuition: Insights from Basic Operations Management Models and Principle*, pages 19–32. Springer, Berlin, Germany, 2008. doi:10.1007/978-0-387-73699-0.

[14] Thomas Bartz-Beielstein. *Experimental Research in Evolutionary Computation: The New Experimentalism.* Natural Computing Series. Springer-Verlag, Berlin, Germany, 2006.

[15] Thomas Bartz-Beielstein, Marco Chiarandini, Luis Paquete, and Mike Preuss, editors. *Experimental Methods for the Analysis of Optimization Algorithms.* Springer, Berlin, Germany, 2010.

[16] Robert D. Behn and James W. Vaupel. *Quick Analysis for Busy Decision Makers.* Basic Books, New York, NY, 1982.

[17] Lawrence Bodin and Aaron Panken. High tech for a higher authority: The place of graduating rabbis from Hebrew Union College–Jewish Institute of Religion. *Interfaces*, 33(3):1–11, May–June 2003.

[18] George E.P. Box, J. Stuart Hunter, and William G. Hunter. *Statistics for Experimenters: Design, Innovation, and Discovery.* Wiley-Interscience, New York, NY, second edition, May 2005.

[19] Stephen Boyd and Lieven Vandenberghe. *Convex Optimization.* Cambridge University Press, Cambridge, UK, 2004.

[20] Stephen P. Bradley, Arnoldo C. Has, and Thomas L. Magnanti. *Applied Mathematical Programming.* Addison-Wesley, Reading, MA, 1977.

[21] B. Branley, R. Fradin, S. O. Kimbrough, and T. Shafer. On heuristic mapping of decision surfaces for post-evaluation analysis. In Ralph H. Sprague, Jr., editor, *Proceedings of the Thirtieth Annual Hawaii International Conference on System Sciences*, Los Alamitos, CA, 1997. IEEE Press.

[22] Bill Branley, Russell Fradin, Steven O. Kimbrough, and Tate Shafer. On heuristic mapping of decision surfaces for post-evaluation analysis. In Jay Nunamaker, Jr. and Ralph H. Sprague, Jr., editors, *Proceedings of the Thirtieth Hawaii International Conference on System Sciences.* IEEE Computer Press, January 1997.

[23] Michelle H. Browdy. Simulated annealing: An improved computer model for political redistricting. *Yale Law & Policy Review*, 8(1):163–179, 1990.

[24] Charles S. Bullock III. *Redistricting: The Most Political Activity in America.* Rowman & Littlefield Publishers, Inc., Lanham, MD, 2010.

[25] Rainer Burkard, Mauro Dell'Amico, and Silvano Martello. *Assignment Problems.* SIAM, Philadelphia, PA, 2009.

[26] Pu Cai and Jin-Yi Cai. On the 100% rule of sensitivity analysis in linear programming. In Tao Jiang and D. T. Lee, editors, *Computing and Combinatorics: Third Annual International Conference, COCOON '97*, pages 460–469. International Computing and Combinatorics Conference, Springer, Volume 1276, Lecture Notes in Computer Science, 1997.

[27] Bruce E. Cain. *The Reapportionment Puzzle.* University of California Press, Berkeley, CA, 1984.

[28] Jeffrey D. Camm. Asp, the art and science of practice: A (very) short course in suboptimization. *Interfaces*, 44(4):428–431, July–August 2014.

[29] F. Caro, T. Shirabe, M. Guignard, and A. Weintraub. School redistricting: Embedding GIS tools with integer programming. *Journal of the Operational Research Society*, 55:836–849, 2004.

[30] Abraham Charnes, William W. Cooper, Arie Y. Lewin, and Lawrence M. Seiford. *Data Envelopment Analysis: Theory, Methodology, and Application*. Kluwer Academic Publishers, Boston, MA, 1994.

[31] F. Chiyoshi and R. D. Galvão. A statistical analysis of simulated annealing applied to the *p*-median problem. *Annals of Operations Research*, 96:61–74, 2000.

[32] Christine Chou, Steven O. Kimbrough, Frederic H. Murphy, John Sullivan-Fedock, and C. Jason Woodard. On empirical validation of compactness measures for electoral redistricting and its significance for application of models in the social sciences. *Social Science Computer Review*, 32(4):534–542, August 2014. doi: 10.1177/0894439313484262 http://ssc.sagepub.com/content/early/2013/04/09/0894439313484262.

[33] Christine Chou, Steven O. Kimbrough, John Sullivan, C. Jason Woodard, and Frederic H. Murphy. Using interactive evolutionary computation (IEC) with validated surrogate fitness functions for redistricting. In Terence Soule, editor, *Proceedings of GECCO 2012 (Genetic and Evolutionary Computation Conference)*, pages 1071–1078, New York, NY, USA, 2012. ACM. http://dx.doi.org/10.1145/2330163.2330312.

[34] William W. Cooper, Lawrence M. Seiford, and Kaoru Tone. *Data Envelopment Analysis: A Comprehensive Text with Models, Applications, References and DEA-Solver Software*. Springer, New York, NY, second edition, 2007.

[35] Gary W. Cox and Jonathan N. Katz. *Elbridge Gerry's Salamander: The Electoral Consequences of the Reapportionment Revolution*. Political Economy of Institutions and Decisions. Cambridge University Press, Cambridge, UK, 2002.

[36] G. Dantzig, R. Fulkerson, and S. Johnson. Solution of a large-scale traveling-salesman problem. *Journal of the Operations Research Society of America*, 2(4):393–410, 1954.

[37] Mark S. Daskin. *Network and Discrete Location: Models, Algorithms, and Applications*. Wiley, Hoboken, NJ, second edition, 2013.

[38] Robyn M. Dawes. The robust beauty of improper linear models in decision making. *American Psychologist*, 34(7):571–582, 1979.

[39] Kenneth A. De Jong. *Evolutionary Computation: A Unified Approach*. The MIT Press, Cambridge, MA, 2006.

[40] Kathryn A. Dowsland. Simulated annealing. In Colin R. Reeves, editor, *Modern Heuristic Techniques for Combinatorial Problems*, pages 20–69. John Wiley & Sons, Inc., New York, NY, 1993.

[41] G. Dueck and T. Scheuer. Threshold accepting: A general purpose algorithm appearing superior to simulated annealing. *Journal of Computational Physics*, 90:161–175, 1990.

[42] Phan Minh Dung. On the acceptability of arguments and its fundamental role in non-monotonic reasoning, logic programming and n-person games. *Artificial Intelligence*, 77(2):321–357, 1995.

[43] Ward Edwards and F. Hutton Barron. SMARTS and SMARTER: Improved simple methods for multiattribute utility measurement. *Organizational Behavior and Human Decision Processes*, 60(3):306–325, December 1994.

[44] A. E. Eiben and J. E. Smith. *Introduction to Evolutionary Computing*. Springer, Berlin, Germany, 2003.

[45] Samual Eilon. Application of the knapsack model for budgeting. *OMEGA International Journal of Management Science*, 15(6):489–494, 1987.

[46] H. A. Eiselt and Vladimir Marianov, editors. *Foundations of Location Analysis*. Springer, New York, NY, 2011.

[47] European Commission. Impact assessment guidelines. http://ec.europa.eu/governance/impact/commission_guidelines/docs/iag_2009_en.pdf, 15 January 2009. Accessed 2010-9-12. See also http://ec.europa.eu/governance/impact/index_en.htm and http://ec.europa.eu/enterprise/policies/sme/files/docs/sba/iag_2009_en.pdf.

[48] M-A Félix and A. Wagner. Robustness and evolution: Concepts, insights and challenges from a developmental model system. *Heredity*, 100:132–140, 2008.

[49] D. B. Fogel, editor. *Evolutionary Computation: The Fossil Record*. IEEE Press, Piscataway, NJ, 1998.

[50] David B. Fogel. *Evolutionary Computation: Toward a New Philosophy of Machine Intelligence*. IEEE Press, Piscataway, NJ, second edition, 2000.

[51] David B. Fogel. *Blondie24: Playing at the Edge of AI*. Morgan Kaufmann, San Francisco, CA, 2002.

[52] L. J. Fogel. Autonomous automata. *Industrial Research*, 4:14–19, 1962.

[53] D. Gale and L. S. Shapley. College admissions and the stability of marriage. *The American Mathematical Monthly*, 69(1):9–15, January 1962.

[54] Michael R. Garey and David S. Johnson. *Computers and Intractability: A Guide to the Theory of NP-Completeness*. W. H. Freeman and Company, New York, NY, 1979.

[55] Michel Gendreau and Jean-Yves Potvin, editors. *Handbook of Metaheuristics*. Springer, New York, NY, second edition, 2010.

[56] Michel Gendreau and Jean-Yves Potvin. Tabu search. In Michel Gendreau and Jean-Yves Potvin, editors, *Handbook of Metaheuristics*, volume 146 of *International Series in Operations Research & Management Science*, pages 41–60. Springer, New York, NY, second edition, 2010.

[57] F. Glover, C. C. Kuo, and K. S. Dhir. Analyzing and modeling the maximum diversity problem by zero-one programming. *Decision Sciences*, 24:1171–1185, 1993.

[58] F. Glover and M. Laguna. *Tabu Search*. Kluwer Academic Publishers, Boston, MA, 1997.

[59] Fred Glover. Future paths for integer programming and links to artificial intelligence. *Computers and Operations Research*, 13:533–549, 1986.

[60] Fred Glover. Tabu search—part i. *ORSA Journal on Computing*, 1:190–206, 1989.

[61] Fred Glover. Tabu search—part ii. *ORSA Journal on Computing*, 2:4–32, 1990.

[62] Fred Glover, John Mulvey, Dawei Bei, and Michael T. Tapia. Integrative population analysis for better solutions to large-scale mathematical programs. In G. Yu, editor, *Industrial Applications of Combinatorial Optimization*, pages 212–237. Kluwer Academic Publishers, 1998.

[63] David E. Goldberg. *Genetic Algorithms in Search, Optimization & Machine Learning*. Addison-Wesley Publishing Company, Inc., Reading, MA, 1989.

[64] B. Golden, S. Raghavan, and E. Wasil, editors. *The Vehicle Routing Problem: Lastest Advances and New Challenges*. Springer, New York, NY, 2008.

[65] Ram Gopalan, Steven O. Kimbrough, Frederic H. Murphy, and Nicholas Quintus. The Philadelphia districting contest: Designing territories for city council based upon the 2010 census. Video on the Web, October 2012. `https://live.blueskybroadcast.com/bsb/client/CL_DEFAULT.asp?Client=569807&PCAT=5277&CAT=5278`.

[66] Ram Gopalan, Steven O. Kimbrough, Frederic H. Murphy, and Nicholas Quintus. Team Fred home page. World Wide Web, November 2012. `http://opim.wharton.upenn.edu/~sok/phillydistricts/`.

[67] Ram Gopalan, Steven O. Kimbrough, Frederic H. Murphy, and Nicholas Quintus. The Philadelphia districting contest: Designing territories for City Council based upon the 2010 census. *Interfaces*, 43(5):477–489, September-October 2013. `http://dx.doi.org/10.1287/inte.2013.0697`.

[68] Harvey J. Greenberg. How to analyze the results of linear programs—Part 1: Preliminaries. *Interfaces*, 23(4):56–67, July–August 1993.

[69] Harvey J. Greenberg. How to analyze the results of linear programs—Part 2: Price interpretation. *Interfaces*, 23(5):97–114, September–October 1993.

[70] Harvey J. Greenberg. How to analyze the results of linear programs—Part 3: Infeasibility diagnosis. *Interfaces*, 23(6):120–139, November–December 1993.

[71] Harvey J. Greenberg. How to analyze the results of linear programs—Part 4: Forcing substructures. *Interfaces*, 24(1):121–130, January–February 1994.

[72] Pietro Grilli di Cortona, Cecilia Manzi, Aline Pennisi, Fedrica Ricca, and Bruno Simeone. *Evaluation and Optimization of Electoral Systems*. Society for Industrial and Applied Mathematics, Philadelphia, PA, 1987.

[73] D. Gusfield and R. W. Irving. *The Stable Marriage Problem: Structure and Algorithms*. MIT Press, Cambridge, MA, 1989.

[74] Christian Haas. *Incentives and Two-Sided Matching - Engineering Coordination Mechanisms for Social Clouds*, volume 12 of *Studies on eOrganisation and Market Engineering*. KIT Scientific Publishing, 2014.

[75] Magnús M. Halldórsson, Kazuo Iwama, Shuichi Miyazaki, and Hiroki Yanagisawa. Improved approximation results for the stable marriage problem. *ACM Trans. Algorithms*, 3(3):30, 2007.

[76] S. Hartmann and D. Briskorn. A survey of variants and extensions of the resource-constrained project scheduling problem. *European Journal of Operational Research*, 207(1):1–14, 2010.

[77] IPPC. Background papers, ipcc expert meetings on good practice guidance and uncertainty management in national greenhouse gas inventories. Pages on the World
Wide Web, 1999. IPPC= Intergovernmental Panel on Climate Change. Accessed
2010-10-31. URL: `http://www.ipcc-nggip.iges.or.jp/public/gp/gpg-bgp.htm`.

[78] IPPC. Good practice guidance and uncertainty management in national greenhouse
gas inventories. `http://www.ipcc-nggip.iges.or.jp/public/gp/english/`, 2000.
IPPC= Intergovernmental Panel on Climate Change. Accessed 2015-07-30.

[79] Robert W. Irving and Paul Leather. The complexity of counting stable marriages.
SIAM Journal on Computing, 15(3):655–667, August 1986.

[80] William James. *Pragmatism: A New Name for Some Old Ways of Thinking*. Project
Gutenberg, `http://www.gutenberg.org/ebooks/5116`, 1907.

[81] Richard C. Jeffrey. *The Logic of Decision*. University of Chicago Press, Chicago, Il,
second edition, 1983.

[82] David S. Johnson, Cecilia R. Aragon, Lyle A. McGeoch, and Catherine Schevon.
Optimization by simulated annealing: An experimental evaluation; part i, graph partitioning. *Operations Research*, 37(6):865–892, Nov.–Dec. 1989.

[83] Goos Kant, Michael Jacks, and Corné Aantjes. Coca-Cola enterprises optimizes vehicle
routes for efficient product delivery. *Interfaces*, 38(1):40–50, 2008.

[84] Ralph L. Keeney and Howard Raiffa. *Decisions with Multiple Objectives: Preferences
and Value Tradeoffs*. Cambridge University Press, Cambridge, UK, 1993.

[85] Hans Kellerer, Ulrich Pferschy, and David Pisinger. *Knapsack Problems*. Springer,
Berlin, Germany, 2004.

[86] S. O. Kimbrough, S. A. Moore, C. W. Pritchett, and C. A. Sherman. On DSS support
for candle lighting analysis. In *Transactions of DSS '92*, pages 118–135, June 8-10,
1992.

[87] S. O. Kimbrough and J. R. Oliver. On automating candle lighting analysis: Insight
from search with genetic algorithms and approximate models. In Jay F. Nunamaker,
Jr. and Ralph H. Sprague, Jr., editors, *Proceedings of the Twenty-Sixth Annual Hawaii
International Conference on System Sciences, Volume III: Information Systems: Deci-
sion Support and Knowledge-Based Systems*, pages 536–544, Los Alamitos, CA, 1994.
IEEE Computer Society Press.

[88] S. O. Kimbrough and J. R. Oliver. Candle lighting analysis: Concepts, examples, and
implementation. In Veda C. Storey and Andrew B. Whinston, editors, *Proceedings of
the Second Annual Workshop on Information Technologies and Systems*, pages 55–63,
Dallas, Texas, December 12-13, 1992.

[89] S. O. Kimbrough, J. R. Oliver, and C. W. Pritchett. On post-evaluation analysis:
Candle-lighting and surrogate models. *Interfaces*, 23(7):17–28, May-June 1993.

[90] Steven O. Kimbrough and Hua Hua. On defeasible reasoning with the method of
sweeping presumptions. *Minds and Machines*, 1(4):393–416, 1991.

[91] Steven O. Kimbrough and Ann Kuo. On heuristics for two-sided matching: Revisiting
the stable marriage problem as a multiobjective problem. Working paper, University
of Pennsylvania, Philadelphia, PA, 2010. `http://opimstar.wharton.upenn.edu/
~sok/sokpapers/2010/matching-marriage-short-generic.pdf`.

[92] Steven O. Kimbrough, Ann Kuo, and Hoong Chuin Lau. Finding robust-under-risk solutions for flowshop scheduling. In *MIC 2011: The IX Metaheuristics International Conference*, Udine, Italy, 2011.

[93] Steven O. Kimbrough, Ann Kuo, Hoong Chuin Lau, Lindawati, and David Harlan Wood. On using genetic algorithms to support post-solution deliberation in the generalized assignment problem. MIC 2009: The VIII Metaheuristics International Conference, conference CD, 2009.

[94] Steven O. Kimbrough, Ann Kuo, Hoong Chuin Lau, Frederic H. Murphy, and David Harlan Wood. Solution pluralism and metaheuristics. In *MIC 2011: The IX Metaheuristics International Conference*, Udine, Italy, 2011.

[95] Steven O. Kimbrough, Ann Kuo, Hoong Chuin LAU, and David Harlan Wood. On decision support for deliberating with constraints in constrained optimization models. Workshop paper, Genetic and Evolutionary Computation Conference (GECCO-2010), Workshop on Constraints Handling, July 2010.

[96] Steven O. Kimbrough, Ming Lu, and Soofi M. Safavi. Exploring a financial product model with a two-population genetic algorithm. In *Proceedings of the 2004 Congress on Evolutionary Computation*, pages 855–862, Piscataway, NJ, June 19–23, 2004. IEEE Neural Network Society, IEEE Service Center. ISBN: 0-7803-8515-2.

[97] Steven O. Kimbrough, Ming Lu, David Harlan Wood, and D. J. Wu. Exploring a two-market genetic algorithm. In W. B. Langdon, E. Cantú-Paz, and others, editors, *Proceedings of the Genetic and Evolutionary Computation Conference (GECCO 2002)*, pages 415–21, San Francisco, CA, 2002. Morgan Kaufmann.

[98] Steven O. Kimbrough, Ming Lu, David Harlan Wood, and D. J. Wu. Exploring a two-population genetic algorithm. In Erick Cantú-Paz et al., editors, *Genetic and Evolutionary Computation (GECCO 2003)*, LNCS 2723, pages 1148–1159, Berlin, Germany, 2003. Springer.

[99] Steven Orla Kimbrough. *Agents, Games, and Evolution: Strategies at Work and Play.* CRC Press, Boca Raton, FL, 2012.

[100] Steven Orla Kimbrough, Gary J. Koehler, Ming Lu, and David Harlan Wood. On a feasible–infeasible two–population (FI-2Pop) genetic algorithm for constrained optimization: Distance tracing and no free lunch. *European Journal of Operational Research*, 190(2):310–327, 2008.

[101] Steven Orla Kimbrough, Ming Lu, and David Harlan Wood. Exploring the evolutionary details of a feasible-infeasible two-population GA. In Xin Yao et al., editors, *Parallel Problem Solving from Nature – PPSN VIII*, volume 3242 of *LNCS: Lecture Notes in Computer Science*, pages 292–301. Springer-Verlag, Berlin, Germany, 2004.

[102] Steven Orla Kimbrough and David Harlan Wood. A note on the deliberation problem for constrained optimization and how evolutionary computation may contribute to its solution. In *Proceedings of the Conference on Information Systems & Technology (CIST 2008)*. INFORMS, 2008.

[103] S. Kirkpatrick, C. D. Gelatt, Jr., and M. P. Vecchi. Optimization by simulated annealing. *Science*, 220(4598):671–680, 1983.

[104] Marc Kirschner and John Gerhart. Evolvability. *PNAS*, 95(15):8420–8427, 1998.

[105] Knowledge@Wharton. A new approach to decision making: When 116 solutions are better than one. `http://knowledge.wharton.upenn.edu/article.cfm?articleid=2850`, 28 September 2011.

[106] Donald E. Knuth. *Marriages Stables*. Les Presses de l'Université de Montreal, Montreal, Canada, 1976.

[107] Donald E. Knuth. *Stable Marriage and Its Relation to Other Combinatorial Problems: An Introduction to the Mathematical Analysis of Algorithms*, volume 10 of *CRM Proceedings & Lecture Notes, Centre de Recherches Mathématiques Université de Montréal*. American Mathematical Society, Providence, RI, 1997. Originally published as [106].

[108] R. Kolisch and S. Hartmann. Experimental investigation of heuristics for resource-constrained project scheduling: An update. *European Journal of Operational Research*, 174(1):23–37, 2006.

[109] P. Korevaar, U. Schimpel, and R. Boedi. Inventory budget optimization: Meeting system-wide service levels in practice. *IBM Journal of Research & Development*, 51(3/4):447–464, May/July 2007.

[110] Ann Kuo. *Unveiling Hidden Values of Optimization Models with Metaheuristic Approach*. PhD thesis, University of Pennsylvania, Philadelphia, PA, May 2014.

[111] Richard C. Larson and Amedeo R. Odoni. *Urban Operations Research*. Prentice-Hall, Inc., Englewood Cliffs, NJ, 1981.

[112] Hoong Chuin Lau and Zhe Liang. Pickup and delivery with time windows: Algorithms and test case generation. *International Journal on Artificial Intelligence Tools*, 11(3):455–472, 2002.

[113] Hoong Chuin Lau, Melvyn Sim, and Kwong Meng Teo. Vehicle routing problem with time windows and a limited number of vehicles. *European Journal of Operational Research*, 148(3):559–569, 2003.

[114] E. L. Lawler, J. K. Lenstra, A. H. G. Rinnooy Kan, and D. B. Shmoys. *The Traveling Salesman Problem: A Guided Tour of Combinatorial Optimization*. Wiley, New York, NY, 1985.

[115] Robert J. Lempert, David G. Groves, Steven W. Popper, and Steven C. Bankes. A general, analytic method for generating robust strategies and narrative scenarios. *Management Science*, 52(4):514–528, 2006.

[116] Robert J. Lempert, Steven W. Popper, and Steven C. Bankes. Confronting surprise. *Social Science Computing Review*, 20(4):420–440, 2002.

[117] S. Lin and B. W. Kernighan. An effective heuristic algorithm for the traveling-salesman problem. *Operations Research*, 21(2):498–516, 1973.

[118] M. Lozano, D. Molina, and C. García-Martínez. Iterated greedy for the maximum diversity problem. *European Journal of Operational Research*, 214:31–38, 2011.

[119] J. Lysgaard. CVRPSP: A package of separation routines for the capacitated vehicle routing problem. Working paper, Centre for Operations Research Applications in Logistics, Aarhus University, 2004.

[120] W. Macmillan. Redistricting in a GIS environment: An optimisation algorithm using switching points. *Journal of Geographical Systems*, 3:167–180, 2001.

[121] Silvano Martello and Paolo Toth. *Knapsack Problems: Algorithms and Computer Implementations*. John Wiley & Sons, New York, NY, 1990.

[122] Zbigniew Michalewicz. *Genetic Algorithms + Data Structures = Evolution Programs*. Springer, Berlin, Germany, third edition, 1996.

[123] Zbigniew Michalewicz and David B. Fogel. *How to Solve It: Modern Heuristics*. Springer, Berlin, Germany, 2000.

[124] John H. Miller. Active nonlinear tests (ANTs) of complex simulation models. *Management Science*, 44(6):820–830, 1998.

[125] Melanie Mitchell. *An Introduction to Genetic Algorithms*. MIT Press, Cambridge, MA, 1996.

[126] Mark Monmonier. *Bushmanders & Bullwinkles: How Politicians Manipulate Electronic Maps and Census Data to Win Elections*. The University of Chicago Press, Chicago, IL, 2001.

[127] M. Granger Morgan and Max Henrion. *Uncertainty: A Guide for Dealing with Uncertainty in Quantitative Risk and Policy Analysis*. Cambridge University Press, Cambridge, UK, 1990.

[128] Frederic H. Murphy, Sidney W. Hess, and Carlos G. Wong-Martinez. Politics. In Saul Gass and M. Fu, editors, *Encyclopedia of Operations Research*. Springer, 2012.

[129] Alexander G. Nikolaev and Sheldon H. Jacobson. Simulated annealing. In Michel Gendreau and Jean-Yves Potvin, editors, *Handbook of Metaheuristics*, volume 146 of *International Series in Operations Research & Management Science*, pages 1–39. Springer, New York, NY, 2nd edition, 2010.

[130] Charles S. Peirce. The fixation of belief. *Popular Science Monthly*, 12:1–15, November 1877.

[131] Howard Raiffa. *Decision Analysis: Introductory Lectures on Choices under Uncertainty*. Addison-Wesley Publishing Company, Reading, MA, 1968.

[132] T. Ralphs, M. Guzelsoy, and A. Mahajan. The SYMPHONY source code. `https://projects.coin-or.org/SYMPHONY`, 2010.

[133] Colin R. Reeves and Jonathan E. Rowe. *Genetic Algorithms—Principles and Perspectives: A Guide to GA Theory*. Operations Research/Computer Science Interfaces Series. Kluwer Academic Publishers, Boston, MA, 2003.

[134] Csar Rego, Dorabela Gamboa, Fred Glover, and Colin Osterman. Traveling salesman problem heuristics: leading methods, implementations and latest advances. *European Journal of Operational Research*, 211:427–411, 2011.

[135] C. S. ReVelle and H. A. Eiselt. Location analysis: A synthesis and survey. *European Journal of Operational Research*, 165:1–19, 2005.

[136] Alvin E. Roth. Deferred acceptance algorithms: History, theory, practice, and open questions. *International Journal of Game Theory*, 36:537–569, March 2008.

[137] Alvin E. Roth. *Who Get What—and Why*. Houghton Mifflin Harcourt, Boston, MA, 2015.

[138] Alvin E. Roth and Marilda A. Oliveira Sotomayor. *Two-Sided Matching: A Study in Game-Theoretic Modeling and Analysis*. Cambridge University Press, Cambridge, UK, 1990.

[139] Andrea Saltelli, Paola Annoni, Ivano Azzini, Francesca Campolongo, Marco Ratto, and Stefano Tarantola. Variance based sensitivity analysis of model output. Design and estimator for the total sensitivity index. *Computer Physics Communications*, 181:259–270, 2010.

[140] Andrea Saltelli, Karen Chan, and E. Marian Scott, editors. *Sensitivity Analysis*. John Wiley & Sons, Ltd., Chichester, England, 2000.

[141] Andrea Saltelli, Marco Ratto, and Terry Andres. *Global Sensitivity Analysis: The Primer*. John Wiley & Sons, Ltd., Chichester, England, 2008.

[142] Andrea Saltelli, Stefano Tarantola, Francesca Campolongo, and Marco Ratto. *Sensitivity Analysis in Practice: A Guide to Assessing Scientific Models*. John Wiley & Sons, Chichester, UK, 2004.

[143] William R. Shadish, Thomas D. Cook, and Donald T. Campbell. *Experimental and Quasi-Experimental Designs for Generalized Causal Inference*. Wadsworth Publishing, New York, NY, second edition, 2001.

[144] Takeshi Shirabe. Districting modeling with exact contiguity constraints. *Environment and Planning B: Planning and Design*, 36:1053–1066, 2009. doi:10.1068/b34104.

[145] Stephen Edelston Toulmin. *The Uses of Argument*. Cambridge University Press, Cambridge, UK, updated edition, 2003.

[146] United States Environmental Protection Agency. Guidance on the development, evaluation, and application of environmental models. http://www.epa.gov/crem/library/cred_guidance_0309.pdf, March 2009. Accessed 2010-9-12. Publication: EPA/100/K-09/003. http://www.epa.gov/crem.

[147] William Vickrey. On the prevention of gerrymandering. *Political Science Quarterly*, 76(1):105–110, March 1961.

[148] Detlof von Winterfeldt and Ward Edwards. *Decision Analysis and Behavioral Research*. Cambridge University Press, Cambridge, UK, 1986.

[149] Yevgeniy Vorobeychik, Steven O. Kimbrough, and Howard K. Kunreuther. A framework for computational strategic analysis: Applications to iterated interdependent security games. *Computational Economics*, March 2014. http://link.springer.com/article/10.1007/s10614-014-9431-1, DOI 10.1007/s10614-014-9431-1.

[150] Andreas Wagner. *Robustness and Evolvability in Living Systems*. Princeton Studies in Complexity. Princeton University Press, Princeton, NJ, 2005.

[151] Harvey M. Wagner. *Principles of Operations Research: With Applications to Managerial Decisions*. Prentice-Hall, Inc., Englewood Cliffs, NJ, 1969.

[152] Douglas N. Walton. *Informal Logic: A Handbook for Critical Argumentation*. Cambridge University Press, Cambridge, UK, 1989.

[153] James B. Weaver and Sidney W. Hess. A procedure for nonpartisan districting: Development of computer techniques. *Yale Law Journal*, 73(2):288–308, December 1963.

[154] Adianto Wibisono, Zhiming Zhao, Adam Belloum, and Marian Bubak. A framework for interactive parameter sweep applications. In Marian Bubak, GeertDick van Albada, Jack Dongarra, and Peter M. A. Sloot, editors, *Computational Science ICCS 2008*, volume 5103 of *Lecture Notes in Computer Science*, pages 481–490. Springer Berlin Heidelberg, 2008.

[155] Justin C. Williams. Political redistricting: A review. *Papers in Regional Science*, 74(1):13–40, 1995.

[156] Peter Winker. *Optimization Heuristics in Econometrics: Applications of Threshold Accepting*. John Wiley & Sons, Ltd., Chichester, UK, 2001.

[157] Wayne L. Winston. *Operations Research Applications and Algorithms*. Brooks/Cole, Belmont, CA, fourth edition, 2004.

[158] Nicklaus Wirth. *Algorithms + Data Structures = Programs*. Prentice Hall, 1976.

[159] Mutsunori Yagiura and Toshihide Ibaraki. Recent metaheuristic algorithms for the generalized assignment problem. In *Proceedings of the 12th International Conference on Informatics Research for Development of Knowledge Society Infrastructure (ICKS'04)*, pages 229–237. IEEE, IEEE Computer Society, 2004. Digital Object Identifier 10.1109/ICKS.2004.1313429, http://dx.doi.org/10.1109/ICKS.2004.1313429.

[160] Mutsunori Yagiura and Toshihide Ibaraki. The generalized assignment problem and its generalizations. http://faculty.smcm.edu/acjamieson/f12/GAP.pdf, ND. Accessed: 2015-01-24.

[161] Mutsunori Yagiura, Toshihide Ibaraki, and Fred Glover. An ejection chain approach for the generalized assignment problem. *Informs Journal on Computing*, 16(2):133–151, 2004.

[162] H. Peyton Young. Measuring the compactness of legislative districts. *Legislative Studies Quarterly*, 13(1):105–115, February 1988.

Index